Gender in Music Production

The field of music production has for many years been regarded as male-dominated. Despite growing acknowledgment of this fact, and some evidence of diversification, it is clear that gender representation on the whole remains quite unbalanced. *Gender in Music Production* brings together industry leaders, practitioners, and academics to present and analyze the situation of gender within the wider context of music production as well as to propose potential directions for the future of the field. This much-anticipated volume explores a wide range of topics, covering historical and contextual perspectives on women in the industry, interviews, case studies, individual position pieces, as well as informed analysis of current challenges and opportunities for change.

Ground-breaking in its synthesis of perspectives, *Gender in Music Production* offers a broadly considered and thought-provoking resource for professionals, students, and researchers working in the field of music production today.

Russ Hepworth-Sawyer is a sound engineer and producer with over two decades' experience of all things audio and is a member of the Association of Professional Recording Services and a former board member and continuing member of the Music Producers Guild, where he helped form their Mastering Group.

Jay Hodgson is an associate professor of popular music studies at Western University, where he mostly teaches songwriting and the project paradigm of record production.

Dr Liesl King is Associate Head of School: Creative Writing, Media and Film Studies at York St John University in York, England.

Mark Marrington trained in composition and musicology at the University of Leeds (M.Mus., Ph.D.) and is currently Senior Lecturer in Music Production at York St John University.

Perspectives on Music Production

This series collects detailed and experientially informed considerations of record production from a multitude of perspectives, by authors working in a wide array of academic, creative and professional contexts. We solicit the perspectives of scholars of every disciplinary stripe, alongside recordists and recording musicians themselves, to provide a fully comprehensive analytic point-of-view on each component stage of music production. Each volume in the series thus focuses directly on a distinct stage of music production, from pre-production through recording (audio engineering), mixing, mastering, to marketing and promotions.

Series Editors
Russ Hepworth-Sawyer, York St John University, UK
Jay Hodgson, Western University, Ontario, Canada
Mark Marrington, York St John University, UK

Titles in the Series

Producing Music
Edited by Russ Hepworth-Sawyer, Jay Hodgson, and Mark Marrington

Innovation in Music
Performance, Production, Technology, and Business
Edited by Russ Hepworth-Sawyer, Jay Hodgson, Justin Paterson, and Rob Toulson

Pop Music Production
Manufactured Pop and BoyBands of the 1990s
Phil Harding
Edited by Mike Collins

Cloud-Based Music Production
Sampling, Synthesis, and Hip-Hop
Matthew T. Shelvock

Gender in Music Production
Edited by Russ Hepworth-Sawyer, Jay Hodgson, Liesl King, and Mark Marrington

For more information about this series, please visit: www.routledge.com/Perspectives-on-Music-Production/book-series/POMP

Gender in Music Production

Edited by Russ Hepworth-Sawyer,
Jay Hodgson, Liesl King, and
Mark Marrington

Routledge
Taylor & Francis Group
NEW YORK AND LONDON

First published 2020
by Routledge
52 Vanderbilt Avenue, New York, NY 10017

and by Routledge
2 Park Square, Milton Park, Abingdon, Oxon, OX14 4RN

Routledge is an imprint of the Taylor & Francis Group, an informa business

© 2020 selection and editorial matter, Russ Hepworth-Sawyer, Jay Hodgson, Liesl King, and Mark Marrington; individual chapters, the contributors

The right of Russ Hepworth-Sawyer, Jay Hodgson, Liesl King, and Mark Marrington to be identified as the authors of the editorial material, and of the authors for their individual chapters, has been asserted in accordance with sections 77 and 78 of the Copyright, Designs and Patents Act 1988.

All rights reserved. No part of this book may be reprinted or reproduced or utilised in any form or by any electronic, mechanical, or other means, now known or hereafter invented, including photocopying and recording, or in any information storage or retrieval system, without permission in writing from the publishers.

Trademark notice: Product or corporate names may be trademarks or registered trademarks, and are used only for identification and explanation without intent to infringe.

Library of Congress Cataloging-in-Publication Data
A catalog record for this book has been requested

ISBN: 978-1-138-61337-9 (hbk)
ISBN: 978-1-138-61336-2 (pbk)
ISBN: 978-0-429-46451-5 (ebk)

Typeset in Times New Roman
by Apex CoVantage, LLC

Contents

List of Contributors   vii

1 Gender in Music Production – An Introduction   1
RUSS HEPWORTH-SAWYER, LIESL KING, AND MARK MARRINGTON

PART ONE HISTORY AND CONTEXT

2 Women in Music Production: A Contextualized History From the 1890s to the 1980s   11
MARK MARRINGTON

3 The Role of Women in Music Production in Spain During the 1960s: Maryní Callejo and the "Brincos Sound"   35
MARCO ANTONIO JUAN DE DIOS CUARTAS

4 The Representation of Women in the Twentieth and Twenty-First Centuries Trikitixa   51
GURUTZE LASA ZUZUARREGUI

5 "Hey boy, hey girl, superstar DJ, here we go . . .": Exploring the Experience of Female and Non-Binary DJs in the UK Music Scene   82
REBEKKA KILL

6 She Plays the Pipe: Galician Female Bagpipers in the Production of Local Tradition and Gender Identity   97
JAVIER CAMPOS CALVO-SOTELO

7 Rare Bird: Prince, Gender, and Music Production   120
KIRSTY FAIRCLOUGH

PART TWO WOMEN IN THE STUDIO

8 Slamming the Door to the Recording Studio – Or Leaving It Ajar? 127
HENRIK MARSTAL

9 Interview With Betty Cantor-Jackson 145
SERGIO PISFIL

10 Twists in the Tracks: An Interview With Singer, Composer, and Sound Producer Aynee Osborn Joujon-Roche 156
LIESL KING

PART THREE PERSONAL PERSPECTIVES

11 Women in Audio: Trends in New York Through the Perspective of a Civil War Survivor 175
SVJETLANA BUKVICH

12 Three-Pronged Attack: The Pincer Movement of Gender Allies, Tempered Radicals, and Pioneers 187
JULIANNE REGAN

13 Gender in Music Production: Perspective Through a Female and Feminine Lens 199
LOUISE M. THOMPSON

PART FOUR INDUSTRIAL EVOLUTION

14 Addressing Gender Equality in Music Production: Current Challenges, Opportunities for Change, and Recommendations 219
JUDE BRERETON, HELENA DAFFERN, KAT YOUNG, AND MICHAEL LOVEDEE-TURNER

15 The Female Music Producer and the Leveraging of Difference 251
SHARON JAGGER AND HELEN TURNER

16 Conversations in Berlin: Discourse on Gender, Equilibrium, and Empowerment in Audio Production 268
LIZ DOBSON

Index 285

Contributors

Jude Brereton is Senior Lecturer/Associate Professor in Audio and Music Technology, in the Department of Electronic Engineering, University of York, UK. She teaches postgraduate and undergraduate students in the areas of virtual acoustics and auralization, music performance analysis, and voice analysis/synthesis. Her research centers on the use of virtual reality technology to provide interactive acoustic environments for music performance and analysis. Until recently, she was Chair of the Departmental Equality and Diversity Committee and was instrumental in achieving the Equality Challenge Unity (ECU) Athena SWAN Bronze Award, which recognizes the department's commitment to gender equality. She is dedicated to progressing gender equality in audio engineering through innovative, creative approaches to teaching grounded in interdisciplinary research. Before beginning her academic career, she worked in arts and music administration, and is still active in promoting research-inspired music and theatre performance events combining art and science for public engagement and outreach.

Svjetlana Bukvich is a Sarajevo-born and NYC-based award-winning composer, producer, and media artist. Her works have been presented widely in the US, including at The Kennedy Center, Tribeca Film Festival, Cal State Fullerton, Ailey City Theater, Lincoln Center, and Berklee Performance Center, among others, and internationally in Beijing, London, Berlin, Johannesburg, Helsinki, Odessa, and Copenhagen. Ms. Bukvich is the recipient of numerous sponsorships and commissions, most notably from the New York Foundation for the Arts (Fellowship in Music/Sound), New Music USA, USArtists International, and the Institute on the Arts and Civic Dialogue at Harvard University. She is featured in the Jennifer Kelly book *In Her Own Words: Conversations with Composers in the United States*, where she is identified as one of the 25 outstanding women composers in America. Her only solo release to date, the genre-busting album *EVOLUTION* (PARMA Recordings), was hailed as "nothing short of spectacular" (Equal Ground). *EXTENSION*, her second collection of works, will come out on PARMA in spring 2020. The Sarajevo Philharmonic recently played a selection from this upcoming release to great acclaim. Ms. Bukvich is currently Assistant Professor at City University of New York, and has previously held positions as Adjunct Professor at Pratt Institute and New York University (NYU).

Javier Campos Calvo-Sotelo holds a double degree in history (Autónoma University of Madrid) and music (Conservatory of Madrid). He earned his doctorate in Musicology in 2008 (Complutense University of Madrid), and has formed part of several research projects on popular music, Celtology, and revival, specializing also in some areas of systematic musicology. The results of his activity have been exposed in several publications and international conferences. A relevant part of Campos's academic work is devoted to Galician culture and the bagpipe as a popular instrument with a high symbolic potential that also conveys gender issues.

Helena Daffern is currently Senior Lecturer in Music Technology in the Department of Electronic Engineering at the University of York. She received a BA (Hons.) degree in music, an MA degree in music, and a PhD in music technology, all from the University of York, UK, in 2004, 2005, and 2009, respectively, before completing postgraduate training as a classical singer at Trinity College of Music, London. Her research utilizes interdisciplinary approaches and virtual reality technology to investigate voice science and acoustics, particularly singing performance, vocal pedagogy, choral singing, and singing for health and well-being.

Liz Dobson is National Teaching Fellow of the Higher Education Academy (HEA), director of the Yorkshire Sound Women Network, composer, and Principal Enterprise Fellow in Music Technology at The University of Huddersfield. In 2012 she completed her PhD entitled "An investigation of the processes of interdisciplinary creative collaboration: the case of music technology students working within the performing arts" (with the Open University). Since 2007 she has been working at the University of Huddersfield, where she teaches modules in sonic arts, sound for media, film music composition, and empirical research for musicians. She has been developing extracurricular practices that foster enterprise creativities and build initiatives from her research, including CollabHub and The Yorkshire Sound Women Network C.I.C. Liz has a commitment to supporting girls in music technology, demonstrated through fundraising and delivering workshops and summer schools, partnering with Sound and Music, and employing internationally established sound artists to teach and work with girls across a wide range of science and music activities. Liz has a number of academic and industry conferences that focus on access, collaboration, and knowledge co-development amongst people from diverse communities. For more information, you can see her research profile here: https://pure.hud.ac.uk/en/persons/elizabeth-dobson

Dr Kirsty Fairclough is Associate Dean: Research and Innovation in the School of Arts and Media at the University of Salford, UK. Kirsty has published widely on popular culture and is co-editor of *The Music Documentary: Acid Rock to Electropop* (Routledge), *The Arena Concert: Music, Media and Mass Entertainment* (Bloomsbury), *Prince and Popular Music* (Bloomsbury), and *Music/Video: Forms, Aesthetics, Media*

(Bloomsbury) and author of the forthcoming *Beyoncé: Celebrity Feminism and Popular Culture* (Bloomsbury). Kirsty's work has been published in *Senses of Cinema, Feminist Media Studies, SERIES* and *Celebrity Studies* journals and has been featured on BBC 4, BBC Manchester, and in *The Guardian* and *Creative Review* amongst others. She is the co-curator of *Sound and Vision: Pop Stars on Film* held recently at HOME, Manchester and has recently organized, *I'll See You Again in 25 Years, Twin Peaks and Generations of Cult Television* (University of Salford, May 2015), *Mad Men: The Conference* (Middle Tennessee State University, May 2016), and the world's first conference on Prince – *Purple Reign: An interdisciplinary conference on the life and legacy of Prince*. Kirsty has lectured internationally on popular culture, most notably at the Royal College of Music, Stockholm, the University of Copenhagen, Second City, Chicago, Columbia College, Chicago, Middle Tennessee State University, Bucknell University, Pennsylvania and Unisinos, Brazil.

Dr Sharon Jagger is a specialist in gender theory and completed her PhD at the Centre for Women's Studies, University of York, a project that focused on the lives of women in the priesthood. Sharon is currently based at York St. John University, researching the university experience of trans and non-binary students and staff. Having also spent more than a decade in the music industry, Sharon combines her passion for supporting women as music makers and her academic research.

Marco Antonio Juan de Dios Cuartas is a PhD in musicology and an audio engineer. He graduated in history and music sciences at the University of Oviedo (Principality of Asturias, Spain) and then graduated in recording arts at the Middlesex University of London. He has been responsible for the academic programs of audio production and music industry at the SAE Institute in Madrid (Spain). He is a member of the Association for the Study of the Art of Record Production (ASARP) and the International Association for the Study of Popular Music (IASPM), associations in which he participates actively. As a member of the SIbE-Ethnomusicology Society, he coordinates the research group in music production. He is currently an associate professor in the Musicology Department of the Complutense University of Madrid.

Dr Rebekka Kill has worked as an academic for over two decades and has held leadership roles for around ten years. Her research interests are: festival performance, disciplinary pedagogy, practice as research, academic identity construction, and social media. Her outputs include performance art, visual practice, and written published outcomes. Her research addresses complex academic identities that are anchored in the dual contexts of discipline-specific pedagogy and practice as research in higher education. Dr Kill's career as an artist is equally varied. She originally trained as a painter and still makes visual art. Alongside this she has worked as a nightclub DJ, taught DJ skills, and has worked at venues and at music festivals.

Dr Liesl King is Associate Head of School: Creative Writing, Media and Film Studies at York St John University in York, England. Her PhD in English from the University of London, Queen Mary focuses on representations of gender and progressive spirituality in women's science fiction and fantasy in the late twentieth century. Through publications, teaching, and public engagement activities, Liesl is particularly interested in exploring the way in which science fiction enables readers to access secular and progressive spiritual perspectives. She is the caretaker of the online magazine *Terra Two: An Ark for Off-World Survival* (https://yorkstjohnterratwo.com).

Gurutze Lasa Zuzuarregui finished her degree in history at the University of Deusto in 2001. She has studied the transmission of Basque culture at the University of Mondragon (2008) and cultural management at the Public University of Navarra (2012). She holds an MA in advanced methods and techniques of historical, artistic, and geographic research from the UNED (2014). In the labor market, she has carried out technical work in scientific libraries and has taught in different Secondary Schools. She is currently a PhD candidate in the Department of Philosophy of Values and Social Anthropology (UPV/EHU), where she is preparing her thesis on the hybridization processes in the Basque culture. From 2015 to 2019 she has been a scholarship recipient to carry out the doctoral thesis of The Mikel Laboa Chair.

Michael Lovedee-Turner received a BSc (Hons) in audio and recording technology at De Montfort University (2014) and anMSc in audio and music technology at the University of York (2015). Michael is currently a PhD student in the AudioLab based in the Department of Electronic Engineering at the University of York, investigating geometry inference for convex and non-convex-shaped rooms using spherical microphone arrays and machine hearing for room acoustic analysis of binaural room impulse response. Michael has previously co-authored an Institute of Electrical and Electronic Engineers (IEEE) paper investigating gender diversity in authorship at audio engineering conferences, "The Impact of Gender on Conference Authorship in Audio Engineering: Analysis Using a New Data Collection Method". Michael's research interests include machine learning, virtual acoustic modeling, acoustic analysis, binaural audio, spatial audio, and room geometry inference.

Mark Marrington trained in composition and musicology at the University of Leeds (MMus, PhD) and is currently Senior Lecturer in Music Production at York St John University. He has previously held teaching positions at Leeds College of Music and the University of Leeds (School of Electronic and Electrical Engineering). Mark has published chapters with Cambridge University Press, Bloomsbury Academic, Routledge, and Future Technology Press and has contributed articles to *British Music*, *Soundboard*, the *Musical Times*, and the *Journal on the Art of Record Production*. Since 2010 his research has been focused on a range of music production topics with a particular emphasis on the role of digital technologies in music creation and production. Other interests include songwriting, music technology pedagogy, the contemporary classical guitar, and British classical music in the twentieth century. His most recent research has been concerned with the aesthetics of classical music recording and a forthcoming monograph on the role of recordings in shaping the identity of the classical guitar in the twentieth century.

Henrik Marstal is a Danish music scholar, writer, composer, and musician (primarily a bass player). He has obtained a PhD in musicology from the University of Copenhagen and holds a chair as an associate professor at the Rhythmic Music Conservatory, i.e. the Danish institute for contemporary popular music. As a musician and producer, Marstal has been involved with several gold-selling rock and pop acts, and he has currently released albums with ambient electronica-project (starchild #2) and dreampop (marstal:lidell). In addition, he is a keen collector of rare effects pedals and has developed a great knowledge about the uses of them for creative purposes. As a scholar, Marstal's main area of research is Western popular music and, to a lesser extent, avant-garde music since 1945. His work has appeared in numerous national and international music research and media outlets. Moreover, he has written numerous books, among these monographies on Arvo Pärt and the story of electronic music since 1900. In addition, he has co-edited a number of songbooks and has been a member of the Danish Arts Foundation. Last, but not least, he is an outspoken feminist and author of the book *Breve fra en kønsforræder* [*Letter from a Gender Traitor*] (2015).

Sergio Pisfil is a postdoctoral research fellow at the Sorbonne Université, France. His PhD, gained at the University of Edinburgh under the supervision of Simon Frith, focused on the history of live sound and its connection to rock music between 1967 and 1973. His research interests include the history, aesthetics, and social aspects of popular music, and he has been published in various edited collections and specialized journals of popular music. He is currently co-editing, with Chris Anderton, *Researching Live Music: Gigs, Tours, Concerts and Festivals* (Routledge, 2021).

Julianne Regan achieved a UK top ten hit as the singer and co-composer of the acoustic ballad, *Martha's Harbour*, in addition to eight top 40 hits with her band All About Eve. She co-wrote and performed on four studio albums, two of which reached the UK top ten album chart. Two major recording and publishing contracts later, she gained a distinction in MMus in songwriting from Bath Spa University, where she has been a lecturer in BA commercial music and MA songwriting since 2015. Choosing to maintain a low profile as a recording artist, she enjoys a collective passing on of the torch to her students.

Louise M. Thompson is an academic, blogger, activist, and accomplished music producer, known professionally as Orthentix. She is currently studying a masters in creative industries at the SAE Institute in Australia. Her practical expertise and scholarship interests include gender and music production, electronic music aesthetics, auto-ethnography, musicology, and DIY modalities of cultural production.

Helen Turner is currently Associate Head (Fine Art) at York St John University. Helen's research is mostly practice based, exploring discourses from a feminist perspective as well as stimulating experimental pedagogy, including attempting to subvert the researcher (performer)-subject power dynamic through participation. She uses ethnographic elements and is researching the creation of artefacts, songs, and rituals through participatory performance.

Kat Young received an MEng (Hons) in electronic engineering with music technology and a PhD in electronic engineering from the University of York in 2015 and 2020, respectively. Their thesis investigated the feasibility of near-field binaural loudspeaker reproduction using computational simulation. They are passionate about gender representation within STEM, particularly non-binary issues, as well as live music, cricket, and hats. They are a member of the Audio Engineering Society Diversity and Inclusion Committee.

1

Gender in Music Production – An Introduction

Russ Hepworth-Sawyer, Liesl King, and Mark Marrington

INTRODUCTION

The *Perspectives on Music Production* (POMP) series was devised to fill a perceived gap in publishing in the academic field of Music Production Studies. In particular, editors Russ Hepworth-Sawyer and Jay Hodgson felt a need for greater industry and professional representation and sought to create a series that would capture and reflect in equal measure both professional practice and academic discourse. We set out with the initial aim of publishing a collected volume on each section of the music production process as we had viewed it. Hence the first book in the series, *Mixing Music*, welcomed chapters by both practitioners and academics and set the blueprint for forthcoming edited volumes. *Producing Music* was released a year or so later, and others in the initially conceived strand, such as *Mastering Music*, are now in production. In addition, our related monograph series was inaugurated in 2019 with *Pop Music Production*, an engaging academically couched reflection on BoyBand culture and record production practice by Phil Harding (former Stock Aitken & Waterman engineer and producer for East 17). Harding's contribution, we feel, successfully epitomizes the connection between industry and academia that the POMP series was created to represent. As the identity of the POMP series has taken shape, however, it has become apparent to us that much of what has been published, or signed for publication, within the series gives little consideration to issues of gender. Only Kallie Marie's chapter, 'Conversations with Women in Music Production', which concluded *Producing Music*, has addressed such matters explicitly, and in doing so served as a significant call to action both for us as series editors and our editorial colleagues at Routledge. A discussion with our then commissioning editor, Lara Zoble, elicited warm encouragement for us to proceed, and the result is the book you now have before you.

In line with all the POMP collected volumes thus far, we have been at pains not to shape the present book with specially commissioned chapters, nor lead the discussion in any way. Our call for contributions simply suggested potential topic areas and themes – for example, gender in the workplace (including bias and stereotyping); gendered modalities of production; gender and intersectionality; case studies, interviews, and reflective pieces. However, we wanted to leave the key terms of the title 'gender' and 'music production' open to interpretation and as far as possible remain flexible regarding the potential scope of the submissions. It has been extremely interesting in these circumstances to have received the breadth of chapters presented herein. Issues concerning specific record production scenarios are here, as one might expect in a book concerning 'music production': the book does contain, for example, the expected historical accounts of female recording engineers and producers and interviews with highly successful industry personnel, as well as academically framed accounts and analysis of gender issues within specific music production practice contexts. However, it also includes contributions that foreground gender issues in the wider context of music practice within the industry – in geographically localized musical performance situations, for example, or within particular socio-cultural contexts. Another important element of this collection is the subjective position piece, which has allowed for uniquely individualized responses that are no less rigorous in their analysis than the more academically framed contributions. This has ultimately made for a diverse 'collection' of differing perspectives and a wider interpretation of the potential meaning behind the title. The book you're about to read, we hope, will consequently provide multiple useful access points to the discussion of matters of gender within the wider context of music production practice.

As with our most recent volume, *Producing Music*, we have organized this book in terms of topic areas within which it was most convenient to group the chapters received: 'History and Context'; 'Women in the Studio'; 'Personal Perspectives' and 'Industrial Evolution'. By way of contextual scene setting, Mark Marrington initiates the present collection with a concise historical sketch of the participation of women in the field from the 1890s to the 1980s. His purpose is to situate women relative to changes in production practice, placing them in close relation to the production aesthetics of particular eras (often relative to genre) so that they can be more conveniently correlated with their male counterparts. While not intending to minimize the impact that gender discrimination has historically had upon the potential for women to succeed in the industry, the chapter nonetheless aims to foreground an optimistic, perhaps even celebratory, stance in regard to the very considerable achievements of those women who were able to make careers in the broadly considered field of music production over the course of the twentieth century.

In a similar but more focused contextual vein, Marco Antonio Juan de Dios Cuartas's chapter, 'The Role of Women in Music Production in Spain During the 1960s,' discusses the first important female Spanish record producer, Maryní Callejo, and her production approach. He provides fascinating insight into the aesthetic context of record production in Spain during the 1960s, which was essentially to defer to the Anglo-American production styles of groups such as The Beatles. Callejo herself participates in this imitative aesthetic but makes an important contribution through her musical and arranging background, as well as her close involvement in the recording process itself. To clarify Callejo's role as a producer, Cuartas also usefully situates Callejo's contribution in relation to Richard Burgess's widely cited producer typologies.

Gurutze Lasa Zuzuarregui's chapter, 'The Representation of Women in the Twentieth and Twenty-First Centuries Trikitixa,' offers a detailed examination of women's representation in the Basque country's historical genre trikitixa, which in the author's words consists of "instrumental, vocal, and literary elements". The author introduces the history of the genre, which spans from the nineteenth century to the present day; outlines the chapter's aims and methodology; and importantly scrutinizes gender in trikitixa in specific musical pieces across a number of topics including romance, sexual expression, motherhood, age, marriage, and in more recent incarnations, non-heteronormative encounters. The study offers a revelatory, in-depth look at the way in which trikitixa serves as a repository for the complex and changing historical attitudes towards women, gender, and sexual identity between the nineteenth and twenty-first centuries.

Rebekka Kill's chapter entitled "'Hey boy, hey girl, superstar DJ, here we go ...': Exploring the Experience of Female and Non-binary DJs in the UK Music Scene" is best explained by Kill herself: "this chapter uses auto-ethnographic, social history and narrative methodologies to explore the experiences of female and non binary DJs working in nightclubs. ... [A]n important observation is that the male to female ratio in DJing is approximately 50:1". The chapter begins by describing the author's historical relationship with music (an obsession) and some of her performance projects, which include a TED presentation entitled 'Facebook is like disco, and Twitter is like punk', and '24/7', where she played seven-inch records in "approximate alphabetical order for 24 hours in a public space in Leeds". Foregrounding her feminist project, she explains that in hindsight she is aware that the latter work was unabashedly about the "size of her collection", her "knowledge" as a female DJ, and "resilience". The author discusses the way her interviews with fellow female DJs explore various challenges of being a woman in an arena dominated by men, highlighting difficulties such as negotiating space harmoniously in the 'tiny' DJ changeover box. The chapter introduces projects and businesses set up to offer women DJs better access to the field, such as Miss Melodie's DJ Academy, Equaliser, and the Keychange program. The writer's key aim is to illustrate the need for more and enhanced spaces for women and non-binary DJs, so that DJs of every gender/sexual identity might have the opportunity to showcase talents, obsessions, and musical know-how.

Next, Javier Campos Calvo-Sotelo's chapter, 'She Plays the Pipe,' provides an engaging historically couched discussion of the evolving situation of female bagpipers who originate from Galicia in the northwest Iberian Peninsula. His chapter begins with an overview of the socially constructed gendered traditions that have surrounded the instrument before proceeding to discuss the considerable transformations that have been brought about within the culture by pioneering female bagpipers since the 1960s. In addition to giving focused coverage to the important contributions made by the various artists profiled to the socio-cultural standing of the female bagpiper, the author also considers their relationship to the music business, the record industry, and the developing global music marketplace.

Concluding the historical and contextual focus of the first part of the book, Kirsty Fairclough's chapter, 'Rare Birds: Prince, Gender, and Music Production,' beautifully encapsulates the way in which Prince – a paradigm-shifting musical icon – not only challenged black and heteronormative concepts of masculinity through his performances, costumes, and artistry, but also created a female-centric vibration in the music production space, one which specifically championed a number of female musicians, singer/songwriters, and music producers, too. As Fairclough explains in her chapter, "Prince's fluid performances of gender as a performer seeped into the studio, creating an environment that fostered inclusivity and equality in a traditionally masculine arena". The first part of the chapter articulates the way in which Prince challenges conventional modes of being and doing (music, performance, gender) in the music business from the 1970s onwards. The second part specifically focuses on Prince's collaboration with a number of female creatives, including Susan Rogers, who "[became] known in Prince fan and critic circles for her years working as Prince's staff engineer in Minneapolis from 1983–87". Finally, Fairclough sums up crucial aspects of the gender-shifting legacy that Prince left to the entire popular music industry, as well as to his fans across the globe, adding important dimensions to our cultural memories of this rare and uniquely dynamic bird.

In the second part of the book, the focus moves to three case studies of individual female practitioners in the studio. Henrik Marstal's chapter homes in on a specific music production encounter in an auto-ethnographic reflection of his experiences of working with Danish alternative pop artist, Where Did Nora Go (the pen name of Astrid Nora Lössl). In doing so, Marstal interrogates gender relationships within the studio and the issues that can potentially arise in a creative environment. After an initial discussion problematizing the studio as a male-gendered environment, Marstal proceeds to an enlightening discussion of the dynamics of his working relationship with Lössl over the course of a series of EP and album projects. His conclusion highlights the importance of maintaining the balance of equal partnership as well as preparing oneself for the "ontological and epistemological tasks at stake" in studio-based gender collaborations.

Betty Cantor-Jackson's career as the 'Betty Tapes' engineer, tracking the Grateful Dead live shows, is represented through a fantastic interview piece by Sergio Pisfil. Through candid discussion, Pisfil enquires about the gender balance in the engineering teams at the Grateful Dead shows. The interview provides insight into the situation Cantor-Jackson found herself in and her particular approach to overcome being in a "totally male dominated situation".

Liesl King's interview with singer-songwriter and audio producer Aynee Joujon-Roche explores a story that takes place between the late 1980s and the present day, across a range of music and audio-production venues in Los Angeles, Denver, and Nashville. During the interview Joujon-Roche describes an early 'Cinderella' story in which she auditions for Star Search and is later telephoned by Glenn Frey's (*The Eagles*) assistant and flown over to Aspen, Colorado to perform – the beginning of a musical and professional journey marked by a variety of complex negotiations and transformative collaborations. The interview exposes examples of particularly hierarchical, male-dominated practices in the popular music and film industries across the period. On the other hand, Joujon-Roche describes the many occasions where she has experienced wonderfully collaborative experiences with men in audio, and offers her own understanding of what contributes positively to creatively productive situations. She explains the ideal in the following way: "Maybe you've experienced this, where you're involved in a creative process, and you're with the opposite sex, where you just become humans, musicians – making music, where I'm not looking at 'Oh, wow, those are nice breasts', or 'Wow, he's so tall and handsome, look at his jaw line'. No, we're just humans, in the same tribe, making music, and that's such a beautiful thing, when those outer things can fall away". Liesl King frames the interview by drawing on theoretical perspectives advanced by Monique Wittig (1992), Adrienne Rich (2003), and J. Halberstam (1998).

Personal perspectives are the particular focus of the third part of the book, beginning with Svjetlana Bukvich, who gives some considerable context to her international experience as a music producer, from her experiences of the Yugoslavian war and those displaced through it, her childhood in Ethiopia and later Scotland, and her eventual settling in New York. This chapter explores the author's rich history of engagement with music and music-making across her experienced geographical spaces. She explains that women's complex, complicated, and difficult histories often enable them to bring resilience to the job of making and producing music in masculine environments. Bukvich's creative, fictocritical essay draws on detailed memories of musical epiphany in order to reveal episodic insight into sonic revelation, gender relations, genre hybridity, and artistic collaborations. Relaying what happens when musicians from a range of cultural and musical backgrounds unite to create music, she says: "to me, what they are sharing ultimately is an understanding of, and a bending of time". Bukvich's narrative makes reference to classical and electronic music, progressive rock, jazz, and a range of other genres, too; yet crucial to the chapter's creative vision is the writer's point that ultimately, genre divisions can prove "stultifying". Through her closing reflections she hopes we can generate a 'win-win' situation where "men and women in the music industry will benefit from creating opportunities for working together".

Next, Julianne Regan provides a revealing and highly personal account of her journey through the music industry since the early 1980s, along the way highlighting some of the key gender-related issues she has encountered when dealing with male colleagues. She then moves to reflect on the current situation, pointing out the important role of what she calls 'gender allies' and 'tempered radicals' (a number of whom are male) in changing the culture of industry from within. At the same time, Regan reminds the reader of those earlier female pioneers – Delia Derbyshire, Kate Bush, Björk, and others – whose work has been pivotal in changing the perception of women within the industry, complementing those profiles outlined by Mark Marrington in Chapter 2.

Concluding this section, Louise M. Thompson draws on interviews with "nine female producers from North America, Mexico, Australia, and the Netherlands," providing a "trans-local perspective on the female producer and an examination of music production through a female lens". She draws on a range of feminist theorists, including Hélène Cixous, Elizabeth Grosz, and Teresa De Lauretis to help underpin her points about feminine, female, and feminist approaches to music production, and to provide a framework for understanding/deconstructing the comments made in reference to women and men by the women who are interviewed. This chapter navigates provocative theoretical terrain, as all the women interviewed here suggest they have a specifically 'feminine' approach to electronic music-making and production, citing (in sum) a preference for minor keys, aesthetically pleasant studio environments, and relaxed, supportive, non-hierarchical production styles. The author acknowledges that not all women producers conceive of themselves as feminine, but it is noteworthy that nearly all the women she interviews celebrate what we might call historically feminized approaches to music-making and recording, which the author terms 'gendered modalities'.

In the final section of the book, the focus turns to gender issues in the wider industry context, opening with a discussion by Jude Brereton and her colleagues, of the reduced uptake of music production as an area of study by females. Anecdotally, here in the UK we cannot fail to observe, each year, the continual weight of male versus female music production students higher education appears to attract. Despite decades of discussion around this disproportion, it sadly never seems to shift, despite concerted efforts by so many of us. Brereton et al. discuss this disproportion and provide us with some insight into reasons why and ways in which our community might redress the balance through a vision for the future in a detailed analysis.

Sharon Jagger and Helen Turner's chapter, 'The Female Music Producer and the Leveraging of Difference,' begins by foregrounding one London-based music studio with male toilets, devoid of tampon machines, as a way into considering whether "the physical and social world of the music production space emphasizes and reproduces sex and gender difference". The writers focus on interviews with two female music producers aged 38 and 22, respectively, which dovetail in their analysis of the way feminine approaches to music production can be used to leverage economic capital and enhance the studio experience. They explore reasons why men may continue to be the majority in educational arenas, and how autodidactic routes pursued by women producers of music often lead to feelings of insecurity and imposter syndrome. Through a deconstruction of the interviews they carry out, the two writers argue that although emotional labor can be seen as an additional burden for women in audio – used to smooth the way for all concerned – successful production depends on social finesse in addition to technological skill, resulting in high-quality artefacts produced and curated by women; therefore, audial artefacts produced and curated by women.

Finally, Liz Dobson's 'Conversations in Berlin: discourse on gender, equilibrium, and empowerment in audio production' considers qualitative data, personal experiences, and narratives that enable readers to understand some of the issues women face in audio production, and provides a path to understand how "personal and systemic sexism may be contributing to a situation where women leave professions in audio engineering". Considering interviews with 18 women in audio based in Berlin, Dobson explores the way in which women are often held to higher standards within the industry, as well as the relationship between some men's stereotyped perceptions of women's technical prowess and the concomitant result on female practitioners. She foregrounds the integral value of access to social networks in terms of facilitating economic, personal, and symbolic capital. Dobson draws on concepts such as Goh and Thompson's (2014) "inspired development of sonic cyberfeminism" to consider ways in which the industry might work from the inside out to increase the numbers of confident, empowered, technically agentic women producers of audio.

In conclusion, we would like to offer a few remarks concerning what we, the editors, regard as being the particular value of the collection of writings published here. In recent years, those closely involved in professional music production have witnessed several heated verbal discussions and debates concerning gender in the music industry, at trade events, academic conferences, and so on, as well as between professional colleagues in various situations, which have not necessarily resulted in the most positive outcomes. Indeed, in some extreme cases a perception has emerged that the foregrounding of this dialogue has set people adrift from their original relationships and collaborations – in other words, from the things that had previously connected them musically and spiritually – precisely at the very juncture when such discussion ought to have been widely inclusive. This is perhaps where a book such as this can provide assistance. Taken as a whole, the work of the writers featured in the current collection represents a potent and prescient cross-section of thought on the subject of gender in music production, available in written form for consultation. As such it provides a convenient vehicle for the more measured contemplation of a body of arguments and ideas painstakingly formulated by a range of individuals, all of whom are deeply concerned with the issues herein. At the very least, it is hoped that such a resource will afford a means of approaching the verbal debates around the various topics broached in this volume with an improved understanding of both the terms in which they are couched and the underlying assumptions that have informed them.

REFERENCES

Goh, A. (2014). Sonic cyberfeminism and its discontents. In CTM Festival & J. Rohlf (Eds.), *CTM 2014: Dis Continuity Magazine* (pp. 56–59). Berlin: CTM.

Halberstam, J. (1998) *Female Masculinity*. Durham: Duke University Press.

Rich, A. (2003) 'Compulsory Heterosexuality and Lesbian Existence'. In: *Journal of Women's History*, Volume 15, No. 3, pp 11–48. Baltimore: The Johns Hopkins University Press.

Part One

History and Context

2

Women in Music Production

A Contextualized History From the 1890s to the 1980s

Mark Marrington

INTRODUCTION

This chapter's purpose is to provide a foundational sketch of the historical participation of women within the mainstream of music production from the late nineteenth century to the early 1980s. The context of the survey is Anglo-American, in keeping with the geographical focus of the bulk of the academic literature concerning music production that has accrued to date. My choice of the 1980s as the upper limit of the time frame is intended to suggest that women have by this point become, if not necessarily a normalized presence in the field of record production, then certainly a visible one, and one that is also quantifiable in terms of a body of recorded work that is regarded as significant. Another indicator of their 'arrival' by this time is the increased commentary concerning women within the recording industry that begins to appear from the early 1990s (see, for example, Jepson, 1991; Philips, 1993; Lont, 1995; Bayton, 1998). A key observation that arises from the present survey is that women were involved in music production practice from a much earlier time than has generally been documented. For example, they were active as field recordists in the acoustic era of recording using some of the first sound capture technologies, were working in recording studios as early as the mid-1930s, and were making significant contributions to the evolution of record production aesthetics from the 1950s onwards. In order to usefully contextualize the activities of the women in question I have aimed, where possible, to situate their careers relative to developments in record production practice as they are commonly articulated in the established histories of the field. I have also made an attempt to distill, where relevant, their particular philosophies of record production and provide some indications of how they were regarded in the critical literature.

A more general aim in undertaking a survey of this nature is to raise awareness of the earlier contributions of women in the field of record production *per se* to address what appears to be a significant gap in historical knowledge. While this is certainly (as far as I am aware) the first academic chapter to attempt a narrative of this scope concerning women's historical presence within the field, it by no means claims to be fully comprehensive within the limited remit of a book section (I again emphasize that the chapter is in the nature of a *sketch*). Rather, it attempts, in the manner of a literature review, to assemble certain facts derived from currently available information in a way that enables connections to be conveniently made, provokes insights, and suggests a basis for future research. In particular my concern has been with foregrounding the creative accomplishments of notable women who have worked in the recording industry, rather than interrogating sociological factors via frameworks deriving from critical theory or gender studies. Having said that, there is no reason why the survey should not usefully inform the perspectives of writers working in these areas in the future.

FEMALE RECORDISTS IN THE EARLY PERIOD OF RECORDING

The idea of 'music production' has evolved considerably since the invention of the first recording devices and has implied a range of practices and processes over the decades. For the purposes of this chapter, music production begins with the appearance of the first sound capture technologies in the 1880s and 1890s, a period commonly referred to as the 'acoustic' era of recording. At this time women played a significant role in exploring the potential of these new technologies through their work as field recordists. Field recording – that is, the practice of using mobile recording equipment (beginning with the phonograph) to capture sound events, musical or otherwise, on location – was essential to the development of the recording industry in the late nineteenth and early twentieth centuries, being by far the most convenient and flexible means of acquiring and amassing a stockpile of recorded material that could be marketed commercially. Field recordists were by turns the first recording engineers, producers, and A&R personnel, and were highly valued for their entrepreneurial outlook and willingness to cast the net widely for material suitable for commercial or documentary use.[1] Also of importance were their interpersonal skills – in an era when recording was still regarded as a novelty, it often took considerable powers of persuasion to encourage musicians to commit their performances to wax. Historical studies of recording typically highlight figures such as Fred Gaisberg and the Sooy brothers (Gelatt, 1977; Fischer, 2012; Burgess, 2014) as early pioneers of commercial recording activity with mobile technology. In the area of documentary field recording, however, women appear to have played a more significant role. American ethnomusicologists Alice Cunningham Fletcher (1838–1923) and Frances Densmore (1867–1957), for example, were among the earliest pioneers of documentary field recording in the 1890s, being noted in particular for their recordings of Native American Indian song.[2] Densmore, who for much of her career recorded using a Columbia Graphophone (a rival machine to Edison's original Phonograph), made more than 2,000 cylinder recordings in her lifetime (Hofmann and Densmore, 1968). One of the most important field recordists of the mid-twentieth century was Laura Boulton (1899–1980), active from the 1930s, who spent 35 years of her life traveling the globe recording musicians in places as far afield as Africa, India, Southeast Asia, Japan, and the Arctic. Her 30,000 or so recordings, many of which were issued by RCA-Victor and Folkways Records, today constitute a substantial contribution to the recorded archives of global musical culture.[3] Boulton's autobiography (1969) reveals that, like Gaisberg, she operated both in an A&R-like capacity in her systematic search for recording opportunities and as a producer in her psychological approach to 'coaching' her recording subjects to give their best performances. Andrew Cordier, in his foreword to Boulton's autobiography, commented that

> she had a capacity to develop a quick and easy rapport with her hosts, whoever they might be, and thus elicited from them not only warm cooperation in the rendition and recording of music, but, as well, a flood of folk habits which gives music a meaningful setting.
> (1969: xiii)

Like many other recordists whose careers straddled the early evolution of the recording industry, she also mastered a range of technological media, from Edison wax cylinders[4] to discs to magnetic tape (Hart and Kostyal, 2003). As Boulton stated in her autobiography, "I seem to have lived through the history of recording for I think I have tried every method and material known" (1969: 27). Field recordists such as Boulton, Densmore, and later contemporaries such as Henrietta Yurchenco[5] (1916–2007), illustrate the synthesis of the creative and technical aspects of music production unique to the early period of recording that were later to become separated as the recording industry became increasingly systematized. There is also an interesting parallel here between their activities and the present context of women engaging in autonomous 'self-production' with current forms of mobile recording media (laptops, DAWs etc.), as pointed out by contemporary writers such as Wolfe (2019), which may merit further exploration.

WOMEN AS STUDIO-BASED PRODUCERS AND RECORDISTS: TWO EARLY INSTANCES

Women first begin to become involved in studio-based record production from the 1930s onwards, both in the capacity of producers and engineers. 'Producer' here refers to a role that had by this time crystallized to entail a range of responsibilities, among the more typical of which were decisions concerning who and what was to be recorded (essentially an A&R remit), the organization of recording sessions, and the employment of musical expertise during the recording process to critique standards of performance and interpretation. In some cases producers also contributed ideas on how a recording ought to sound, although the achievement of any particular objectives in this regard usually required collaboration with specialist recording personnel – the engineers (or recordists) – who possessed the relevant technical know-how. The demarcation between these two quite specific territories of music production practice remained pronounced until the 1960s.

The routes by which women entered the field as producers were varied and often the result of quite specific circumstances. For example, Toronto-born Helen Oakley Dance (1913–2001), arguably the first female jazz producer, worked as a journalist for *Down Beat* and as a concert promoter for jazz artists. In 1934 she had moved from Canada to the United States to seek out the live music scene, establishing the Chicago Rhythm Club as a vehicle for promoting public concerts with such luminaries as Benny Goodman, Duke Ellington, and Fletcher Henderson. Her recognized expertise within the field coupled with her immersion in the live scene were instrumental in securing her first recording sessions at the Brunswick label's studios in Chicago, with artists such as Paul Mares and his New Orleans Rhythm Kings (Placksin, 1982). Many of these sessions were funded by the income generated from the concerts Dance promoted. However, her big career break as a record producer came when she moved to New York in 1936 to work in A&R for Irving Mills' short-lived Master and Variety labels. In this role Dance instigated and produced numerous recording collaborations between many different jazz artists of the era. Ward and Huber (2018: 84) state that her remit lay in "assembling experienced, often extremely well established recording artists, supplying them new material and new session-mates to inspire their creativity, and allowing them to record in new combinations that rarely threatened prior contractual obligations".

In regard to the engineering context, it was much more unusual to find women working in technical areas such as sound capture, mixing, or disc cutting during this period, mainly due to the nature of the hierarchical systems that governed employability in studios at this time. This is summarized by Kealy (1979), who uses the expression 'Craft-Union mode' to refer to an ethos of engineering practice that crystallized during the 1940s in which recording was undertaken by specialists and governed by particular rules and regulations. The Craft-Union context of professional engineering can be seen to have directly informed the career of Mary Shipman Howard,[6] a notable female recordist active from the early 1940s until the early 1950s.[7] Howard was a classically trained musician who from a young age had also been fascinated by electronics and sound: "since I always loved acoustic mechanical things the process of translating a sound wave into an electrical impulse and back into sound I got into recording" (Perlis, 1974: 209). Eager to pursue a career in recording, Howard moved to New York in 1940 and applied for an engineer's position at NBC (National Broadcasting Company) but was barred from accessing such a role on the grounds of her gender. As Howard recalled: "it was unusual for a woman to be a recording engineer, particularly as far as the union was concerned" (Perlis, 1974: 209). Instead, Howard was hired as a secretary, but this was a short-lived role – with the increased need for manpower overseas following the entry of the United States into the war, the union reversed its decision and engaged her as a disc cutter. This enabled Howard to gain valuable studio experience through, for example, the opportunity she had to assist on the groundbreaking series of recordings made by Toscanini with the NBC Symphony Orchestra for RCA Victor (Perlis, 1974).

Howard appears to have become quickly dissatisfied with the employment ethos of NBC, however, and shortly after the war established her own recording studio at 37 E. 49th St., New York. By the end of the 1940s, she had left NBC to make this the main focus of her recording activity. To complement her own disc-cutting skills, she hired sound recordist Don Plunkett (1924–2005), later a founding member of the Audio Engineering Society, as Chief Recording Engineer (Benzuly, 2005).[8] The typical 'Services Offered' by Howard's studio, as listed in *Radio Annual* 1949, were:

> Off-the-air and off-the-line recordings. Commercial records, transcriptions, all studio facilities. Package shows and spots. Tape recording and editing facilities. (Tape To Records – Records To Tape.) Recording all audio ends of TV shows.
> (Alicoate, 1949: 765)

In addition to offering general recording services to the industry, Howard also produced and released records by a small number of popular musicians on her Mary Howard Recordings (MHR) label. These were cut to 78 rpm disc, the dominant format until the early 1950s, and featured the company's distinctive music-themed logo. A flavor of Howard's recording work can be heard on her first commercial release, the Chittison Trio's *Album No. 1*, recorded in 1947, whose six sides capture the vibrancy of the trio performing jazzed-up arrangements of classical pieces. In the same year, she also recorded the popular African-American singer Ethel Waters (1896–1977), performing standards by Gershwin, Berlin, and others to piano accompaniment.[9] Howard is also notable for the informal recordings she made in 1943 of the composer Charles Ives playing excerpts from his *Concord* Sonata for piano, which have since come to be regarded as an important historical document.[10]

Howard's uniqueness as a female recordist brought her to the attention of *Newsweek* (Anon, 1947) and the trade literature, such as *Audio Record* (1948), which provide a revealing document of her views on the recording profession. In particular, she advocated for the improvement of recording techniques in the USA, and was keen to raise awareness of the importance of the recording engineer to the success of the production process. In a 1948 interview for *Audio Record*, she argued for an holistic approach to production in opposition to the prevailing Craft-Union set-up:

> Unfortunately, the interest and ingenuity of the recordist has often been overlooked. Recording is not a dull craft at all if engaged in all its technical phases. There seems to be a prevalence in large organizations for specialization cutting technicians, studio technicians, maintenance, etc. which often results in poor recording because of lack of interest or information in all phases of the recording operation. If

interest and enthusiasm were carried all the way through the recording organization, and management, perhaps time might he found to raise the general recording standards in America.

(*Audio Record*, 1948: 4)

Howard's work as an independent recordist can be understood in relation to the expansion of the recording industry in the United States during the 1940s and 1950s. As increasingly affordable technology allowed smaller operations to compete with the majors, these decades saw the appearance of many independently run (and often short-lived) labels that enjoyed success at a local level. In New York these included names like Musicraft, Allegro, and the Spanish Music Center (SMC), all of which contributed in their own small way to the heritage of recorded music during this period. In such an environment, entrepreneurial women like Howard were able to sidestep the hierarchical constraints of the large companies such as NBC and pursue their careers with considerable freedom. Another female beneficiary of this climate, who also established an independent recording studio in New York, was Bebe Barron (1925–2008). Barron's introduction to the field was sparked by the acquisition of an early Stancil Hoffmann tape machine in the late 1940s at a time when the technology was still not widely available (Holmes, 2015). Together with her husband Louis, she established one of the earliest American electronic music studios in Greenwich Village, among whose many pioneering projects were the recording and editing of the material for John Cage's landmark collage-based electronic composition, *Williams Mix*, in 1952, and the creation of the entirely electronic music score (the so-called electronic tonalities) for the film *Forbidden Planet* in 1956. The field of electronic music continued to provide opportunities for women to work within studio environments during the 1960s.

HIGH-FIDELITY RECORDING AESTHETICS IN THE 1950S AND THE INNOVATIONS OF WILMA COZART

The emergence of Wilma Cozart (1927–2009) in the late 1950s signals the increasing involvement of women at the cutting edge of commercial record production. Through her collaboration with engineer-husband C. Robert Fine, Cozart made significant and innovative contributions to the development of classical recording aesthetics with Mercury, a small American independent label founded in 1945. Cozart studied music and business at North Texas State University before embarking on a career in orchestral management as a secretary to conductor Antal Dorati (best known for his work with the Minneapolis Symphony). She joined Mercury in 1950 and initially worked in an A&R capacity, augmenting the classical music profile of the label through her brokering of recording deals with unknown or neglected US orchestras of the period, including the Chicago Symphony, Minneapolis Symphony, and Detroit Symphony. By 1956, in recognition of her marketing achievements, Cozart had been made a Vice President of Mercury (unique for a woman at the time), which provided a vehicle for her to expand her remit into the area of record production and participate directly in the recording sessions. From this point she became a key contributor to the evolution of the celebrated Mercury 'Living Presence' classical recording aesthetic,[11] whose principal concern was with fidelity – namely, to duplicate (in recording) and replicate (in cutting the consumed recording) as accurately as possible what was played in the recording session (Ranada, 1991: 96).

Central to the Living Presence sound was the 'minimal' miking technique developed by C. Robert Fine, which involved the careful positioning of a single omnidirectional Telefunken microphone (a Neumann U-47, and later the Schoeps 201M) above the sound source, augmented in the stereo era by two additional microphones on either side. The Cozart-Fine team also placed an emphasis on capturing the interaction of the natural ambience of the room in which the recording was taking place with the sound source in question. Cozart commented that "the room is also very much a part of the performance. . . . We always thought that that was a very, very important part of the naturalness of sound" (Ranada, 1991: 96). This approach was aligned with a general trend in the 1950s away from recording classical music in the studio, which often produced lifeless and unrepresentative results, towards the use of concert halls and similar venues prized for their acoustic properties (Swedien, 2009; Schmidt-Horning, 2013). To this end the Cozart-Fine team focused on securing acoustically desirable venues around the US and in Europe, including Chicago Hall, Carnegie Hall, Old Orchestra Hall (Detroit), Cass Technical High School Auditorium (Detroit), and Watford Town Hall (in the UK). The overt concern of the Mercury production team with audio quality was also reflected in the presentation of its discs, whose sleeve notes were always partially devoted to explaining in technical terms how each particular recording had been achieved.[12]

In the studio, Cozart was designated an 'Executive Producer', a role which involved coordinating recording sessions, monitoring the progress of takes, and discussing aspects of the recording process with musicians. In addition, she worked closely with C. Robert Fine on matters of microphone placement and participated at the console in the sound mixing process (Gray, 1989a). In an interview in 1991, Cozart recalled her approach to mixing Mercury recordings in the early stereo era, revealing both her appreciation of the technical process and the importance of her musical background to its success:

> We started to record in stereo in 1955 because we knew that was coming. We quickly decided we needed three mikes to get the full sound stage and the spread of the orchestra. All the microphones were omnidirectional. The centre was our monaural mike. When I determine placement I always go to the centre first, because if the centre microphone is in focus and in place you're home free. Then you can add the two sides. Then what you do is 'ghost' the centre back to each side. This is not an engineering task, you have to have a musical person

doing this. At the sessions I always monitor all three channels with no mixing. Then in structuring the final result I have to have the score in front of me and put the illusion of the centre back in. Which means that the clarinets have to sit where they sat, that the flute be heard where he sat, the bassoon and the percussion and double basses have to be where they were. If you don't get that right the whole thing falls apart.

(March, 1991: 1485)

Cozart has arguably received the most sustained critical recognition for her achievements over the decades, and as a result has become highly regarded within audio circles. For example, during the heyday of the Mercury label in the 1950s and 1960s, she maintained a consistently high profile in the industry and was frequently mentioned in audiophile literature, particularly the US publication *High Fidelity*[13] magazine. Then in the 1990s, the project to remaster the entire catalogue of Mercury Living Presence recordings for CD (overseen by Cozart until her death in 2009) led to a revival of interest in the production techniques of the Mercury team.[14] During this period Cozart re-appeared in numerous features and interviews (March, 1991; Ranada, 1991), which served to further solidify her reputation (along with that of C. Robert Fine) as a major innovator in record production.

FEMALE PRODUCERS IN GREAT BRITAIN AFTER 1950

In Great Britain women first began to achieve recognition as record producers in the 1950s, as illustrated by the work of Olive Bromhall (1909–2002) and Isabella Wallich (1916–2000), both of whom, like Cozart, worked in the field of classical recording. Bromhall was a trained musician, graduating from the Royal Academy of Music as a piano teacher in 1930,[15] and her career in the recording industry began when she joined the staff of the Education Department at EMI in the early 1950s. Established in 1919 in the days of the earlier Gramophone Company, the Education Department was primarily concerned with promoting the gramophone as a pedagogical aid through the development of audio resources designed to support the teaching of musical appreciation (Wimbush, 1969). Many of the classical production assignments that Bromhall undertook for the organization over the next few years were historical in nature, as reflected the remit of the Education Department. A major project with which Bromhall was involved in the late 1950s was the *History of Music in Sound* collection, a long-running collaboration between HMV and Oxford University Press. Beginning in 1957 this involved the creation of a series of LPs charting the evolution of music from ancient times to the present, each accompanied by a volume of notated music. Bromhall is identified in the Foreword to the first volume of the series, in which she is thanked for the 'gathering of materials', suggesting that repertoire research was a key part of her production remit (Wellesz, 1957). She subsequently worked on a series of similar historical and education-related productions thereafter, including *Instruments of the Orchestra* (1962), *Music of Shakespeare's Time* (1963), *From Plainsong to Polyphony* (1966), and selections for the five-volume *Treasury of English Church Music* (1966).[16] Among Bromhall's earliest production credits are three recordings of the counter-tenor Alfred Deller performing music of the English Baroque composer Henry Purcell (1659–1695), suggesting this may perhaps have been one of her musical specialisms. These were undertaken at Abbey Road Studios in 1951, and as was typical of the Craft-Union style hierarchy of Abbey Road at this time, Bromhall was assisted on these sessions by various EMI 'house' engineers, including Francis Dillnutt, Robert Beckett, and Harold Davidson.[17]

Isabella Wallich, who can be regarded as Cozart's counterpart in Great Britain, was undoubtedly the most significant British female classical music producer of the mid-twentieth century. Surprisingly, despite a career that lasted into the 1980s, she is hardly mentioned in contemporary accounts of record production history, even where classical recording is concerned,[18] and her detailed autobiographical account published shortly after her death in 2000 (*Recording My Life*) provides the main source for her life's work. Wallich was, like Bromhall, a trained musician, and had a brief career as a professional concert pianist performing recitals and touring during the pre-war years. Significantly she was also a niece of Fred Gaisberg, the aforementioned pioneer record producer who was pivotal in the rise of the Gramophone Company in the early twentieth century. She was therefore immersed from an early age in the recording world and exposed to many influential musicians and industry personalities. Wallich's first encounters with the studio occurred through the opportunities she had to sit in on Gramophone Company recording sessions overseen by Gaisberg in the 1920s. Referring to these sessions, Wallich wrote in her autobiography:

> My greatest joy was to be invited to the studios, when Uncle Fred would allow me into the engineer's room to watch the delicate operation of cutting the grooves into the recording wax. . . . Even at a very young age, I was absolutely fascinated by the process of recording, and was proud to be allowed to attend the sessions.
> (Wallich, 2001: 22).

Later she formed an association with Walter Legge, the pre-eminent British classical record producer of the 1940s and 1950s, who had also been a protegé of Gaisberg. Legge was instrumental in progressing Wallich's career in the early 1950s when he appointed her as the manager of his Philharmonia orchestra, leading to a tour of Europe with conductor Herbert Von Karajan. During this period Wallich also received an offer of employment from EMI as a classical producer. She turned this down, however, and in 1954 established her own record label, Delysé, which became the main focus of her record production work. Like Cozart, Wallich took an innovative approach to developing an identity for the Delysé label by focusing on untapped areas of repertoire and unknown artists. In the early period of the label's history, for example, she made Welsh and Irish music the focus of her recording strategy, while in her later classical music endeavors, she undertook a pioneering series of recordings of Gustav Mahler's music at a time when the composer was little known to the general public. Among the musical artists that Wallich was instrumental in bringing to wider public attention through her recordings were the baritone Geraint Evans, the mezzo-soprano Janet Baker, the conductor Wyn Morris, and the classical guitarist John Williams.

While Wallich undoubtedly had a good understanding of the intricacies of the recording process, she was not a recording engineer herself and instead enlisted specialist engineering personnel to work alongside her. By the late 1950s, Wallich had established a fruitful working relationship with Allen E. Stagg, an engineer at British independent recording studio IBC.[19] The choice of Stagg reflected Wallich's concern to work with technical personnel who were not affiliated to any of the major labels, for reasons explained in her autobiography:

> I could have placed the technical side of the operation into the hands of either Decca or EMI, but I decided against this because I wanted to be in complete control and entirely responsible for the atmosphere which I wanted to achieve.
>
> (2001: 146)

Wallich's desire for 'complete control' reflected an awareness of the important role that engineering processes played in defining the sonic character of a recording – by avoiding the established methods of the competing major labels, she would be more likely to produce something unique. As regards the 'atmosphere' she wished to obtain, in an echo of the Cozart-Fine philosophy, Wallich was adamant she did not want 'a studio sound' and instead took pains to locate suitable halls for all of her recordings. For her first released disc (*Welsh Folk Music*, EC 3133), Wallich chose Conway Hall in Red Lion Square, which became a regular venue for Delysé recording sessions thereafter. Her affiliation with Stagg, a purist who was opposed to doctoring classical recordings with EQ or compression, was also key to the shaping of the Delysé aesthetic,[20] and Wallich worked closely with him to set up recording sessions to achieve the sound she was looking for. Wallich's approach to production is well documented in her autobiography, particularly regarding the sessions for two of her most acclaimed Mahler orchestral song recordings, *Des Knaben Wunderhorn* (1966) and *Das Klangende Lied* (1967). Both records were made in Watford Town Hall, a venue still prized for its fine acoustic characteristics, which was purposefully chosen by Wallich for *Das Knaben Wunderhorn* because of the "free, almost out-of-the-door sound that the music required". Wallich recounts her detailed work with Stagg in the control room "to fine-tune the orchestral balance" for this recording, with numerous 'trial takes' to test microphone positioning. This painstaking approach to pre-production ensured that little further editing was necessary after the recording sessions. The recording of *Das Klagende Lied* was a somewhat more complex affair, which involved careful positioning of the singers "so that the listener would be able to imagine the position of the voices in relation to one another and the text" (2001: 200). In addition, there was the problem of positioning of a band in relation to the singers to give the impression that the sound was emanating from the interior of a castle. Rather than solve this problem through artificial means – for example, by splicing the music in later from a separate recording – Wallich and Stagg took the decision to record the band live with the orchestra, "positioning this group of musicians behind the conductor, as far away as possible within the hall" to achieve a "naturally distant sound". This attention to detail in 'staging' of the recording to enhance the listening experience immediately brings to mind the innovative production strategies of John Culshaw when recording the highly celebrated Wagner Ring Cycle with Georg Solti at Decca during the same period.

Wallich was extremely well regarded in the heyday of her career, and her productions were reviewed frequently in the British trade literature. In a 1972 *Gramophone* retrospective of Wallich's achievements, Roger Wimbush wrote:

> in her work as a producer she has accepted the most daunting challenges and has been responsible for many remarkable recordings. It is not easy to catch the essential frisson of massed Welsh choirs, military bands and children's voices. In the solo field it required the faith born of experience to launch John Williams on record long before the great guitar boom. Her recordings of Mahler have been acclaimed in the face of the fiercest competition in the business, and here her inherited and cultivated flair saw the potential of Wyn Morris, who directed Janet Baker and Geraint Evans for her in songs from *Des Knaben Wunderhorn* and then went on to record *Das Klagende Lied*.
> (Wimbush, 1972: 1199)

WOMEN IN MUSIC PRODUCTION IN THE 1960S AND 1970S

The climate of the recording industry as regards its accessibility to women by the late 1960s is summarized, somewhat bleakly, by Lont (1995), who notes that although in this period more women were trying to develop the relevant skills to enter the music business – namely, production, engineering, and management – they remained largely 'unwelcome'. She also asserts that during the 1970s, "major recording companies kept women out of the business end of the industry", with any woman who wanted to work in the "business" "most often placed as a secretary or in publicity" (1995: 326). In reaction to this, some women were prompted to form their own independent collectives, as Lont explains:

> In these early years, a group of women musicians and women engineers unable to get work in the mainstream music industry joined

> political activists in the lesbian-feminist movement to form a "women's music" independent recording industry. The industry included women's music recording labels (Olivia), performers and musicians (Meg Christian, Cris Williamson, and Margie Adam), engineers (Joan Lowe, Leslie Ann Jones), producers (June Millington), album designers (Kate Winter), and photographers (JEB). The albums were distributed via mail order or through women's bookstores to thousands of women who wanted music that realistically portrayed their lives.
> (Lont, 1995: 326)

Lont (writing in 1995) suggests that the 'women's music industry' in effect began with these activities in the early 1970s and "continues today as a successful alternative form of music in the United States". There was also another 'fringe' field of studio-based activity in which women were beginning to make headway at this time – electronic music, a technology-intensive area of experimentation that was more typically the province of public broadcasting companies, academic institutions, and private inventors. Among the women pursuing careers in this field during the 1960s were Alice Shields, who composed tape music at Columbia-Princeton Electronic Music Center (a University-sponsored hotbed of avant-garde electronic music experimentation), and Delia Derbyshire, who made important contributions to the recorded output of the BBC Radiophonic Workshop (Brend, 2005, 2012).[21] Also at this time composers Linda Fisher, Suzanne Cianni, and Pauline Oliveros were independently evolving innovative and highly personal uses of the newly available Buchla and Moog synthesizers (Trocco and Pinch, 2004). Part of the appeal of electronic music to women, perhaps, was that it afforded the individualistic pursuit of creative goals through experimentation with technology whose potential was not yet fully codified in practice. To put it another way (alluding to the appeal of 'self-production' discussed by Wolfe), rules delineating technological specialism had not become entrenched in the way they had in the commercial recording industry, meaning likely fewer constraints on creativity.

While there is much truth in Lont's observations regarding the outlook for women wishing to enter the mainstream recording industry during this period, it is nonetheless the case that a small number were successful in pursuing careers within this environment from the 1960s onwards. These women joined the industry at a time when the division between specialist handlers of recording technology (engineers) and those who directed them in its creative use (producers) was still very much in place. This meant that women generally tended to enter the business from the traditional A&R producer perspective, rather than as engineers. However, it was also during this period that the boundaries between the technical and the creative aspects of production were being gradually eroded, particularly in the field of popular music. On the one hand, recording artists were now becoming interested in participating in the production process (see Kealy, 1979), while on the other hand, producers desired to expand beyond their traditional A&R remit to become more involved in areas usually reserved for the technician. This blurring of boundaries can be seen to have informed the work of women working in both popular and classical recording production.

In the popular music sphere, the 1960s saw the emergence of Helen Keane (1924–1996), an important jazz record producer, and Ethel Gabriel (b. 1921), who produced a wide range of popular and easy listening music for RCA. According with Lont's account of the employment situation for women in this field, Keane's first job in the music industry was as a secretary for MCA in the 1940s. However, the organization possessed enough hierarchical flexibility to permit her to work her way up to the position of 'female talent agent', in which capacity she was responsible for a number of major signings, including Andy Williams and Harry Belafonte. After undertaking a similar role with CBS Television acquiring talent for the Ed Sullivan and Gary Moore shows, she left to set up her own management agency, and from around 1962 began a long-term association with jazz pianist Bill Evans (Pettinger, 1998). From 1966 Keane was both manager and full-time producer of Evans's work, taking both an A&R approach in her persuading him to expand the scope of his recording projects beyond what was typical for jazz artists at the time (Schroeder, 2013), and in becoming increasingly more involved in the recording and production process itself. Such was Keane's technical knowledge of the recording process that Nat Hentoff, who observed her working with pianist Joanne Brackeen, remarked that she "possessed the audio engineering skills to take over the control room" (Placksin, 1982: 275). In an interview (Dahl, 1984: 247), Keane has qualified this, acknowledging the demarcation between herself and the engineer in the studio and recognizing the relationship as essentially collaborative in that the engineer is directed by the producer towards achieving the sound she is after, but "*his* ears are going to be working if yours get tired". Among the most valuable accounts of Keane's career as a manager-producer for Bill Evans are Pettinger (1998) and Shadwick (2002), while interviews in Dahl (1984) and Gourse (1996) are more revealing as regards her production philosophy. On the role of the producer-manager synthesis, for example, Keane has remarked that:

> The ideal way to function as a manager is to be the producer. They are two separate functions, but the manager really knows more about the artist than anyone else – his or her creativity, life, habits, how disciplined or undisciplined they are when they work, what music they like best, how they choose their material, how they like to record. Therefore the manager can obviously be the best producer.
> (Dahl, 1984: 247)

Ethel Gabriel enjoyed a 43-year career as a producer for RCA Victor, leaving a substantial legacy of recordings, and serving as an inspiration for women to succeed in an environment that was often hostile to their presence. The practical difficulties surrounding Gabriel's career, who, in her capacity as A&R producer, "invaded one of the most hallowed male areas of the music business" are discussed in revealing detail in Lucy

O'Brien's book *She Bop II* (2002). Gabriel joined RCA in 1940 as an A&R manager and record inspector working at the label's plant in Camden, New Jersey, and in an echo of Bromhall's career path, was later transferred to New York to work in the educational record department. Her move into production occurred in 1955 when she was entrusted by RCA's then president, Mannie Sacks, with producing Mexican bandleader Perez Prado's US hit 'Cherry Pink and Apple Blossom White' (Mayfield, 2001). On the strength of this success, she went on to produce recordings for key artists of the era such as Chet Atkins, Cleo Laine, Perry Como, and Roger Whittaker. One of Gabriel's most notable RCA projects was the 'Living' series of easy listening albums, released on RCA's economy label Camden, during the 1960s (Lanza, 1994; Aldinolfi and Pinkus, 2008). These were pioneering productions in the Muzak idiom, designed, in Gabriel's words, "for a person who loves music, but at the very moment you get too classy with them, it's over their head" (Lanza, 1994: 90). Her approach on the *Living Strings* albums (which won her a Grammy in 1968) was essentially to adopt pop production approaches to the processing of the classical string orchestra sound, through the use of echo. Gabriel commented that:

> Echo was important back then. Before we got the German echo chambers in the early 'sixties, we found that the best echo was through the men's room on RCA's studios on 24th Street. When Toscanini recorded at places like Manhattan Center, they would channel his music through it and pipe the sound into the studio.
> (Lanza, 1994: 90)

As regards the specifics of Gabriel's approach as a record producer, while interviews and features have offered some sense of her detailed and perfectionist attitude towards the recording process (see, for example, McDowell, 1983), in comparison to what we know of Cozart and Wallich, this has unfortunately received only sporadic attention.

Women also continued to make a significant impact in the field of classical music during this period, as exemplified by the important triumvirate of producers that emerged in the early 1970s: Patti Laursen (1927–2013), Joanna Nickrenz (1936–2002), and Eleanor Sniderman (b. 1920). Laursen, who played a key role in developing the classical music profile of Angel Records (the classical division of Capitol Records), began her career in classical record retail, before being employed as a distributor for the Vox classical series. She also spent a period working as a station librarian and programmer for KFAC-AM, a major classical music broadcaster based in Los Angeles. These early experiences were pivotal in building the extensive knowledge of the classical repertoire that formed the bedrock of her later production expertise. In 1963 she joined Capitol Records, working with the Angel label as an assistant to producer Robert E. Myers, and within only four years had risen to become a producer herself (Dexter, 1974). As a producer Laursen worked with many iconic classical artists, including Christopher Parkening, Leonard Pennario, Itzhak Perlman,

Angel Romero, and Yehudi Menuhin, helping them to establish identities that would enable them to succeed in the classical music marketplace. For example, she assisted guitarist Christopher Parkening (with whom she worked on many recordings) in developing a distinctive profile at a time when the marketplace for classical guitar music had reached saturation point. In his autobiography (2006), Parkening states that Laursen's knowledge of classical music made her an excellent adviser on the repertoire choice for his albums, encouraging him to move away from well-established potpourri programs to musical projects of a more conceptual nature. Laursen received two Grammy nominations for her work with Parkening – specifically for *Parkening and the Guitar* (in 1977) and a later album *Pleasures of their Company* (1986).

Joanna Nickrenz was a classical pianist by training who had distinguished herself as a recording artist in the late 1960s.[22] She entered the world of record production when she was hired by engineer Marc Aubort to work for Elite Recordings, a small freelance recording outfit best known for its work for the Vox and Nonesuch labels.[23] This launched a long and fruitful career in classical recording, which brought her widespread recognition and many awards. She received her first Grammy ('Best Engineered Classical Recording') in 1974 for her work with Aubort on the album *Percussion Music* with the New Jersey Percussion Ensemble (Nonesuch, H71291). Eleanor Sniderman (née Koldofsky) was a key figure in the establishment of the classical recording scene in Canada. She began her career at Boot in the early 1970s, a label associated with country music, which was then branching out to explore the marketplace for Canadian classical music. Her first production (for the Boot Master Concert Series) was the debut album of the classical guitarist Liona Boyd, which was recorded at the iconic Manta Sound Studios in Toronto, a venue in which Sniderman undertook much of her production work during her career (Boyd, 1998). In 1975 she left Boot to establish Canada's first classical label, Aquitaine (inspired by her idol, twelfth-century queen Eleanor of Aquitaine), and proceeded to build a roster of Canadian classical artists, including Gisela Depkat, Arthur Ozolins, Victor Schultz, and Alan Woodrow.

Like Cozart and Wallich before them, all three women involved themselves to varying degrees in the recording process, often being initiated into this area via close collaboration with male engineering colleagues. Laursen, for example, formed a long-term relationship with Robert Norberg until she retired from Capitol in the late 1980s, which influenced her particular 'purist' classical recording aesthetic of "keep it simple. No plethora of microphones spread all over the place; no multitrack gimmicks" (Smith, 1991). Laursen's musical background also informed her detail-oriented approach to recording: as Smith, who observed Laursen undertaking a recording session for Harmonia Mundi, commented, she "followed every note of the score. No intonation slip escaped her notice; no weak attack or blurred passage got past her" (Smith, 1991). Laursen's career also encapsulated the introduction of digital recording in the late 1970s, a development which she embraced wholeheartedly (Terry, 1979). Nickrenz became closely involved with the intricacies of studio practice early in her career when engineer Marc Aubort undertook to train her in recording techniques. This enabled her to successfully straddle the line between a more conventional musically focused producer role and that of the studio technician (Fremer, 2010). In particular, Nickrenz was highly skilled in the area of editing – that is, the splicing and cutting together of multiple takes of performances – a process requiring acute aural awareness and considerable precision. Remarking on her editing ability, William Bolcom, a pianist who worked with Nickrenz on a number of recordings for Nonesuch, described her as an "absolute whiz . . . probably the best in the business" (Harvith and Harvith, 1987: 306). Nickrenz's consummate skill as an editor was a reflection of her general view that recordings stood as a permanent record of an artist's capabilities and therefore ought to be as technically perfect as possible. Joan Morris summed up Nickrenz's production philosophy in the following terms:

> They really go for technical perfection. "Well," Jo says, "you know you're going to be listening to this for the next seventy-five years. If that note's a little sharp and you don't bother about it – 'Oh, I'm too tired to do it' – that's going to bother you for seventy-five years. You might as well, if you can, have it as perfect as possible".
> (Harvith and Harvith, 1987: 296)

While such a detail-oriented approach to the construction of recordings can be seen (relative to the era) as a more progressive recording philosophy, at least where classical music was concerned, in other respects the Nickrenz-Aubort team erred on the side of tradition in their audiophile recording aesthetics. For example, the Nonesuch team took a similar approach to Mercury in their insistence upon minimal miking and the use of concert halls and churches with high-quality acoustics to make their recordings (Horowitz, 1973; Fremer, 2010).

In contrast, Sniderman took a more progressive attitude towards recording classical music, preferring the studio environment to location recording and willing to use the technological resources of the studio (such as artificial reverberation and tape splicing) to shape her productions. Her most significant (and controversial) recording, which was also a benchmark in large-scale classical music production for the era, was the 14-LP set of Beethoven's complete piano sonatas, recorded (between 1975 and 1977) by the Vienna-born Canadian pianist Anton Kuerti (b. 1938). For the recording, Sniderman close-miked the piano,

> in an acoustically dead studio just large enough to contain Kuerti and his piano. The dry sound was passed through an echo chamber on its way to the tape, then fed back to Kuerti through the headphones as he played.
> (Hathaway, 1977: 253)

This unusual approach, which had more in common with popular music recording aesthetics, was seized upon by some critics as a travesty. For example, reviewing the discs in 1977, Thomas Hathaway claimed that the dead studio environment robbed the instrument of the acoustic resources necessary to accurately portray its colors, while the effects distorted the balance of frequencies and made the instrument sound too close. He also argued that the set-up influenced Kuerti's performance style in a negative manner, producing dynamic emphases where they were not necessary. This was a minority perspective, however and elsewhere Sniderman's innovative production approach was held in high esteem by the audiophile recording community, as evidenced by a cable she received from the preeminent European classical recording label of the era, Deutsche Grammophon, congratulating her on her achievement (Dzeguze, 1979). Sniderman also received a Juno Award for the Kuerti recordings in 1978.

DISCUSSION: HISTORICAL FACT AND CONTEMPORARY OPINION

My aim in the preceding commentary has been to foreground the careers of some of the most widely accomplished, but largely undocumented, female record producers of the last century, with a view to providing a foundational narrative for contextualizing the writing on the subject that has emerged since the 1990s. In one sense, this is intended to serve the general purpose of augmenting the existing accounts of women's presence within the narrative of music production history. For example, the information in this chapter might usefully complement Barbara Jepson's 1991 article on the situation of women in the classical recording industry, bridging the careers of Cozart, Laursen, and Nickrenz (who are also acknowledged by Jepson) with their younger contemporaries, such as Judith Sherman, Elizabeth Ostrow, and Elaine Martone. Or it may function to expand on the tidbits of information concerning the earlier history of record production, as exemplified by Susan Schmidt-Horning's pioneering 2013 history *Chasing Sound*, which, within its own remit, can only give brief attention to Mary Howard's activities in the 1940s or the achievements of Wilma Cozart in the 1950s.

It may also reveal the shortcomings of the extant historical accounts of music production, which in general have not taken the broad approach to the field necessary to capture the presence of its female participants. Part of the problem relates to the fact that much of the writing on the subject, whether couched in historical or theoretical terms, has tended to delimit the field to the territory of popular music, which itself is defined narrowly in reference to a mostly male canon of iconic producers. Hence, as will have been observed in the cases of those women working in specialist or fringe areas such as jazz and easy listening – namely Helen Oakley Dance, Helen Keane, and Ethel Gabriel – it has been necessary to consult a range of disparate sources in order to piece together the circumstances of their careers. Classical recording has also been largely excluded from the music production studies literature, with the exception of historical accounts such as Gelatt (1977) or Day (2000)[24] and autobiographical commentaries such as Gaisberg (1942), Culshaw (1982), and Schwarzkopf

(1982). In general, there has been little systematic documentation of the work of classical producers per se, which may reflect the common assumption that in classical recording the producer role, by contrast to the 'producer-as-creative-agent' perspective that is widely recognized in popular music (see Moorefield, 2005), is not usually elevated above the artist and the musical interpretation.

The survey has also drawn attention to particular themes that might be profitably explored in further research. In particular, as this chapter has shown, a number of the women discussed – Olive Bromhall, Isabella Wallich, Wilma Cozart, Eleanor Sniderman, Joanna Nickrenz, and Patti Laursen – all worked as producers in the field of classical music. This has been an area of record production in which women have clearly been able to flourish and make important contributions, but this fact that has so far received little acknowledgement. As Patti Laursen noted, in her response to Chuck Philips' 1993 article highlighting the ongoing problem of the accessibility of the record business to women:

> In Chuck Philips' excellent article on the growing number of women in executive positions in the recording industry ("You've Still Got a Long Way to Go, Baby," April 18), and thank goodness for that, he did not mention, nor perhaps know, that the leader in this area has been the classical recording industry. . . . Perhaps our abilities and professional dedication helped top label executives understand that women have made, and are making, a significant contribution. The recording industry will be the richer for it.
>
> (1993: 83)

The reasons for the apparent success of female record producers in the classical field clearly merit further exploration, and there are certainly indications in the earlier literature as to the directions this potentially might take. Comments made by certain interviewees in Jepson (1991), for example, suggest that classical recording environments may have generally been more hospitable to female producers. Judith Sherman, while acknowledging that the industry is competence-oriented, states that: "The only time I feel a rub is when I go into a pop studio – in the pop world, women are a commodity" (1991: 345), while Alison Ames comments that "We're better off in the classical side than rock and roll" (1991: 344). These are qualified statements, in that Jepson's article acknowledges that the industry ethos at the time of writing was generally challenging for women. However, it indicates that there are potential differences in studio culture based upon genre that obviously should not be ignored when considering these issues.

On a related point, it is also hoped that the foregoing survey might usefully inform the reading of gender studies accounts (usually focusing on popular music) that have tended to problematize male-gendered studio culture (see, for example, Bayton, 1998; Leonard, 2007)). The culmination of this critique in the recent writing of Wolfe (2012, 2019) is to advocate a re-location from such culture into situations of individualized 'self-production' in order "to disrupt the gendering that, historically, has taken place within the field" (2019: 24). What the historical record demonstrates, however, is that women have been able to function successfully within the traditional studio context, many of them working in productive collaborative relationships with their (usually) male engineering colleagues, whose expertise they harnessed in the service of their production visions. Furthermore, the various women discussed in this chapter have not indicated (at least as far as what they have put on record about their careers is concerned) that they were intimidated by the technological context within which they were working. Rather, like their male counterparts, they were stimulated by it and made a success of their careers by using the available technological resources as a vehicle for achieving their creative production goals. Certainly no one would deny that circumstances of their music production activities were to varying degrees conditioned by the gender-based distribution of studio roles that obtained during the era in which they were working, and it would be naive to suggest that entrenched values concerning gender did not in particular cases throw up significant obstacles to career progress (as directly experienced by Mary Howard with the NBC union example). At the same time, however, an over-reliance on reductionist hegemonical interpretations of gendered power to evaluate the nature and scope of their contributions may obscure the importance of their very considerable achievements. As the successes of these women clearly illustrate, the recording industry cannot have been an entirely unassailable male fortress.

To summarize, women were active in the recording industry in a range of capacities as early as the 1890s, and this was by no means a peripheral or tentative development: the women in question were highly successful and influential in the field. In several of the cases discussed, they contributed significantly to the evolution of recording practice with their ideas on production and their willingness to innovate with emerging technologies for recording. Given the current high-pitched rhetoric that continues to problematize the recording industry's attitude towards women, one could be forgiven for assuming that the pioneering work to establish women within the field was just now beginning. As this survey has demonstrated, however, this is far from the case, and it is hoped that the preceding attempt to re-write the aforementioned women into the received history of record production will provide a useful basis for future research.

NOTES

1. See Western (2018) for an insightful essay on the ways in which field recording can be regarded as a unique area of early music production practice.
2. Fletcher was assisted in her recording projects by Francis La Flesche (1857–1932), a notable professional Native American ethnomusicologist.
3. Major collections of Boulton's recordings are housed at Columbia and Harvard Universities.

4. The use of wax cylinders remained popular with field recordists working in the 1930s due to the portability of the recording equipment. In her autobiography (1969: 17), Boulton wrote: "To have preserved anything at all with that early equipment was something of a miracle".
5. Henrietta Yurchenco pioneered field recording in Mexico in the 1940s. Originally trained as a pianist, Yurchenco had begun her career as a broadcaster on New York radio station WNYC, where she had gained a reputation for 'weird' programming, in reference to her playlists of American folk and world music. In 1941 she left this position to take up residence in Mexico with her husband Basil Yurchenco, and over the next few years toured that country and Guatemala for recording opportunities with local tribes (Hart and Kostyal, 2003; Yurchenco, 2003).
6. Later sources refer to Howard by her married name, Pickhardt.
7. See Sutton (2018) for a succinct summary of Howard's career.
8. See Schmidt-Horning (1999) for a fascinating interview in which Plunkett discusses his time working with Howard.
9. The Chittison and Waters recordings can currently be heard in remastered versions on *The Chronological Classics: Herman Chittison 1945–1950* (Classics 1334) and *The Chronological Classics: Ethel Waters 1946–1947* (Classics 1249).
10. They were commercially released in the *100th Anniversary* box set of Ives' works in 1974 (Columbia Masterworks M4 32504).
11. This expression came from a New York critic's comment on the first released Mercury LP that experiencing the recording was like being "in the living presence of the orchestra".
12. See, for example, the sleeve notes for two of the label's most innovative recordings, the Tchaikovsky *1812 Overture* (1959) and *The Civil War, Its Music and Its Sounds* (1958).
13. See, for example, a lengthy *High Fidelity* article by Shirley Fleming (1961) documenting Mercury's stereo re-recording of the *Civil War* album.
14. The most recent remastered box set compilation, *Mercury Living Presence: The Collector's Edition 3*, was issued by Decca in 2015.
15. Confirmed by a brief mention in the *Musical Times* dated June 1, 1930.
16. Little has been written about Bromhall from a critical perspective, but it has been possible to discern the circumstances of her career from liner notes, occasional references to her activities in periodicals, and informal discussions with EMI archival personnel. In particular I am grateful to Lester Smith, Ken Townsend, Tony Locantro, and Malcolm Walker for the assistance they provided in enabling me to confirm certain facts concerning Bromhall's work at EMI.
17. The recordings are 'Retired from after any mortal's sight'; 'Thus to a ripe consenting maid' and 'Hark, how all things'. They were re-issued on The HMV Treasury series in 1982 as HLM 7234.
18. She is noticeably absent, for example, from Timothy Day's *A Century of Recorded Music*, other than a brief citation in reference to Fred Gaisberg.
19. Stagg became General Manager of Abbey Road Studios in 1967. For further discussion of IBC and Stagg's activities, see Massey (2015). See also Wimbush (1967) and Vinyl House UK (2016).

20. Stagg's particular attitude toward classical recording is outlined in his technical notes reproduced on the sleeve of the Mahler *Des Knaben Wunderhorn* LP, which state that "no artificial echo, equalisation or screens were used at any time, the aim being to record as natural a sound as possible".
21. Established in 1958 at Maida Vale, West London. The composer Daphne Oram (1925–2003) was a co-founder of this organization.
22. Nickrenz specialized in performing contemporary music. See, for example, her 1968 recording with the Claremont Quartet of works by Schoenberg, Webern, and Stravinsky (Noneusch H71186).
23. Nonesuch was also notable for its female director, Teresa Sterne, who was instrumental in transforming the label from a small European outlet to a leading exponent of progressive music.
24. Day mentions Isabella Wallich briefly, but this is only in reference to her uncle Fred Gaisberg.

REFERENCES

Adinolfi, F. and Pinkus, K. (2008) *Mondo Exotica: Sounds, Visions, Obsessions of the Cocktail Generation*. Durham: Duke University Press.

Alicoate, J. (ed.) (1949) *The 1949 Radio Annual*. New York: Radio Daily Corp.

Anon (1947) 'Woman with a Disk' In: *Newsweek* 30 (26) 29/12/1947 p. 42.

Audio Record (1948) 'The War Gave Mary Howard Her Big Chance to Make Good in Recording; She Did – And How!' In: *Audio Record* 4 (2) p. 1, 4.

Bayton, M. (1998) *Frock Rock: Women Performing Popular Music*. Oxford: Oxford University Press.

Benzuly, S. (2005) In: *Rememberance*: Donald J. Plunkett, 1924–2005. At: www.mixonline.com/recording/rememberance-donald-j-plunkett-1924-2005-377355 (Accessed 23/08/2019).

Boulton, L. (1969) *The Music Hunter: The Autobiography of a Career*. New York: Doubleday.

Boyd, L. (1998) *In My Own Key: My Life in Love and Music*. Toronto: Stoddart.

Brend, M. (2005) *Strange Sounds: Offbeat Instruments and Sonic Experiments in Pop*. San Francisco, CA: Backbeat.

Brend, M. (2012) *The Sound of Tomorrow: How Electronic Music Was Smuggled into the Mainstream*. New York: Bloomsbury Academic.

Burgess, R. J. (2014) *The History of Music Production*. Oxford: Oxford University Press.

Culshaw, J. (1982) *Putting the Record Straight*. New York: Viking Press.

Dahl, L. (1984) *Stormy Weather: The Music and Lives of a Century of Jazzwomen*. New York: Pantheon.

Day, T. (2000) *A Century of Recorded Music: Listening to Musical History*. New Haven and London: Yale University Press.

Dexter Jr, D. (1974) 'Laursen: Angel's Angel – Fem Producer Runs Her Own Artists Stable' In: *Billboard* 05/10/1974 pp. 4, 82.

Dzeguze, K. (1979) 'Eleanor, for the Record' In: *MacLean's* pp. 14–18.

Fischer, P. D. (2012) 'The Sooy Dynasty of Camden, New Jersey: Victor's First Family of Recording' In: *Journal on the Art of Record Production* 7. At:

www.arpjournal.com/asarpwp/the-sooy-dynasty-of-camden-new-jersey-victor%E2%80%99s-first-family-of-recording/ (Accessed 30/09/2019).

Fleming, S. (1961) 'Gettysburg in Stereo' In: *High Fidelity* pp. 56–59.

Fremer, M. (2010) *Elite Recordings: A Conversation with Freelance Recording Engineer Veteran Marc Aubort*. At: www.analogplanet.com/content/elite-recordings-conversation-freelance-recording-engineer-veteran-marc-aubort-0 (Accessed 22/08/2019).

Gaisberg, F. W. (1942) *The Music Goes Round*. New York: The Macmillan Company.

Gelatt, R. (1977) *The Fabulous Phonograph, 1877–1977*. New York: Collier Books.

Gourse, L. (1996) *Madame Jazz: Contemporary Women Instrumentalists*. Oxford: Oxford University Press.

Gray, M. (1989a) 'The Winged Champion: Mercury Records and the Birth of High Fidelity (Part I)' In: *The Absolute Sound* (60) pp. 47–59.

Gray, M. (1989b) 'The Winged Champion: Mercury Records and the Birth of High Fidelity (Part II)' In: *The Absolute Sound* (61) pp. 46–56.

Hart, M. and Kostyal, K. M. (2003) *Songcatchers: In Search of the World's Music*. Washington, DC: National Geographic.

Harvith, J. and Harvith, S. E. (1987) *Edison, Musicians, and the Phonograph: A Century in Retrospect*. New York: Greenwood Press.

Hathaway, T. (1977) 'New Recordings: Kuerti, Schnabel and Toscanini' In: *Queen's Quarterly* 84 (2) pp. 252–257.

Hofmann, C. and Densmore, F. (1968) *Frances Densmore and American Indian Music: A Memorial Volume*. New York: Museum of the American Indian, Heye Foundation.

Holmes, T. (2015) *Electronic and Experimental Music: Technology, Music, and Culture*. New York and London: Routledge.

Horowitz, I. (1973) 'To Elite, Quality Speaks for Itself' In: *Billboard* 27/01/1973 p. 42.

Jepson, B. (1991) 'Women in the Classical Recording Industry' In: Zaimont, J. L. et al. (eds.) *The Musical Woman: An International Perspective Vol. 3: 1986–1990*. Westport, CT and London: Greenwood Press, pp. 337–352.

Kealy, E. R. (1979) 'From Craft to Art the Case of Sound Mixers and Popular Music' In: *Work and Occupations* 6 (1) pp. 3–29.

Lanza, J. (1994) *Elevator Music: A Surreal History of Muzak, Easy-Listening, and Other Moodsong*, Revised and Expanded Edition. New York: Picador.

Laursen, P. (1993) 'Women in Music' In: *Los Angeles Times* 02/05/1993 p. 83.

Leonard, M. (2007) *Gender in the Music Industry: Rock, Discourse and Girl Power*. Aldershot: Ashgate.

Lont, C. M. (1995) *Women and Media: Content, Careers, and Criticism*. Belmont, CA: Wadsworth Publishing Company.

March, I. (1991) 'Mercury Returns' In: *Gramophone* 02/1991 pp. 1484–1485.

Massey, H. (2015) *The Great British Recording Studios*. Milwaukee, WI: Hal Leonard Corporation.

Mayfield, G. (2001) 'Bob Jamieson and Jack Rovner' In: *Billboard* 12/06/2001 p. 64.

McDowell, E. (1983) 'The Art of Fine-Tuning a Recording' In: *New York Times* p. H18.

Moorefield, V. (2005) *The Producer as Composer: Shaping the Sounds of Popular Music*. Cambridge, MA and London: The MIT Press.

O'Brien, L. (2002) *She Bop II: The Definitive History of Women in Rock, Pop and Soul*. London: Continuum.

Parkening, C. and Tyers, K. (2006) *Grace Like a River*. Carol Stream, IL: Tyndale House Publishers.

Perlis, V. (1974) *Charles Ives Remembered: An Oral History*. New York: W.W. Norton & Co.

Pettinger, P. (1998) *Bill Evans: How My Heart Sings*. New Haven and London: Yale University Press.

Philips, C. (1993) 'You've Still Got a Long Way to Go Baby' In: *Los Angeles Times* 18/04/1993 p. 9.

Placksin, S. (1982) *American Women in Jazz: 1900 to the Present: Their Words, Lives, and Music*. New York: Seaview Books.

Ranada, D. (1991) 'Art and Technology: Passing the Torch' In: *Musical America* 111 (3) p. 96.

Schmidt-Horning, S. (2013) *Chasing Sound: Technology, Culture, and the Art of Studio Recording from Edison to the LP*. Baltimore: JHU Press.

Schmidt-Horning, S. (1999) *Interview with Don Plunkett* 9/02/1999. At: https://nunncenter.net/ohms-spokedb/render.php?cachefile=2016oh215_chase045_ohm.xml (Accessed 23/08/2019).

Schroeder, D. (2013) 'Produced by Helen Keane' In: *IAJRC Journal* 46 (2) pp. 14–17.

Schwarzkopf, E. (1982) *On and Off the Record: A Memoir of Walter Legge*. London: Faber & Faber.

Shadwick, K. (2002) *Bill Evans: Everything Happens to Me, a Musical Biography*. San Francisco, CA: Backbeat.

Smith, T. (1991) 'Philharmonic Wraps Up Recording Session' In: *South Florida Sun Sentinel* 03/11/1991. At: www.tribpub.com/gdpr/sun-sentinel.com/ (Accessed 27/12/2018).

Sutton, A. (2018) *American Record Companies and Producers 1888–1950: An Encyclopedic History*. Denver, CO: Mainspring Press.

Swedien, B. (2009) *Make Mine Music*. Milwaukee, WI: Hal Leonard.

Terry, K. (1979) 'Digital process still limited classical producers claim' In: *Cashbox* 27/10/1979 pp. 16, 52.

Trocco, F. and Pinch, T. J. (2004) *Analog Days: The Invention and Impact of the Moog Synthesizer*. Cambridge, MA: Harvard University Press.

Viny House UK (2016) 'Linda Stagg On Her Father Allen Stagg's Career & Legacy' At: www.youtube.com/watch?v=kbZ2qYRfzYM (Accessed 30/08/2019).

Wallich, I. (2001) *Recording My Life*. London: Sanctuary Publishing, Limited.

Ward, B. and Huber, P. (2018) *A&R Pioneers: Architects of American Roots Music on Record*. Nashville, TN: Vanderbilt University Press and Country Music Foundation Press.

Wellesz, E. (ed.) (1957) *The History of Music in Sound Vol. 1: Ancient and Oriental Music*. Oxford: Oxford University Press.

Western, T. (2018) 'Field Recording and the Production of Place' In: Bennett, S. and Bates, E. (eds.) *Critical Approaches to the Production of Music and Sound*. New York: Bloomsbury Publishing USA.

Wimbush, R. (1967) 'Allen Stagg to EMI' In: *Gramophone* 10/1967 p. 198.

Wimbush, R. (1969) 'Here and There: Education' In: *Gramophone* 09/1969 p. 377.

Wimbush, R. (1972) 'Delysé' In: *Gramophone* 01/1972 p. 1199.

Wolfe, P. (2012) 'A Studio of One's Own: Music Production, Technology and Gender' In: *Journal on the Art of Record Production* 7. At: www.arpjournal.com/asarpwp/a-studio-of-one%E2%80%99s-own-music-production-technology-and-gender/ (Accessed 12/09/2019).

Wolfe, P. (2019) *Women in the Studio: Creativity, Control and Gender in Popular Music Sound Production.* (s.l.): Routledge.

Yurchenco, H. (2003) *Around the World in 80 Years: A Memoir.* Point Richmond, CA: MRI Press. At: www.music-research-inst.org/html/pubs/yurchenco.htm (Accessed 16/08/2019).

3

The Role of Women in Music Production in Spain During the 1960s

Maryní Callejo and the "Brincos Sound"

Marco Antonio Juan de Dios Cuartas

INTRODUCTION

In the flourishing economic environment of the late 1950s, an authentic restructuring within the Spanish music industry took place. Record catalogues would now no longer focus exclusively on folklore, copla, or classical music and instead began to take advantage of the new political and social developments that would soon encourage investment in this sector. The musicologist Celsa Alonso (2005) highlights in her study on Spanish beat the importance of the popular music of the 1960s in the social and cultural changes of "developmentalism". The author raises the interpretation of the Spanish beat as a gamble of the "incipient record industry that saw an interesting market among young middle-class Spanish people" (Alonso, 2005: 229). Although this phenomenon of Anglo-Saxon musical importation could be approached from an exclusively commercial perspective, we cannot ignore the fact that, beneath this new music for the youth, a set of "institutional strategies whose objective was to build an image of normality, modernity and economic, social and cultural renewal" were also hidden (Ibid.). It is an indisputable fact that the new musical proposals at the beginning of this decade were a response to the need for expression, rebellion, and leisure of a youth markedly influenced by their Anglo-Saxon referents and the institutional use of this phenomenon. More importantly, Spanish beat music represented a new market for a record industry that had to respond to this phenomenon in two directions: adapting human resources on the one hand and technical resources on the other.

Within this new environment of social and cultural renewal, in which the incursion of rock and pop music took place, as in other countries such as England or the United States, the figure of the musical producer in Spain is consolidated and linked to the processes of record production within recording studios, which themselves acquire an increasing prominence as creative agents, with evident consequences for the music's sound.

The music industry in Spain during the last years of Franco's rule took advantage of the international opening raised by the political regime in order to establish contacts with recording studios, engineers, and producers from different European capitals, particularly London, Paris, and

Milan. One of the names associated with the foundations of the figure of the producer in Spain is Alain Milhaud. He developed his work with the Columbia label in Madrid and is associated with productions of Spanish bands made in English studios. This reflected a dissatisfaction on the part of some producers with Spanish studios, which were still far from possessing the technological means of achieving the new sound of rock emerging from abroad, which was being made in studios that were no longer simply spaces to fix a musical interpretation, but creative tools in themselves. During this process of change, not only the technical resources of the studio had to be updated, but an adaptation of the different professional profiles – namely, audio engineers, producers, or session musicians – was also necessary. If at the beginning of the 1960s it was unusual to find the figure of the music producer clearly defined during a recording process, the fact of finding a woman developing this role in a recording studio was nothing short of groundbreaking. According to the study conducted by Gil Calvo (1989), from 1962 to 1966 the labor insertion in Spain of women between 20 and 24 years was reduced to 29.3% of the total.

It is within this special social and cultural situation in Spain that María de las Nieves Callejo Martínez-Losa, known artistically as Maryní Callejo, began her professional activity. An excellent musician, precocious discoverer of talent through her work in A&R for the record company Zafiro, and arranger and musical director, we must consider Callejo the first female producer in the history of record production in Spain. Callejo represents the case of the musician of academic training who accesses the recording and publishing industry, evolving and finally fixing her position as a record producer. The study of the professional role of Maryní Callejo entails a great difficulty because there is a complete lack of information concerning her professional career. For example, neither the writer and guitarist Salvador Domínguez, in the chapter devoted to the first Spanish producers from his book *Bienvenido Mr. Rock* (2002), nor the specialized music press have documented the important work of Callejo within the Spanish music industry of the 1960s. Furthermore, the absence of information about Callejo on the Internet is equally significant, and it is reduced to a few commentaries by some of the musicians who worked for her. This fact confirms the need to definitively address her professional contribution to the Spanish music industry and gives a great relevance to the personal interviews that have been made for the preparation of this article.

MARYNÍ CALLEJO: A HISTORIC MILESTONE IN THE INCORPORATION OF WOMEN IN THE SPANISH RECORD INDUSTRY

Maryní Callejo was born in Madrid in 1943 to a middle-class family with no previous tradition of musicians. She studied piano, composition, and orchestral conducting at the Royal Conservatory of Music in Madrid, acquiring a solid classical education. She finished her piano studies with honors and was granted, by the director at that time, Jesús Guridi, a scholarship to continue her studies in Italy and Germany. When her father became ill, she decided to renounce her scholarship and stay in Madrid. During her adolescence, when she lived in Ferraz Street in Madrid, she often went to the nearby Pintor Rosales Street, where she met other groups of teenagers with whom she played and, thus, began to have her first taste of popular music. At that moment, these groups were beginning to play in the so-called society parties that took place among the bourgeois families and the aristocracy of the time. Maryní remembers her first performance at Guadalupe School in Madrid and what this first performance meant for her professional beginning within pop music:

> On the day of Guadalupe they had a party at the school . . . and I went with my friends. There was a grand piano there and, before the party started, I begun to play the songs that we performed on the street. At that moment, Augusto Algueró and Carmen Sevilla appeared, who were to be the stars of the party. When they heard us, Augusto suggested that we should go to see his father the following day. So, the next day I went to see Augusto Algueró's father and, from that moment, he became my second father.[1]

Augusto Algueró and Carmen Sevilla represented the most successful artistic couple in Spain during that time. Augusto was a composer of recognized talent and projection within the Spanish music industry and Carmen, one of the most successful singers since the decade of the 1950s. The compositions of both Algueró father and son had a significant impact on the publishing world, becoming part of the repertoire of important Spanish artists of the time, and appearing in advertising jingles for radio and television. The result of this first contact with the Algueró family was the consolidation of what would be the first professional musical project by Callejo as a musician: Los Brujos, one of the first Spanish groups of "modern music" at the dawn of pop music. From Los Brujos, recognized singers like Luis Gardey or Tito Mora were to emerge, who subsequently went on to have important careers. Working with Augusto Algueró senior's publisher, Callejo's first group recorded a dozen works whose national impact served to consolidate her professional career. This contact with the Algueró family allowed Callejo to begin working in the publishing world, cataloguing scores and making musical arrangements for different artists. It was also at this time that she began to work as an orchestral director in the recordings of string sessions organized by the publisher for music for advertisements. Callejo's contact with recording studios became constant from the time of her experience with Los Brujos, whose work ranged from recording everything from jingles to choirs for the films of successful singers and actresses of the time, such as Marisol.[2] These jingles were recorded in Estudios Celada at América Avenue, later renamed as Estudios Kirios in a new location on the outskirts of Madrid, which today constitutes one of the most important studios in the history of record production in Spain. Other studios in which Callejo worked on both Los Brujos recordings and the advertising jingles were the modest studio located at Ramiro de Maeztu School and Columbia studios on Barco Street.

However, the beginning of Callejo's professional career as a producer is also closely related to the Zafiro record company and its sub-label Novola. Her access to the record company was motivated after listening to the Spanish band Los Brincos, the subject of her debut as a musical producer. The meeting with Los Brincos involved a firm commitment to a band and a sound, which had unquestionable repercussions for music in Spain:

> At that time, I worked in the mornings with Augusto Algueró (junior) in the publisher that his father had in Carmen Street. And I was offered to go to Zafiro in the afternoons to see if I liked it. I started to classify things in the offices of the record company and, then, one of the directors, Esteban García Morencos, proposed me to meet a group of young boys. I, who came with the ideas of classical music, didn't have high hopes about what I was going to find. We went to the place where they rehearsed, which was in Iberofón factory, next to the Manzanares River. That group was Los Brincos. They started to play and at that moment I said to Mr. Esteban: I'm going to leave Augusto; I've liked the group a lot and I think we can do wonderful things. I worked there for 3 years.

This first album by Los Brincos marked the inception of the beat sound in Spain, under the direction of Callejo, who was chiefly responsible for the recording.

1964: RECORDING OF THE FIRST ALBUM BY LOS BRINCOS AND BEGINNING OF CALLEJO'S CAREER AS A MUSIC PRODUCER

The relationship of Callejo with Los Brincos and the widespread use of the term "the fifth Brinco" leads us to a necessary comparison between this case and that of The Beatles, whose model it was also intended to imitate. Within the functional classification of producers carried out by Burgess (2013), the case of Callejo is adapted, like that of Martin, to that of the producer-collaborator. Callejo is at the forefront of the production process, although exhibiting a flexibility that allows her to extract the maximum value from the artist's ideas. On the back cover of the first album by Los Brincos, Callejo wrote the following words about the experience of working in the studio with the band:

> There are many groups dedicated to 'ye-ye' but, unfortunately, very few that make music in this genre and not just electronic noise accompanied by distorting movements. Except for the 'vedette' groups, which take care of the instruments, repertoire, assembly of songs, etc. down to the last detail, there are many who are uneducated, who scratch discs and discs of these until they manage to copy them! Now

we are finally lucky: a Spanish group, Los Brincos, has decided not to copy anyone. All the songs they perform are their own compositions, with a frankly good commerciality and style. It is difficult to be impartial in a commentary when those who are being commented turn out to be great friends; however, in this case it is very easy for me, because the truth is obvious. . . . Because of the position that I occupy within the Novola Company, I have had to take charge of the musical direction of this recording. To be honest, I must say that, at first, I thought this was going to be just another group, but I changed my opinion after I listened to them. In truth, what they play and sing is in perfect harmony, there are no 'strange' chords (the so-called 'by ear'); everything is adjusted to a very correct technique. From the first moment I listened to them, I grew very fond of this recording, to the point of looking like another Brinco. Today, after finishing this LP, we are proud that in Spain we have such an exceptional group that, hopefully, will mark a pattern that many others will follow. For now, go ahead Brincos! Triumph awaits you because you deserve it.[3]

In this text, Callejo defines her work as "musical direction of the recording", leaving aside the technical part for which an audio engineer would be responsible. Indeed, her position within Novola was that of musical director, and the term musical producer was not yet regularly applied to describe this figure during the first half of the 1960s. At this time the term "producer" is associated exclusively with the "phonographic producer" and affiliated to the company in the recording credits. Callejo is not mentioned as a producer in the credits of any of the two albums by Los Brincos produced for Novola, although in her later stage at Philips (as the producer of Fórmula V, another successful Spanish band from 1968), the singles as well as the EPs and LPs begin to be labelled with the roles: "Arrangements, direction and production: Maryní Callejo".

The role of the producer is from that moment linked, at least in Spain, to the musical concept, to the arrangement and, above all, to the authorship of the music rather than the sound. The limited technical knowledge of our first producers and their limited experience in professional studios explains their lack of prominence in technical actions, where this function was delegated to an engineer who had to translate his sound concept at the mixing console. Callejo's text alludes to some stereotypes that confront pop music with rock, such as "commerciality" versus "authenticity" (not copy), which in this case is objectively not very credible considering the pursued admired English sound. Musical perfection is shown through the absence of "strange" or "by ear" chords. In spite of her training, there is an approach to the musicians, highlighting a "perfection" at the harmonic level that conforms to the patterns of Callejo's classical training. The fact of being considered another Brinco brings Callejo's work in the recording studio closer to the role of a producer-collaborator, her role at the musical level being unquestionable, although in no case does it eclipse the intentionality and the aesthetic values of the band. But perhaps we should also analyze the figure of Callejo using the functional typology of Burgess

(2013), within the producer-consultant. Callejo empathizes with the band to the point of living the experience of recording the album with the same enthusiasm as those four young people who were facing the adventure of entering a recording studio for the first time. In this sense, she remembers:

> We all got in my Seat 600: the four Brincos, the 'electric monkey', who was really called Miguel Ángel but everyone knew him by this nickname and was the technician Los Brincos had, a boy who was 16 or 17 years old and who helped with the cables (that's why they called him "the electric monkey"). As we did not have much money, we went first to the fast food restaurant Rodilla in Callao Square and there we collected the money we had to buy some sandwiches and a soda and pay the gas needed to go to RCA.

The album is developed, therefore, within an atmosphere of cordiality and fraternity between "the musicians" on the one hand and "the person responsible for the production of the record" on the other, taking into account that, unlike the case of The Beatles and George Martin, musicians and producer belonged to the same generation, and both had the same inexperience in the recording studio. The fact that a woman faced the leadership of a project of these characteristics in the Spanish society of the first half of the 1960s is highly significant and unique in the music industry of the time. There are very few examples of women who have been at the forefront of relevant positions in the Spanish music industry: Myriam Von Schrebler (mother of the music producer Carlos Narea) in RCA, Carmen Grau in Zafiro within the record companies, and Rosa Lagarrige in management or Daniela Bosé in the world of publishing.

As regards the recording of the first album by Los Brincos in RCA studios, Callejo comments: "We recorded in stereo, but the stereo was composed of a track with bass, drums and guitars, and then we mixed everything into a track and, therefore, there was again a free track for voices and choirs". Callejo objectively adapts to the profile of producer-musician and producer-arranger who delegates the technical work to a trusted engineer: "Although I'm very interested in what the recording is in itself, I've always had my engineers". The first album by Los Brincos would be recorded under the technical supervision of José María Batlle and in accordance with an organizational chart made by engineer, assistant, and music producer.[4] The album by Los Brincos represents a unique case in which a Spanish band has the opportunity to tackle a recording that emulates the English sound but with original musical ideas, against the usual tendency in which, as musical critics like Jesús Ordovás point out, record companies demanded bands to make versions of The Beatles, The Animals, or The Rolling Stones – this was the situation of bands like Los Mustang, Lone Star, or Los Salvajes: "Under the pressure of their companies, Spanish groups will only be able to record a song of their own between cover and cover version" (Ordovás, 2010: 11).

In the case of Los Brincos, promoted by Fernando Arbex as part of an artistic evolution that begins with the pioneering rock-and-roll band in

Madrid Los Estudiantes, the idea was to create a group in the image of The Beatles.[5] From the perspective of musical production, the evolution of Los Brincos over the course of their albums is very significant and worthy of analysis: a first LP recorded in Madrid, a second LP recorded in Milan, and the third and fourth LPs recorded in London. Unlike The Beatles, whose company maintained an infrastructure that made the Abbey Road studios available to the band, Los Brincos belonged to a company that did not have its own recording studios and that forced – or perhaps we should say, allowed – them to record in different studios inside and outside our borders.

The selection was made not only based on the technical qualities of the recording studio, but also by the audio engineer working in it: in the case of Callejo's productions, the change from Saar to Fonit Cetra studios – where the first single by Fórmula V was produced – was determined because the engineer Plinio Chiesa, in whom Maryní trusted the technical part of her recording productions, moved there. In any case, the decision to record the second album at Saar studios in Milan responds to the search for technical parameters that were not found in Spain:

> In Milan, we went to record on 4 Telefunken tracks, and thus we could put the battery and the bass on a track. Being able to independently raise [the volume of] the battery and the bass seemed incredible to us. On another track, we put the two guitars. The voices were in another track and the soloists, in another.

The decision to produce the second album by Los Brincos in Italy was therefore determined by what has been called the multitrack war and the search for full-range frequency quality. The Radiotelevisión Española historical archive possesses images of the recording of the first album by Los Brincos in the old studios of RCA record company – which later would become a property of Philips – in Madrid.[6] Manolo González, bassist in Los Brincos, recalls in an interview his experience of working on this recording with the sound engineer and with the music producer:

> The studio hired by Zafiro, which did not have its own studios, was that of Fonogram, which belonged to Polydor [both labels were business-related with Philips, which was the one who acquired the old recording studios that belonged to RCA]. It was placed in Madrid, on América Avenue, very close to CEA studios, in what was then called Barajas Motorway. The engineer who had to record us was José María Batlle, a nice guy as well as an excellent professional who, from the very first moment, took the job as if he were one of the group. The production was carried out by Maryní Callejo. She could be considered as the fifth Brinco, given her great enthusiasm and dedication to the project.[7]

The record companies that did not have their own recording studio, as in the case of Zafiro, had to necessarily reach an agreement with other record companies in order to make their recordings. The recording of the album takes place, therefore, in the studio of another company, in a moment in which professional studios belong, in general, to the record companies. The recording made in RCA[8] studios, placed on América Avenue in Madrid, takes place during a period of transition in which the studios of this company are being acquired by the record department of the Philips firm. The magazine *Billboard* documented on September 22, 1962, in an article signed by the music journalist Raúl Matas, the signing of the agreement in the following terms:

> Philips Records has acquired the RCA studio and pressing facilities here in a move to increase its importance in Spain. Negotiations have been top secret, but it is believed the terms will say RCA will continue to use the setup for the present, and Philips will take over the large installation on Airport Highway, only a few miles from here, in the near future. Philips has also completed negotiations with Siemens in Germany and will handle Polydor, Coral and Brunswick and DGG labels in Spain. Julio Sampedro will continue as general manager with José María Quero and Ricardo F. de La Torre continuing as A&R execs. Philips will also increase promotion and distribution of Mercury and affiliated labels, and its jazz titles acquired through the Interdisc organization. These include such American jazz independents as Blue Note, Riverside and Contemporary.

The acquisition of RCA studios by Philips determined the transfer of the facilities and staff from Delicias Street to América Avenue in a clear commitment of the company for the development of its record department. Curiously, the studio in which the first project directed by Maryní Callejo was recorded would end up being the administrative location of her future productions, although the latter would be mostly made abroad.

Manolo González points out how the technology used in this studio was far from what was used in other countries and that, in Spain, it would take yet a few years to appear: "Abroad, you could already record with sound qualities practically unknown in our country" (Ibid.). The old RCA studio – in a stage of transition in which it had already been acquired by Philips – had Philips two-track recorders, so the instrumental base had to be recorded live through "an extensive set of microphones dedicated to each and every instrument" (Ibid.). In the images that follow, we can appreciate two different stages of the recording process. The first shows the recording of the musical backing made by drums, electric bass, and two electric guitars; and in the second, we can see the recording of the voices.

The recording, at this stage, is technically carried out live with all the instruments within the same room, which is acoustically conditioned with a large curtain designed to absorb the reflections of the room. The microphonic pickup is made through a set of condenser microphones intended for the recording of the four sound sources: a large diaphragm condenser microphone at the top for the overheads, two for each guitar amplifier, and another for the bass amplifier. An analysis of these images allows us to determine that all of the microphones share the same characteristics (identical transduction method and diaphragm, etc.), without taking into account the characteristics of the sound source, something that indicates that there was no real sense of the production beyond trying to make the recording as faithful as possible to the sound of the instruments of the band.

The microphone technique used is based on distant miking, possibly following the instructions that the recording studio gave to the engineers in order to guarantee the durability of the technical gear. Although the "emulation" they tried to do of the sound of The Beatles was an approach to the "color" of their recordings, the recording studio itself was a limitation in terms of technical means. The separate location of the components of The Beatles in Studio 2 at Abbey Road or the use of panels between instruments – mainly for the drums – minimized the effect called "spillage" or "leakage", that is, the unwanted sound that inevitably reaches a microphone from a different source to the one that it is intended to capture, for example, the sound of the snare or the cymbals that leaks through the microphone of the bass drum. Although the "spillage" or "leakage" among the different elements of the drum set is something inevitable even in the current musical recording, the setting of the microphones in the recording room in the case of Los Brincos contributes to this phenomenon on the rest of the instruments as well as the drum set.

Figure 3.1 Microphone Setting and Disposition of the Musicians in the Recording Room

Source: RTVE Archive.

In a second sequence of the images taken from the archive in Radiotelevisión Española, which documents the recording of the song *I Can't Make It*, we can appreciate how the voices were recorded in a second phase, monitoring the musical base through headphones, in what is one of the first instances in Spain of the overdubbing technique.

In this image, we can appreciate a closer microphonic pickup of the voices ("close miking") through some condenser microphones that seem to be used in a standard way in both instruments and voices and the absence of a "pop filter".[9] Another significant fact is the setting of the microphone with the capsule pointing upwards in the case of the main voices of Juan and Junior, and the capsule pointing downwards in the case of the voices of Fernando and Manolo, in a search for different tone colors in the capture and with a different position "off-axis" of the microphone in both cases. This fact leads us to think that there actually was an awareness of microphone techniques and their repercussions for the final sound of the production, starting from the premise that the mixing process begins to be "visualized" at the moment of placing the first microphone.

The recording of the first album by Los Brincos was made using almost exclusively Neumann U47 and U67 condenser microphones, a method of acoustic-electric transduction that implies certain precautions when placing the microphones – these do not generally support the same level of sound pressure (SPL) as other transduction methods such as "moving coil" microphones – conditioning their distance from the source. The original U47 – actually distributed by Telefunken[10] – appears in 1948. It is the first interchangeable polar pattern condenser microphone; that is, it allowed capturing the sound from different angles, allowing a more or less directional and even omnidirectional recording of the source, which would entail the capture of a greater number of room reflections. The omnidirectional polar pattern could be interesting in order to add a natural reverb that would glue all the elements in the mix, taking into account that the possibilities of applying mechanical methods of reverberation through plates or springs were really limited and reduced to their application to the main voice and/or choirs. The U47 is a tube microphone that incorporates a VF-14 amplifier and that provides a characteristic color to the recording. The U47 only allowed the use of cardioid (directional) or omnidirectional patterns, which justifies the use of a U67 for the capture of the voices. The U67 already included substantial improvements compared to its predecessors by incorporating a high pass filter[11] that reduced the "proximity effect" when doing "close miking", or a greater versatility when selecting polar patterns. The use of a bidirectional pattern – also known as "figure 8" – allowed the simultaneous recording of two opposing voices while cancelling the lateral leakage of the other two that, located in the same room, were recorded simultaneously using another condenser microphone with the same polar pattern (as can be seen in Figure 3.2).

The voices were recorded in a smaller room, taking into account that in this case the reverb would be applied by artificial means and not acoustically, achieving a greater presence and definition in the mix. In contrast, the distant miking technique can be appreciated when listening to the sound

Figure 3.2 Microphone Position in the Recording Process of the Voices
Source: RTVE Archive.

file because the drum set sounds distant and with the acoustic reflections of the room itself. Manolo González points out the aesthetic intention of musicians and technical team to enrich the recording through overdubs on the initial instrumental base:

> From our very first recordings, we wanted to enrich the musical base with contributions from other instruments, so we had to reproduce in playback the recorded base and send it to a track in another recorder. Later on, the new sounds were added to its second track. Then everything was mixed and we had the instrumental part ready. Next, and with the same technique that I've just described, we added the voices while listening to the music with headphones.
>
> (Ibid.)

The monitoring of the recording project together with the live sound being recorded was, therefore, a usual practice in the recording studio, laying the foundations of a work dynamic that will be consolidated in the 1970s with the expansion of the number of tracks available.

Although we can find certain parallels between the work dynamics in Abbey Road and the recording process in RCA, the "emulation" of the "Beatles sound" necessarily involved accessing the latter's own backline.

The Vox guitar amplifiers were one of the hallmarks of the characteristic sound in the first albums by The Beatles, so this was a basic aspect in order to achieve the goal. In this regard, Manolo González comments:

> The contacts that Fernando [Arbex]'s family had, helped us to import from England the guitars and amplifiers that we used for a while. Novola also gave us a hand in this matter. We changed the guitars several times during the time we were active. It did not happen the same with the amplifiers, which were always Vox.
>
> (Ibid.)

Juan and Junior used a Vox AC50 to amplify their guitars, whereas Manolo González used a Vox T60 for his electric bass. The AC50 guitar amplifiers were a revision of the legendary AC30 with more power but equally amplified by valves, while the Vox T60 was one of the first solid-state Vox amplifiers. This constitutes a significant fact when analyzing the frequency characteristics of the mix, taking into account the different harmonic content generated by one or another amplification system.

An analysis of the sequence recorded by Radiotelevisión Española also allows us to perceive some group working dynamics that were the prelude to the methodology that will eventually be introduced into Spanish recording studios from the 1970s. A young Callejo sits at the mixing console supervising the recording and alternatively looks at the "control room"

Figure 3.3 Maryní Callejo With José María Batlle at the "Control Room" and Los Brincos in the Recording Room

Source: RTVE Archive.

Figure 3.4 Maryní Callejo With Los Brincos Receiving the Sant Jordi Award, Barcelona, 1965

Source: Maryní's personal archive.

(in the hands of José María Batlle) and the band that is on the other side of the glass, indicating through the "talkback" if the recording is valid or not. In what constitutes an historical image, we can appreciate the main roles of a record production within the recording studio: producer, audio engineer, and tape operator.

CONCLUSION

As Wolfe, as recently as 2012, has pointed out, in many cases it is still assumed that the female artist is "only the singer", and being "only the singer" is considered to not require any of the skills associated with the mastery of an instrument or the technology of recording and music production. The decision-making autonomy of a woman in a relevant role within the music industry in the middle of the 1960s in Spain therefore finds a unique and isolated case in Maryní Callejo, as a musical director of the production process, who determined what was or was not valid. The image of a young woman giving indications through the talkback system to a team comprised entirely of men also underlines this important paradigm shift. Although Callejo's formation was predominantly musical and her influence on her record productions is mainly determined by her work as an arranger, she was also closely involved in the mixing process, sitting next to the engineer to supervise the actions that were carried out from the mixing console. Her main objective was to find the musical and technical means for her artists (who often lacked the necessary musical knowledge) to succeed in realizing their creative goals. Callejo's autonomy within Zafiro, although limited if we take into account that she had to obligatorily accept the artists that the company "imposed", allowed her to make important decisions regarding the recording of Los Brincos' second album in Saar studios in 1966, and the first album by the successful Spanish band Fórmula V in Fonit Cetra studios. In the various conversations held with Maryní Callejo during the writing of this chapter, she always indicated her total freedom to make decisions, making it clear at all times that her condition as a woman did not inhibit her career: her leadership ability and her solid training elicited great admiration from the artists with whom she worked, including Los Brincos: "all the artists I've worked with had confidence in me, it's a confidence that I've earned". While Callejo's first albums were made in Madrid with José María Batlle as the audio engineer, her goal was to record in the European studios that she considered references to achieve a specific sound. For this reason, during the second half of the 1960s, Callejo chose the Italian Plinio Chiesa as her main engineer, with whom she achieved some of her greatest successes.

Although her contribution to the consolidation of the foundations of pop music production in Spain is unquestionable, Callejo has not received recognition for her work from either the music industry or the academic sector. This is despite her impact, in a career spanning the decades from the 1970s to 2000 (when illness intervened), being unquestionable (Nino Bravo, Mari Trini, Bertín Osborne, Massiel, and others). The limited recognition of the impact of Callejo on music production practice in Spain during the 1960s arguably relates to a generational gap between current music producers and their predecessors. In Spain, it is less common for one producer to end up becoming a reference for another (in comparison to Anglo-Saxon producers, whose recordings are part of their sonic identities), and this is one of the biggest obstacles when we try to claim the relevance of proto-producers such as Alain Milhaud, Rafael Trabucchelli, or Maryní Callejo. Primarily, the source of inspiration for Spanish producers and engineers – the mirror in which they were reflected – was the Anglo-Saxon music that was slowly seeping into the country. In the same way that the goal of many artists was to make music similar to that of their English and American "idols", producers and engineers sought to imitate the sound of their records: in essence, the Spanish approach to production has always been through technical and musical "imitation". Hence, in Callejo's case the first thing Los Brincos did when arriving at the studio was to listen to albums by The Beatles in order to extract the characteristics of the sound of the bass drum, the snare, or the guitars. However, Callejo's close involvement with the project, her friendship with the band, as well as her leadership in the decision making, represents a unique case in musical production in an era when the approach of other proto-producers towards the artists was not so generous. Callejo's entering into the world of production, even taking into account certain fortuitous aspects that led her to run into some relevant protagonists of the music industry of the time such as Augusto Algueró or Esteban García Morencos, represented a unique opportunity to claim a leadership role, which was recognized by the artists she worked with as something natural. In conclusion, an appreciation of the contribution of Maryní Callejo is an important and necessary first step in changing the perception of the low presence of women in the historical evolution of music production in Spain. Furthermore, her example should offer valuable inspiration to those women contemplating a career in the field of music production in the future.

NOTES

1. All statements by Maryní Callejo have been transcribed by the author of this article during various telephone conversations made in 2018 and 2019.
2. We should take into account that Augusto Algueró (junior) was the usual composer of the successes of these artists.
3. Maryní Callejo (1964). *Los Brincos*. [LP Vinilo] Madrid: Novola.
4. Maryní Callejo remembers that the recording assistant was a young man whom everyone called "Paquito".
5. As Ordovás points out, even the band's name has a direct relationship with the band from Liverpool: "At the beginning, the group was going to be called The Black Sheep, but Fernando Arbex proposed the one of Brincos, more similar to Beatles".
6. Although they could be "fictitious" scenes re-creating a deliberated recording situation for the visit of the cameras of Radiotelevisión Española, we understand that it is a graphic document equally valid for the analysis.
7. Pop Thing (2008). *Pop español de los 60: Brincos en el estudio*. [online] Available at: www.popthing.com/zona_pop/pop_espanol_de_los_60_brincos_en_el_estudio.php [Last checked: 22/04/2019].
8. In the 1970s, RCA opened new recording studios in Madrid at Maurrás Street.
9. Nylon screen around a hoop or hollow metal disc that is placed in front of a condenser microphone in order to avoid possible distortions on the diaphragm caused by the air generated by the occlusive consonants, mainly 'b', 'p' and 't'.
10. Neumann had a distribution agreement with Telefunken, which was dissolved in 1958, the year after which Neumann started to distribute its own products under its own name.
11. A high pass filter (HPF) attenuates the low frequencies of the spectrum to the maximum, allowing only the high frequencies to pass through. The extent to which the frequencies will be attenuated depends on the slope of the filter and it is expressed in dB per octave.

REFERENCES

Alonso, Celsa (2005). "El beat español: entre la frivolidad, la modernidad y la subversión". *Cuadernos de Música Iberoamericana*, Vol. 10, pp. 225–253. Instituto Complutense de Ciencias Musicales. https://revistas.ucm.es/index.php/CMIB/article/view/61202/4564456547870

Burgess, Richard James (2013). *The Art of Music Production*, 4th ed. New York: Oxford University Press.

Domínguez, Salvador (2002). *Bienvenido Mr. Rock. Los primeros grupos hispanos 1957–1975*. Madrid: Fundación Autor.

Gil Calvo, Enrique (1989). "Participación laboral de la mujer, natalidad y tamaño de cohortes". *Revista española de investigaciones sociológicas REIS*, no. 47, pp. 137–175.

Ordovás, Jesús (2010). *Los discos esenciales del pop español*. Barcelona: Lunwerg Editores.

Wolfe, Paula (2012). "A Studio of One's Own: Music Production, Technology and Gender". *Journal on the Art of Record Production*, Vol. 7. http://arpjournal.com/a-studio-of-one's-own-music-production-technology-and-gender/

4

The Representation of Women in the Twentieth and Twenty-First Centuries Trikitixa[1]

Gurutze Lasa Zuzuarregui

INTRODUCTION

Music is an important expression of the political, social, and cultural context of contemporary society. Furthermore, musical lyrics have a significant influence on the construction, perpetuation, and/or the transformation of gender structures and roles. Nevertheless, little research has been carried out on Basque music from this perspective.

This work intends to contribute to this emerging line of research and favor the critical transmission and reception of music. To this end, the final results are presented of a semantic and socio-cultural study of the audio heritage left by a deeply rooted musical expression of the Basque culture[2]: the trikitixa.

In the first section of this article, a definition of "trikitixa" will be provided to introduce the particular characteristics that constitute this musical expression and to establish the definition that will be used throughout this work. The history of the trikitixa will then be explained in order to contextualize the focus of this study.

The second section will specify the objectives, criteria, corpus, and methodology used to carry out the semantic and socio-cultural analysis.

The third part of the work presents the final results of the study. The most significant themes and songs will be highlighted, and it is proposed that they are actively listened to via current applications. It should be noted that reference will be made to the artist or musical group that includes, within its record production, the musical composition, and that this does not always coincide with the authorship of the literary texts.

Among the texts mentioned are the *kopla zaharrak* or couplets. Numerous artists and musical groups from the last 40 years have reused this type of literary text, resulting in hybrid compositions that combine musical innovation with texts that, from a semantic point of view, reflect previous socio-cultural contexts. However, musical groups of the new millennium have tended to write their own lyrics. Today, newly created groups where female members predominate show a clear tendency not only to compose their own texts but also to transform previous discourses from a gender perspective. Thus, the fourth part of this study aims to raise, from a gender perspective, this discussion of the reuse of old couplets in the current socio-cultural context and, finally, to extract the conclusions of the study on the representation of women in the trikitixa of the twentieth and twenty-first centuries.

TRIKITIXA: DEFINITION AND HISTORY

The term "trikitixa" is a polysemic concept whose meaning continues to be debated. The evolution that it has undergone since the end of the nineteenth century and the meanings attributed by specialists, interpreters, and citizens make it difficult to reach a consensus on the specific use and meaning of the concept in question. According to the ethnomusicologist Juan Mari Beltran, it is an onomatopoeia derived from the sound "trikiti-trikiti" produced by hitting the *pandero*[3] or tambourine with the fingers. However, Kepa Perez Urraza uses the term to refer to the combination of the diatonic accordion and the tambourine.[4] Thus, "trikitixa" is often used to describe the musical group formed by the person who plays the *soinu txiki* or diatonic accordion (*soinujole*) and the person who plays the *pandero* or tambourine (*panderojole*) (Aguirre 1992: 21). According to Rafael Aguirre, one of the four pieces traditionally played by this group is known as "trikitixa",[5] and the same concept is also used for the dance associated with the sound of the diatonic accordion and the tambourine (Aguirre 1992: 21–22). Currently, "trikitixa" is often used to refer to the diatonic accordion.[6]

According to Rafael Aguirre, in addition to the accordion and the tambourine, the third element of the trikitixa is the voice (1992: 152). Through the vocals, the old *koplas*[7] the *bertsos*,[8] and the *irrintzi*[9] are transmitted, as well as, as of the 1980s, songs of diverse nature.

Considering the focus of this study, the way that the term has been used in the past and the validity of its use by the population, throughout this work the term "trikitixa" will be understood as the expression or musical genre resulting from the variable combination of the previously mentioned instrumental, vocal, and literary elements.

The instruments and their use, the training, the melodies, the themes, the places of performance, the socio-economic relevance, and the profile of the person linked to the trikitixa have evolved alongside Basque society from the end of the nineteenth century to the present day.

Together with the *alboka*,[10] the tambourine was used by musical groups for the *romerías* (religious festivals) that were held in the vicinity of rural hermitages. It was the women who usually played the tambourine and, therefore, sang the couplets (Aguirre 1992: 52, 152). According to Aingeru Berguices, the conduction of the public dance allowed these women (*panderetereak*) to exert the only managerial role socially accepted to women in the public sphere (Berguices 2019: 55). Aingeru Berguices adds that they fed and reproduced the traditional literature of this musical expression, sustained and minimized the internal tensions of the population through the irony they used in the improvised couplets, and their public exposure encouraged the collective participation of other women (2019: 55, 57).

The introduction of the diatonic accordion at the end of the nineteenth century constituted an important change in the music of the time, especially in rural areas of the Basque Country. Although the theory that it was introduced through the port of Bilbao is widespread in the world of the trikitixa, Rafael Aguirre believes it is more likely that it originated from the arrival of workers from Piedmont and Valle de Aosta, hired in 1860 for the construction of the Beasain-Olazagutia section of the Irun-Madrid railway line (1992: 13, 41).

Due to its sound, it allowed a wide range of individuals to play many different melodies, the diatonic accordion was immediately accepted by popular musicians and, above all, by the population of rural Basque-speaking towns in Gipuzkoa and Bizkaia. Thus, although initially it was introduced to accompany the tambourine, it acquired greater prominence, relegating the tambourine to a secondary instrument (Harana 1986: 32). According to Mielanjel Harana, it also replaced the *txistu* as a dance instrument (1986: 33) and relegated the *alboka* (Aguirre 1992: 97).

A meeting point for a dispersed population and a focus of leisure and courtship, the trikitixa was an indispensable element of *romerías* in hermitages and festivals. In a society governed by a strict education in Catholic doctrine, the isolation of the hermitages facilitated an escape from the control not only of the family but also of civil and ecclesiastical authorities, and at the same time the music fostered relationships between couples (Aguirre 1992: 68). This led the Church to consider the diatonic accordion as *Inpernuko auspoa* or the Bellows of Hell. Rafael Aguirre adds that the social stigma of *soinu txikia* was also contributed to by factors such as its acceptance by subordinate social groups, its foreign origin, its opposition to the established order, and the discredit of its music (1992: 77–78).

However, despite this stigmatization and the decadence of the *romerías* of the hermitages, the trikitixa, in general, and the diatonic accordion, in particular, continued to evolve prior to the Spanish Civil War (1936–1939), adapting to a Basque society which was increasingly urban. According to Rafael Aguirre, groups of trikitixa were often formed by two or more *soinujoles* and several *panderojoles* (1992: 97), and thus enlivened mountain marches, Basque fiestas and other celebrations organized by political parties and societies (Aguirre 1992: 90). According to Joxean Agirre, it was also an indispensable element in the *batzarres* or gatherings that were organized in the neighborhood bars; once the beloved had been accompanied to her house, the men returned to continue singing some of the most picaresque couplets (Agirre 1983: 2). In this way, "schools"[11] of trikitixa were formed, such as the Zegama school, which, according to the testimony of Miguel Urteaga,[12] influenced the Trikitixa of Zumarraga.

The *panderojole* Joxe Oria and the *soinujole* (or *trikitilari*) Joxe Larrañaga *Etxasakorta* would be the first to record a Trikitixa album in 1924. According to Miguel Urteaga,[13] the recording was made in the studios that the Regal record company (later absorbed by Columbia) had in the Kursaal in Donostia. This slate record consisted of four pieces, and Regal published it under the title "Triki-triki de Zumarraga". This album was the first work of the 13 Trikitixa albums recorded between 1924 and 1936 (Figure 4.1).[14]

Figure 4.1 The record "Jota vasca" played by Serafín y Antonia Aranceta Compañía del Gramófono Odeon S.A.E., [1941][15]

Source: ERESBIL – Basque Archive of Music

After the Spanish Civil War, the changes introduced by the *trikitilari* Jacinto Rivas *Elgeta* in the diatonic accordion would continue with his pupil Faustino Aspiazu *Sakabi* (Harana 1986: 33–34). Likewise, the musical heritage of the trikitixa would begin to gain pace again with the Trikitixa from Zumarraga. As Miguel Urteaga remembers,[16] the brothers Urteaga-Oria and Joxe Miguel Ormazabal recorded the first post-war album in 1957. The record company Columbia published the work under the title "Bailables vascos populares" (popular Basque dance tunes).

The trikitixa was acquiring social relevance, becoming a source of income in a society where the expansion of radio meant a revolution for popular music as the television appeared and foreign cultural trends arrived. From 1961, the radio station Loiola Irratia became the ideal media for recording and broadcasting the trikitixa. The use of the diatonic accordion also continued to evolve, especially in the hands of performers such as Iñaki Garmendia *Laja* and Martin Aginagalde and the accordionist

Pepe Yantzi (Harana 1986: 34). Mielanjel Harana affirms that within a few years a marked difference in technique developed between traditional and innovative "schools" and "styles" (1986: 34).

Together with the performances in the villages, from the 1960s onwards the trikitixa championships, promoted by the speaker Irratia Joxe Mari Iriondo from Loiola Irratia, reflected the evolution of social relevance and the trends that were predominating within the trikitixa. The death of dictator Franco made way for the so-called Spanish Transition (1975–1982). Basque society experienced a political, social, and cultural emergence. As far as trikitixa, the 1986 and 1988 championships demonstrated the draw of this musical expression and highlighted the existing debate between those who backed the traditional trikitixa and those who supported experimenting with the diatonic accordion, as in the performances of the innovative Tapia eta Leturia and Kepa Junkera.

The latter did not participate in the last championship held in 1991, and each embarked on musical paths that opened up new territory in the trikitixa scene, based on former traditions. In this way, they would set a precedent for future generations by introducing new instruments and musical expressions and fusing the inherited musical and instrumental tradition with other rhythms, languages, and cultures. The old couplets, *bertsos*, and songs passed on orally, the existing audio heritage and secret recordings of the *romerías* of the time, did little to reflect the worldview of the new generation that burst onto the trikitixa scene in the 1980s. Thus, from a desire to express their own vision of contemporary Basque society, new texts were created by *bertsolaris* and contemporary authors.[17] The musicians themselves also gradually began to learn how to shed their complexes caused by previous social undervaluation of trikitixa and to write the lyrics for the melodies they created. Tapia eta Leturia even introduced a new concept of the musical group and transformed the traditional *soinujole* and *panderojole* ensemble into a live band.

This new paradigm quickly became established and transformed the world of the trikitixa from a formal and semantic point of view. This is evidenced by the boom in bands that emerged from the 1990s onwards. The trikitixa not only continued to demonstrate its ability to coexist, adapt, and expand but also became one of the favorite musical expressions of female Basque artists. Among the numerous bands of so-called Triki pop and Triki rock, there were mixed-sex and all-female bands who would become creators of a previously almost non-existent female tradition.[18] Likewise, the appearance, in the "Post Pum"[19] age (Matxain 2017: 46–47), of the "Millennials" would impact on the formation and transmission of new musical discourses, which would contribute to the shaping of new identities.

ETA EMAKUMEAK, ZER? [WHAT ABOUT THE WOMEN?]

This article presents the final results of a study carried out in order to elucidate how women have been represented in the musical heritage of the trikitixa of the twentieth and twenty-first centuries. It also aims to discern an answer to the question "What about the women?" with which the triki rap group Lin Ton Taun ended the song "Hodei ilunak", included in their first album of the same name (Esan Ozenki 1993).

Thus, the aim of this research project is:

- To understand the ways in which women are represented.
- To elucidate the gender roles represented.
- To draw attention to the gender stereotypes transmitted.
- To favour, from a gender perspective, the critical transmission and reception of music.

To this end, a series of criteria have been established to define and shape a corpus of the most representative records of the musical heritage of the trikitixa from the "Diskografia" section of the website of the Euskal Herriko Trikitixa Elkartea association. The records with the following characteristics have been analyzed:

- Individual and unpublished works: collective albums and compilations have been discarded. It is worth mentioning the exception of the project "Haziak" as this is an initiative of the Euskal Herriko Trikitixa Elkartea association, which aims to raise awareness of the new values of the trikitixa world and, therefore, is of interest when researching contemporary discourse on the topic of this research project.
- The albums of the bands[20] that form the mainstream of the twenty-first century Trikitixa: the complete list in the "Diskografia" section of Euskal Herriko Trikitixa Elkartea has been included, with the works of the most influential groups included in the special collection "Triki" published by the record company Elkar.
- The recordings of the most representative bands of the so-called New Trikitixa: the records of Tapia eta Leturia, Kepa Junkera, Esne Beltza, Gose, Huntza, and Koban have been included and thoroughly analyzed.
- Stability of the bands: we have analyzed the work of bands that have more than one disc on the market, except in the case of artists and groups formed in part or in their entirety by women.
- We have analyzed the songs that make express mention of women or where women are involved.

On the basis of these criteria, we have actively listened to 1,584 songs included in 110 records of 43 bands that make up a universal set as diverse as trikitixa. Tapia eta Leturia, Kepa Junkera, Esne Beltza, Gose, Huntza, and Koban are responsible for 41 of the total recordings analyzed. For practical purposes, it should also be pointed out that 20 of the 43 bands mentioned previously comprise both men and women, and six bands are made up solely of women.

The data was recorded using the Drive application. To this end, 13 categories were pre-established, defined according to the particular characteristics of the musical compositions of Trikitixa and the object of study of semantic analysis: the masculine figure, woman-morality (negative), woman-morality (positive), woman-homage, woman-romantic love, woman-sexuality, romantic love, sexuality, empowerment, gender-based violence, instrumental, various, and others.

It should also be noted that throughout the transcription process certain aspects have had an impact that should be taken into account when assessing the results of the study:

- The subjectivity and cultural background of the author of the study.
- The historical and therefore cultural context to which texts are ascribed (discourses).
- The comprehension difficulties derived from the poor quality of the recordings from the 1920s and 1930s.
- The diverse nature of the texts of the trikitixa: some bands opt for original subjects written by the group in question and/or created specifically by *bertsolaris* and/or authors. In their compositions, others combine new lyrics, old couplets, *bertsos*, and instrumental pieces. Likewise, there are also artists who often reuse and reinterpret old couplets and/or combine instrumental pieces with old couplets and new couplets written in a similar way to the old ones.
- The discs of some female bands contain lyrics written by both men and women.

THE REPRESENTATION OF WOMEN IN TRIKITIXA MUSIC: FINDINGS

Traditions and, by extension, the discourses that derive from them have a specific function in the historical, social, and cultural contexts in which they are created and constructed, respectively. Therefore, their preservation can be an indication of the continued existence of certain functions. Notwithstanding, Pilar Ramos reminds us that postmodernism has not only called into question the stability of concepts such as music, identity, and objectivity but also abandons the essentialist discourse of concepts such as race, ethnicity, class, and gender (Ramos 2013: 37). As it is impossible to truly deconstruct or de-learn elements that make up who we have become, due to the range of cultural and familial influences that influence who we are, we are left with our ability to learn, provided that we are able to overcome the fear of what we might find. Knowledge enables comprehension, which leads to acceptance. That acceptance enables transformation, which, ultimately, enables us to move forward. Thus, as pointed out by Marisa Manchado, the key is not simply to remember, but to recognize and to gain knowledge (Manchado 2011: 17).

The definitive findings of the semantic analysis of the most important musical heritage of the trikitixa of the twentieth and twenty-first centuries, presented as follows, aim to demonstrate how women have been represented and, ultimately, how those representations have contributed to transforming the discourses transmitted through this musical expression. According to Laura Viñuela, musical practices are not only a way of expressing a specific context, but they also contribute to defining, disseminating, confirming, and deconstructing stereotypes (Viñuela 2003: 25).

As we begin to analyze the results of the study, we can observe that only 31.6% of the included compositions address or include references related to the pre-established categories. Table 4.1 shows us that other issues are raised and transmitted in 20.5% of the songs. Meanwhile, instrumental compositions constitute 52% of the total. While there is a sector within the world of the trikitixa that has clearly opted for another type of musical pieces (for example, Gose or Koban), the findings of this study leave no doubt about the importance that instrumental compositions have had in the world of the trikitixa.

But what concerns, ideas, etc. do musical groups express through the trikitixa? And, more importantly, how do they represent women? What stereotypes do they transmit and/or try to transform, and how?

Let us find some answers in the analyzed songs.

Others

The lyrics in this category address topics beyond the object of study. In short, they constitute an essential source of information for studying the social, economic, political, cultural, and existential evolution that took place in the Basque Country in the last century. Such transformations are clearly visible in the chosen discography, which covers what are considered paradigmatic models.

Table 4.1 Semantic content of musical texts by category.[21]

Subjects

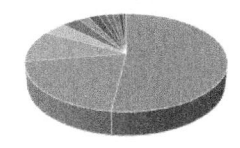

- Instrumental
- Romantic Love
- Sexuality
- Woman-Romantic Love
- Woman-Morality (Negative)
- Masculinity
- Woman-Morality (Positive)
- Others
- Various
- Empowerment
- Woman-Sexuality
- Woman-Tribute
- Gender-based violence

Source: Gurutze Lasa Zuzuarregui, 2019.

As a reflection of the political, social, and cultural changes that they experienced from the 80s onwards, Tapia eta Leturia started to include texts where they expressed their vision of the Basque society, the construction of which they actively contributed to through their music. Songs such as "Trikitilariak omen gara (trikitixa) (1987), "Trikitixaren sua" (1992), and "Zeruko hauspoa" (2002) reflect their respect towards the inherited musical tradition, as well as the changes taking place in the world of the trikitixa as a result of experimentation and the market society. The group, originally from the Guipuzcoa province, alludes to the traditional rural world with irony in "Txapelau" (1992) and breaks with the religious zeal of the past in "Gipuzkoarrok, errosayo santuba! (1995). In songs such as "Begi nini ñimiñoa" (1992) and "Arraun" (2002), they examine the political situation and assert the right of self-determination of the Basque Country and other stateless nations, while "Gezurtientzat" (2002) explores the concept of historical memory. Likewise, they express their opinion regarding Basque language (Euskara) in "Gurasoak euskaraz" (1987), the richness of multilingualism in "Oihan Jaia" (1997), compulsory military service in "Histori okerra" (1987), the topic of stress in "Stressa" (1990), provide a critique of capitalism in "Amway" (1992), and sing about football in "Buruball" (1995). However, as demonstrated by songs such as "Galdurikan nago" (1995) and "Supositorioa" (1995), they do all this without losing their festive edge, so strongly linked to the world of the trikitixa.

The discography of Kepa Junkera is strongly characterized by a musical search and, as a result, it has a strong instrumental element. In terms of semantics, Kepa Junkera displays a strong tendency to reuse old couplets, as we can see throughout this research project. Thus, songs such as "Haika mutil" (2008), "Markosen txarria" (2010), and "Cançó dels traginers" (2017) take us back to traditional rural settings of the Basque Country and other cultural spheres, while others such as "Hara nun diran" (2008), "Barku sinple batean" (2009), and "Primi, Romualda, Andresa, Leona, Mikaela, Pantxika eta Martzelina" (2014) remind us of the migratory past of the Basque people. However, some new topics are also included, such as the sea and its people, in songs such as "Itsasoa belztuta" (1994) and "Marea Zumaian" (2016). In "Akelarretik josapatera" (1990), he challenges the Catholic religion, while criticizing colonialism in "Billy from overseas" (1994) and praising the Basque language in "Egun da Santi Mamiña" (2010). In "Espetxea" (1990), he alludes to the political situation in the Basque Country, "Els segadors" (2017) is a cry out in favor of Catalonia, and "The way that the wind blows" (1994) is about the socio-economic crisis. He praises the beauty of "Bilbo" (1994) and the island of "Izaro" (2001), and explores cultural hybridization in "Ari naizela" (1994). He sings about sorrow in "Sodade" (1995) and happiness in "Fali-Faly" (1998), while encouraging people to be themselves in "Ny Hirahira" (2001).

Criticism and political claims are highly present throughout the discography of Esne Beltza, in songs such as "Lehen" (2009), "Posta zaharra (errefuxiatuei eskeinia . . .) (2007), "Sonidero" (2010), "Mugitu harriak" (2016), "Gure askatasuna (2018), "Distantzia" (2011), and "Sutzen"

(2013). Resilience is also an important element, for example in songs such as "Astindu" (2008), "Ibilian" (2011), and "Sometimes" (2013). Both traditions and contemporary worldviews are present throughout their entire discography, in songs such as "Esnesaltzailearena" (2008) and "Hommes et femmes de couleur" (2010).

As we shall see later, Gose presents a new way of looking at the topics of love, desire, and female sexuality. They make a conscious break with tradition in "Bye, bye, Maurizia" (2005), "Jangura" (2007), "Beaucoup de police" (2007), and "Portaloian" (2009), while invoking the notion of historical memory in "Amets gorria" (2007) and "Errua" (2014). Their punk spirit comes out in the social, political, and religious critique displayed in "Non-noiz" (2005), "Surik bai?" (2009), "Banku atrakatzaileak" (2014), "Hil da Europa" (2012), "2004-XII-09" (2009), "Kamarada" (2012), and "Mantis atea" (2007), while people are encouraged to live their lives and to be themselves in "Alegrantziaren aldeko konjura" (2014) and "Bang!" (2012).

Huntza explores a diversity of topics in their discography; they "reconcile" the rural world with the urban one in "Aldapan gora" (2016), while in "Herri Unibertsitatea" (2016) they advocate the creation of a people's university. They explore the topic of immigration in "Hautsetatik" (2016) and question orthodoxy in "Deabruak gara" (2018), while "Buruz behera" (2018) is an invitation to life and "Olatu bat" (2018) to protect the environment.

As for Koban, their only album critiques the current socio-cultural model through songs such as "Nekropolia" (2018) and "Panpin puskatuaren fabrika" (2018), and materialism in "Daltonikoak" (2018), while encouraging people to enjoy the simple pleasures of life in "Plazer txikiak" (2018).

As with other musical formations, representations of women are also present in the analyzed models. It is mainly the maternal figure, women suffering in the political sphere, and the stereotyping of prominent political symbols that stand out within the multiple and diverse representations of women (as athletes, musicians, rebelling housewives, alcoholics, charitable or beautiful characters).

In "Bihotzeko Ama" (2009) by Laja eta Mikel, for example, a baby pleads for its mother's protection. In "Itsasoa laino dago" (2010) by Kepa Junkera and "Amasemeak oherakoan" (1995) by Tapia eta Leturia, mothers protect and enjoy their baby's company. At the same time, in "Haurtxo txikia" (2008) by Kepa Junkera, a mother nursing her baby expresses her discontent at her husband's gambling addiction, and in "Lua lua" (2009), also by Kepa Junkera, a father tries to get his baby to sleep, telling them how he has sold the baby's mother. Likewise, the blind Solferino expresses his sadness at not being able to see his mother again in a song with the same title performed by Tapia eta Leturia (1990) and Kepa Junkera (2010). In "Anaren mundu arraroa" (2014) by Gose, Ana becomes a mother soon after losing her own and a prostitute in order to maintain her child. In "Pello Joxepe" (2010) by Kepa Junkera, Pello Joxepe questions his paternity, and in "Gure aitak amari" (2009) by the same author, a father gives a red skirt to the mother, who also acts as a transmitter of language.

She acts as the sole transmitter of language in "Euskaraz eta kitto" (1994) by Lin Ton Taun, while doing it together with the father in "Kantari hasterakuan" (1996) by Maltzeta aita alabak. Through the examples given, both classic and innovative groups represent selfless, suffering, and, in some cases, morally questioned women. They also perpetuate the roles of care, protection, and cultural transmission traditionally assigned to them. In this sense, it is remarkable that it is a classical formation Maltzeta aita alabak who equally represents the latter role.

The gendering of political symbols should also be highlighted. According to "Oihuka" (2002) by Urgabe, one's land is transmitted by grandfathers, while the family house is built by a father in "Herri hau" (1999) by Joseba Tapia. However, in general, mothers are seen as symbolic of the homeland of the Basque people, and Esne Beltza call for a defense of the house of the father and mother in "Nafar erreinua" (2010). The Basque provinces are referred to as brothers and sisters in "Kondaira" (1998) by Etzakit, while in "Zazpiak oihal batetik" (2002) by Laja eta Mikel and "Zortzigarren taupada" (2003) by Kontrairo they are referred to as sisters. In several songs, solidarity, peace, political power, and, above all, the figure of the militant activist are linked to men. Women appear alongside other vulnerable subjects that need protection, acting as supporting elements for male fighters in songs such as "Eldu da gure ordua" (2001) and "Châteauguayko bataila" (1999), recorded by Joseba Tapia. According to "Amatxo laztana" (1975) by Sakabi eta Egañazpi and "Urruntasuna" (2001) by Triki ta Ke, mothers are remembered and missed during hardship and suffering. Regarding socio-political activism, Lin ToTaun emphasizes the absence of women as active subjects in "Hodei iluna" (1993), and, despite the fact that Etzakit vindicates the presence of women in "Gurekin" (1998), it is only 25 years later that Irati Gutierrez Artetxe responds adequately to the question of what happens to female political activists, in "Laztanen preso" (2017).

Romantic Love

This category consists of songs with lyrics that describe the different phases, conditions, and, above all, emotional attitudes linked to romantic love. A cultural model of love that, according to Mari Luz Esteban, emphasizes the idealization, mystery, and exaltation of emotions and sexual desire.[22] Romantic love, adds the anthropologist, places on the top of affective interactions the relationship of heterosexual couple and being monogamous, which in turn is linked to marriage. This is an idealized form of loving interaction that, according to Mari Luz Esteban, also sustains an unequal order of gender, class, and ethnicity, as it serves to "unite" people who are also constructed as unequal.[23]

In "Maitemindu kontuak" (1999) and "Zaharrak berri" (2001), Epelde eta Larrañaga and Triki ta Ke describe what they view as some positive changes that have taken place in the historical evolution of how love and sexual relations are experienced. Regarding games of seduction, the tendency to glorify the physicality of women is clear. It is also conspicuous how musical groups, such as Esne Beltza, allude to the ancient tradition of accompanying women to their farmhouse in "Baltxekua" (2009). Likewise, seduction is linked to economic factors and social status in "Baserritarrak gera" (1924) by Trikitixa de Zumarraga, "Tristerik nago" (1975) by Sakabi eta Egañazpi, in the old couplets "Din don" (2008), "Madalen Buzturingo" (2009), "Habanara joan ginan" (2009), "Begiak barrez" (2010), and "Jota del Campello" (2017), recorded by Kepa Junkera, as well as in more recent songs, such as "Ezpada" (2000) by Gozategi and "Oilo ipurdia" (2003) by Kontrairo. It is not only fathers who stand in the way of those in love, as in "Maitia nun zira?" (1998) and "Eperrak" (2010), but also mothers put an end to the love affairs and arrange marriages against their daughters' will in old couplets such as "Amak ezkondu ninduen" (2009) and "Markesaren alaba" (2010), recorded by Kepa Junkera. Meanwhile, Laja eta Mikel allude to the advantages of marriage in "Maitia nahi badezu" (2009). There are also many songs that express the desire to be with a loved one, the expectance of a beloved's return and goodbyes. Maltzeta aita-alabak describe the great fortune of being together with a loved one in "Damatxo" (1996), "Eman neri eskua" (1996), and "Lehengo erromeritan" (1996), while Sakabi eta Egañazpi struggle with rejection in "Tristerik nago" (1975). Likewise, there is a tendency to describe women's singledom in negative terms. Gozategi views the status of so-called spinsters (*neska zaharrak*) as something negative in "Zenbait" (1995), an attitude shared by a suitor in the old couplet "Jota de Sella" (2017), recorded by Kepa Junkera. In the same manner, "Ala kinkiri ala kunkuru" (2009) encourages men and women to get married; however, a woman speaks against marriage due to bad experiences in "Tbilisi" (2006) by Kepa Junkera. This attitude is shared by the male protagonist in "Prima eijerra" (2008) due to having been cheated on by his beloved.

The idea of possession also appears in several descriptions of love and sexual desire. Notwithstanding, and despite expressing the desire for mutual possession in "Txikita" (2008), in "Sherezada" (2008), Esne Beltza voices the need to construct a common space, whilst Kepa Junkera alludes to accepting the personal emancipation of our loved ones in "Txoriak txori" (2009). Meanwhile, the male figure in the couplet "Ene izar maitea" (2010), recorded by Kepa Junkera, describes emotional dependence, and in "Vals de tardor" (2017), recorded by Josep Maria Ribelles and Kepa Junkera, a person in love urges his beloved, Griselda, to stop crying, while Tapia eta Leturia generally encourage people to live love to the sound of music.

In old couplets, there is a clear tendency to specify the gender and/or the name of loved ones and/or others involved. In more recent songs, there is a shift to more neutral language, with a clear tendency to refer to loved ones with expressions such as 'love' (*maitea*). However, gender is also specified in songs, such as "Txikita" (2008) by Esne Beltza and "Maddalen" (1990) by Tapia eta Leturia. It is also worth noting that when gender is specified, heterosexual relationships are the most dominant type, both in old couplets and more recently published lyrics. As Mari Luz Esteban points out, in the last 30 years there has been a great conceptual and experiential transformation of gender, the body, sexuality, and emotions (love) (Esteban 2009: 33); these are transformations, adds the anthropologist, that can be backfeeded as they are spheres with their own entity but deeply connected to each other (2009: 34). Thereby, in the words of Maria Luz Esteban, reflection on romantic love implies reflection on (one's) heterosexuality (2011: 175), a revisiting or transcendence of love that, according to the anthropologist, generates what she calls *disconfort* (Esteban 2011: 176). In this respect, Teresa del Valle affirms that the processes of disidentification and configuration of new identities generate tensions and negotiations (del Valle 2002: 34), which in this case, denote the relevance of inherited forms that hinder the emergence and/or visibilization of diverse identity models.

Various

This category has been defined taking into account the structure of old couplets in order to register the diverse contents and, by extension, representations that such compositions present; the following section will provide some examples.

Trikitixa de Zumarraga alludes to the beauty of the women of Zumarraga in "Al entrar a Zumarraga" (1962), while expressing reluctance to go looking for "spinsters" (*neska zaharrak*) in "Oraingo mutikuak" (1962). Sakabi eta Egañazpi describes the power of the Sanctuary of Urkiola to bring boys and girls together, and subtly reminisces about sexual relations played out in the same place in "Durangon bazkalduta" (1975). In "Ederregia zera zu" (1979), Maurizia, León eta Fasio encourages a beautiful woman to find a man and thus be able to free herself of work in the countryside. We can also observe different representations within a single musical piece; for example, in "Txerri jana" (1999) by Imuntzo ta Beloki, Pello Joxepe denies his fatherhood, a woman is reluctant to engage in physical contact with a suitor, San Cristobal mistrusts the carpenter's wife, and a woman encourages her husband to lose weight. We can also find similar compositions by more recent musical groups: a courting couple fears being caught by the mother, a grandmother buys a red apron, needed to marry off the suffering Maddalen de Busturia, and a woman is advised to marry a blacksmith instead of a basket weaver in "Andrakaseko errementaria" (2002) by Urgabe. Meanwhile, in "Triki koplak I" (1999), Ene Bada encourages "spinsters" (*neska zaharrak*) to look out of the window in order to find a boyfriend. However, there is no doubt that Tapia eta Leturia and, above all, Kepa Junkera have used old couplets throughout their discography the most. With regard to the old couplets recorded by Kepa Junkera in his various musical formations and projects, one can observe diverse, and sometimes contradictory, representations of gender roles and women; for example, "Ederra baina mizkea" (1987) describes the gleam in the eyes of a person who is loved, criticizes the chin of Mari Andres, and refers to the joy of not receiving her affection. "Kaixo lagunak" (1987) is about the life of the wife of a sailor and Saint Anthony's difficult task of uniting men and women, while "Nere sentimendua+Orain kantatuko dut+Madalenara" (2008) describes how the "spinsters" (*neska zaharrak*) selflessly accept that Saint Magdalene will not grant them a boyfriend due to their age. "Atxiketan potxiketan" (2009) alludes to the great butts of the old women who go to Elorrio on donkeys, to the boyfriends of a lady of Donostia, and discourages marriage to the old bachelor and spinsters. In "Ipiñaburu Leku Aituan" (2010), a suitor criticizes the decision of the mother of his beloved, which is of an economic nature, not to give him her daughter's hand. In "Aitite Merkurio" (Txotxongiloa) (2010), women who like to party and to drink txakoli wine are considered lightheaded, while "Kantetan Biher dot Has (Zortzinangoa) (2010) gives an account of the infamies committed by spinsters (*neska zaharrak*) towards the so-called old bachelor in the festivities of San Blas of Abadiño. "Maletak" (2016) alludes to the red cheeks of a barmaid after dancing, and a Romany woman Luisa is also told not to weep for the beloved, etc. As for Tapia eta Leturia, "Abadiñoko Karmentxu" (1998) describes the beauty of Karmentxu, while a suitor awaits the arrival of his beloved. In "Itsasoan ure lodi" (1998), they allude to the threat that shy women pose to men, and a widow, Mari Juana, is advised to look for a man. "Santa Luzietan" (1999) describes the sexual desire of a woman who is married to an old man to whom she is not attracted, the preference for money over girls, the women of Azkoiti, Azpeiti, and Arratia, and suspicions regarding the whereabouts of a husband and who he is with. Meanwhile, "Eguzkie joan da ta" (1999) describes a woman wearing a red skirt travelling on donkeyback to the village Ermua, as well as an old man flirting with young women, etc. The examples given show that both the old couplets reinterpreted in a contemporary way and those newly created transmit diverse and sometimes opposed gender roles; for example, women are praised and satirized according to their physical characteristics, or their well-being is looked after and they are subjected to moral judgment in the few cases in which they move away from the private sphere and the traditionally assigned roles. But the institution of marriage steals the limelight. This is demonstrated by the constant references to so-called bachelors and, above all, spinsters. These women express a desire to marry and/or they are socially placed to leave singledom behind. Nevertheless, in specific cases they reject and/or undervalue male suitors. Likewise, marriage often appears linked to the economic factor that not only leads to the unhappiness to woman but also the male suitor who is rejected by the woman on the basis of his purchase power. Therefore, we can conclude that the well-being of the woman appears directly linked to "a good" marriage and, in certain cases, it is she who stands in the way of an emotionally positive and healthy relationship.

Sexuality

This category focuses on lyrics that relate to sexuality in its broadest sense.

In this vein, "Ezin leike hola" (2001) by Epelde eta Larrañaga describes the Catholic moral that prevailed in past times. In "Alkar maitatzen ez digute uzten" (1977), Laja eta Landakanda describe what they see as positive changes in the ways of loving and experiencing sexuality. However, in "Kulebrin kulebron" (1995) by Gozategi, a woman continues to be a passive subject with regard to the sexual relations within her marriage.

At the same time, Etzakit and Kepa Junkera assert the need for sexual liberty in "Zergatik ez" (1998) and "Kalejira al-buk" (1994), respectively. By contrast, in "Tandremo" (1998), Kepa Junkera gives boys and girls a moral warning with regard to sex. Meanwhile, Esne Beltza mainly links the desire to engage in casual sexual relations to the idea of developing stronger feelings towards the desired person, who, in the case of "Eskuekin" (2010), we know is a woman because she is described as a 'brunette' (*morena*). In "Oh Pello Pello" (2008), Kepa Junkera describes the case of a man who avoids intimacy with his wife, while the protagonist in "Aita San Antonio" (2009) reminisces about his sexual relations with Ramona at a *romería* celebration at the Urkiola sanctuary. Meanwhile, in "Cant de batre del comtat" (2017), recorded by Kepa Junkera, the protagonist feeds his libido by seducing a woman.

Also, sexual diversity can be observed in some of the songs; sexual games are described as a source of pleasure in "Bexamela eta pastela" (1997) by Maixa ta Ixiar, while in "Barruti honen larretan" (1997) and "Sexonet" (1997) by Tapia eta Leturia Band, sexual encounters and games take place both in the toilet of a bar and on the Internet. In "Pekatamundi" (1996) by Imuntzo eta Beloki, Donibane asserts her right to sexually desire and love other women, while "Maiu-Maiu" (2004) by Urgabe describes the shock experienced by the protagonist in an encounter with a transsexual person. Despite relating the encounter sympathetically, it should be pointed out that the protagonist of this last song alludes to the embarrassment he feels telling about the event and the subsequent trauma caused by this non-heteronormative encounter.

Lastly, it is worth mentioning the presence of the figure of a bluesman and a blonde woman with red lipstick (represented as a *femme fatale*), who awaits him at the bar in "Azken aurreko notak" (1999), by Imuntzo eta Beloki.

Although the latter case shows the perpetuation of certain stereotypes, in the last 25 years there has been a clear tendency to vindicate freedom and make sexual diversity visible. The inherent tensions of this sociocultural process are not identified in the most classic musical duos but in some relatively recent songs of groups such as Gozategi or Urgabe. In our opinion, this circumstance denotes the great rootedness of the traditional cosmovision in the most local areas both on the Basque coast and inland.

As for reused old couplets, these compositions transmit diverse, and sometimes contradictory, gender roles. However, the old couplets of the Basque and Catalan cultural traditions also constitute the metaphorical expression of sexual desire and relationships consummated within societies strictly controlled by the Catholic Church's doctrine until a few years ago.

Empowerment

This category focuses on texts about women who decide to be themselves and live their sexuality, love, and music freely.

The contribution of Imuntzo eta Beloki in this category is considerable, for example, the female protagonist in "Markesaren alaba" (2007) rejects the concept of love linked to economic factors, and she is described as being in control of her love life due to her academic training. "Amona Inaxi" (2007) describes the fulfilling life of an older woman, and "Bizi pasioak (Mexikana)" (2015) encourages self-care, socialization, and experiencing sexual desire freely.

Likewise, in "Urruti dagoen alabari" (2002) by Laja eta Mikel, a father accepts his daughter's search for her own path, while "Beldurrak ahaztuta" (2001) by M-Ezten asserts the need for sexual empowerment. "Ahopeka" (1995) by Trikitixa Kontrairo and the Big Band constitutes the first statement in favor of lesbians' right to freedom of sexual expression, and in "-ak (Tximosak)" (1996) and "Gogoak emana hala" (1997), Maixa ta Ixiar assert the right of women to decide the way they want to look and to dance freely, while in "Zezenak" (1996), women take the initiative in love.

The discographies of musical groups formed in the last decade include more texts that relate to this category. In "Ni" (2018), for example, Esne Beltza asserts the right of transgender people to express their sexuality freely, while they promote women's self-care and personal well-being in "Mi revolución" (2018), and highlight a victim of bullying's capacities for resilience in "Paretaren kontra" (2018).

Regarding groups that are composed mainly of women, such as Gose, Huntza, and Koban, we can also observe a shift in their discourse. These groups use attitudinal resources and literary figures to reinforce the narrative element of their musical compositions. Gose, for example, uses direct language accompanied by body language that transmits security and strong-mindedness in order to reinforce its discourse in favor of transgender people and of building new languages in "Naizena izateko remix (DZ with Patrol Destroyers)" (2012), and to express the importance of making one's own choices in "Ezetz" (2012). Meanwhile, Koban calls on us to reject capitalist models and all forms of female slavery in "E.E.I.E" (2018) and "Congo Square" (2018). In addition, they use their characteristic ironic touch to reject conservative stereotypes in "Señorita bat Akelarre batian" (2018). Similarly, Huntza encourages people to reject traditions and asserts women's right to freedom of expression in the world of music in "Zer izan" (2017).

This type of discourses conveyed from a woman's body not only allows to transform semantically the representation of women in the trikitixa but also makes visible other ways of being (woman) (musician) that attract and configure other worldviews, desires, eroticisms, and sexualities.[24] Ultimately, they create a fundamental (feminine) frame of reference as far as it affects not only the expression and reconfiguration of gender and sexual identities but also the renewal and/or invention of the (feminine) musical tradition.

Women and Romantic Love

Women adopt and live according to different, and often contradictory, attitudes regarding love. In "Gaurko eskon-berrientzat" (1981), Maltzeta aita-alabak present traditional marriages as a source of emotional well-being. By contrast, "Mariñelaren zain" (2011) by Imuntzo eta Beloki describes the sorrow of a girl whose father arranges her marriage. Likewise, loneliness acts as a source of sorrow for the protagonist of "Marigorri" (2002) by Urgabe. In "Neska zaharren koplak" (2018), Koban uses irony to transform the traditional view regarding so-called spinsters (*neska zaharrak*). Meanwhile, "Kalabazak" (2016) by Huntza represents a woman who is active in love and not afraid of failure, and in "Ilusioz bete" (2018), one that is able to leave her beloved in a positive way and continue on her own path. Gose describes a similar type of determination and love towards a loved one in "Au revoir" (2012). However, many songs still display the desire to be with and/or be reunited with a loved one. "XXX" (2012) by Gose is about the fear of falling in love, while "Milaka zati" (1999) by Alaitz eta Maider explores jealousy and other feelings of discontent. At the same time, while "Jonvals" (2002) by Alaitz eta Maider describes the benefits of travelling the solitary path, "Damu" (1999) by Ene Bada is about the impossibility of living one's life without a loved one. On the contrary, "Bat baina bi" (2003) by Triki ta Ke suggests that love can be healthy as long as one stops idealizing toxic love. According to "Ametsa naiz" (2003), by the same group, healthy love also brings stability.

Women and Sexuality

"Inshala" (1999) by Alaitz eta Maider represents the first invitation for women to use their bodies to dance freely with no strings attached, an idea that is repeated in another song of theirs, "Eta besterik ez" (2002), three years later. Meanwhile, M-Ezten encourages persons who are the object of someone's desire to enjoy sexual pleasures in "Zain nago" (2001) and "Egin dezagun" (2001). In "larru truk" (2003) by Triki ta Ke, the content regarding sexual relations is more explicit. Doubtless it reflects the changes in attitude and discourse that were taking place in society.

These changes are also reflected in "Rimmel" (2009), where Gose rejects the typical princess narrative, both with regard to semantics and body image, and replaces it with an independent and confident woman who expresses her thoughts, desires, and sexual experiences freely. In "Lokartu arte" (2009), a woman expresses her sexual preferences to a person that she is intimately involved with, while in "Bondage" (2005), "Andrë" (2005), "Album" (2005), "16 eskailera" (2008), and "Gose" (2008), women enjoy sexual games and relations, which are explicitly described in the songs. "Hey boy!" (2008) refers to the classic prostitute-straitlaced binomial, which has traditionally been used to divide women into those that have and/or indulge in (or let others indulge in) sexual fantasies and an active sexual life, and those who don't. The above, especially

"Rimmel" (2009), is contrasted by "Gezurrez" (2005), which describes the desire of a woman to be loved by her sexual partner.

Nonetheless, Gose breaks, both semantically and attitudinally, with the subtlety and sweetness with which the previous trikitixa bands expressed their desire to enjoy their sexuality to the full. That plenitude, diversity, and attitude are later reflected in "Klimax gaiztoa" (2018) by Koban.

Women and Morality (Negative)

Some old couplets, but also newer songs, represent women in a negative way. We can find adjectives such as 'spinster' (*mutxurdin*), 'frigid' (*itxi*), 'nun' (*monja*), 'crone' (*atso zaharra*), 'old, crazy and crippled' (*vieja, loca, coja*), and 'birdbrained harlot' (*buscona casquivana*) that are used to describe women. Likewise, both in old couplets and newer songs from the 90s, one can observe a general tendency to morally judge and parody women based on their clothes, physical characteristics, and/or lifestyle; in "Neskatxa gaztea" (1975), Sakabi eta Egañazpi warn young women about the potential harm to their reputation if they go out to dance halls, while in "Matxalen Busturiko (Aurreskuarena)" (2010), recorded by Kepa Junkera, Matxalen de Busturia is forced to leave her home village because of getting a bad reputation. Meanwhile, in "No te cases con la Luchi" (1963) by Trikitixa de Zumarraga, and "Paparra sabalduta" (1933) by Trikitixa de Elgoibar, it is suggested that a woman has not fulfilled her traditional role as a married woman.

At the same time, the musical pieces "Abadiño San Blasetan" (1979), sung by Maurizia Aldeiturriaga, "Amonaren mingaina" (1996) by Imuntzo eta Beloki, and "Tapatata" (1996) by Gozategi demonstrate that sometimes women themselves judge and transmit such representations and negative ideas regarding the female figure and the role assigned to women.

However, the witches and the wife of Pello firmly confront such insulting views in the couplets "Anbototik Oizera" (1987) and "Miru zuria+Ikusi nuenean" (2008), recorded by Kepa Junkera.

Tributes to Women

Imuntzo eta Beloki pay tribute to female *trikitilaris* or trikitixa players and female family members in "Zuei ohore" (1993), "Soinu ttikia" (1996), "Lazkanon amona" (2011), "Nahia (Kunbia)" (2015), "Martxea albokeagaz" (2015), and "Juliana" (2004). Meanwhile, in "Harro gaude" (2016), Huntza acknowledges the *pandereteras* or female *pandero* players of the past. However, Maurizia Aldeiturriaga is unquestionably the *panderetera* that receives the most public recognition, her name being referred to in song titles, such as "Bye, bye, Maurizia" (2005) by Gose, "Aupa Maurizia!" by Korrontzi, and "Nondik jo Maurizia" by Kepa Junkera, who also recognizes the importance of the trikitixa players "Primi, Romualda, Andresa, Leona, Mikaela, Pantxika eta Martzelina" (2014) in the history of the trikitixa, pays tribute to his mother in "Roman eta Kontxa Urraza Zollon" (2014), and dedicates one of his albums to Maren (2001).

Meanwhile, in "Etxekoandre" (2000), Ene Bada highlights the invisibility of the work carried out by housewives. "Oroitzen zaitudanean ama" (2011) by Esne Beltza is based on maternal memories, while mothers also receive their children's recognition in "Amak egin ninduben" (1924) by Trikitixa de Zumarraga and "Adorea" (2003) by Triki ta Ke. Meanwhile, Maixa ta Ixiar praise the qualities of "Mikele" (1998), while Tapia eta Leturia commend the figure of the Palestinian Wafa Idris in "Wafa" (2002).

Masculinity

We can observe a great diversity of representations of the male figure in terms of worldview, attitudes, and experiences. While the single man in "Mutil zahar biziotsua" (1981) by Laja eta Landakanda is reluctant to get married, "Lur giro" (1999) by Epelde eta Larrañaga presents a man who is willing but unable to get married. "Artzai bat" (1991) by Imuntzo ta Beloki describes the sorrow of a lonely shepherd, while in "Bertsuak jarri dizkat" (1982) by Miren eta Bingen a man justifies his lack of commitment through the physical defects that he assigns to his former suitresses. However, in "Gizontxo gizentxo" (1991) de Imuntzo eta Beloki, a woman alludes to her husband's weight problem.

Toxic love – according to Mari Luz Esteban, a colloquial expression of great popular roots linked to the concept of romantic love[25] that, as Coral Herrera point out, associates the love to the suffering, sacrifice, submission, and renunciation (Herrera 2018: 33) – also makes men emotionally dependent in "Berriz ihes egin didazu" (2001) by Triki ta Ke and "Baina baina" (1996) by Maltzeta Neym, unlike the peasant that is represented in "Ikusten duzu goizean" (2008), recorded by Kepa Junkera. In the latter case, the peasant feels like a true man in the village that was chosen by the father of his father for him to settle down in with his farm, work, house, and the family that he has formed. Here it is worth highlighting the traditional gender roles represented by the peasant's family: the wife that takes care of the house and family, the shepherd son and the beautiful daughter, destined, in the future, to fulfill her role as a wife.

Gender-Based Violence

From the mid-1990s onwards, trikitixa music groups started to include texts related to rape, physical/psychological violence, and the murdering of women in their songs. This is a tendency that, in our opinion, constitutes an expression of the increase in the social and, progressively, institutional response, in the face of sexual aggressions, social inequalities, machismo at work, and gender violence observed by Jose Antonio Martin Matos (Martin 2013: 251). According to the scholar and musical journalist, this mobilization was encouraged by the advance of the feminist movement and the work carried out by social collectives (2013: 251). Thereby, in "Asetuta" (2001), M-Ezten encourages women to free themselves from psychological violence and to stand up to the person who is producing it. "Muxu urrunak" (2003) by Triki ta Ke describes the heartbreaking case of a child who is robbed of his innocence through witnessing a situation of domestic violence experienced by his terrified and suffering mother. In "Hil ezazu aita" (2007), Gose describes the case of a father who murders almost his entire family. Meanwhile, in "Don Francisco" (2017), recorded by Kepa Junkera, a man kills his wife and her lover.

Xabier Solano and Jon Mari Beasain have also denounced gender violence throughout their musical trajectory. In "Bihotz zatitua" (2000) by Etzakit, only the lonely night bears silent witness to the killing of a woman in the middle of the street. In "Zauriak" (2013), Esne Beltza highlights the solitude and resilience of battered women to the rhythm of ska. Likewise, the group encourages respecting the affective-sexual needs of women in "Ez da ezetz" (2015).

Women and Morality (Positive)

There are only a few songs in this category, which focuses on texts that represent (and value) the female figure in a clearly positive light. However, it should be noted that there are several positive references to women in the texts included in the previous categories. It is also worth mentioning that the use of gender-neutral language makes it difficult to be sure about the exact amount of positive references corresponding to women.

As demonstrated by the texts included in the category *Others*, there is a tendency to highlight the physical characteristics of women in the lyrics of old couplets. The couplet "Kale barrendik-asita" (1963) by Trikitixa de Zumarraga presents a clear example of this tendency.

The song "Azpeitiko neskatxak" (2010), recorded by Joseba Tapia, also falls under this category, as it welcomes the fact that the girls of Azpeitia decided to dance neither with the Carlists nor the liberals.

The few texts included in this category verify that the positive representation of women is directly related to their physical characteristics and social behavior when this is in accordance with the gender roles traditionally assigned to them.

ANCIENT COUPLETS: A HERITAGE AND/OR A TRADITION THAT SHOULD BE CHANGED ?

As pointed out at the beginning of this work, artists and musical groups have often re-used traditional couplets. From the 1980s onwards, Tapia eta Leturia and Kepa Junkera decided to explore new directions and laid the foundations for a new model of trikitixa, which was different in its musical style as well as its concepts and lyrics. As Joseba Tapia states, the scarce discography that exists and the circumstance that the traditional couplets of Bizkaia were forgotten and almost completely lost during the Franco dictatorship (1939–1975) led to the creation of new lyrics by the *panderojole* (tambourine players) who wished to sing in the Basque language again and to a Basque society that was in the middle of a political, economic, and socio-cultural transition.[26] For example, Juanito Maltzeta created, for the first time, his own rhythms and lyrics, which he included, together with his daughter, in the album "Maltzeta aita-alabak" (Soinutek 1981). However, Tapia eta Leturia really reversed the widespread lack of textual and aesthetic research in the world of trikitixa. This artistic couple expressed the worldview of this new Basque generation of rural origin that had now become urbanized.

According to Joseba Tapia[27], trikitixa performers had to commission *bertsolaris* to write lyrics for them because few other resources were available. Only a small number of literary texts existed, written by previous *bertsolaris* and referring to outdated socio-cultural contexts and discourse (such as bachelors, spinsters, and chauvinists). According to Tapia, the remoteness of the lyrics, a lack of identification with this type of text and discourse, and the low value that trikitixa performers placed on their ability to write their own lyrics motivated them to search for new texts. The Guipuzcoan *trikitilari* states that they went to *bertsolaris* who, through form, aesthetic, imagery, and original metaphors, transmitted other concerns that were relevant to them. Tapia highlights the contribution of Xabier Amuriza and Jon Sarasua. Xabier Amuriza, an expert in old couplets, created compositions with forms that resembled traditional texts, but that were new to Tapia eta Leturia. On the other hand, the texts created by younger *bertsolaris* such as Jon Sarasua would "topple" the previous worldview of rural areas. According to Tapia, the reading of literary works also led trikitixa performers to opt for writers who provided texts in the form of couplets, but with a different brushstroke.[28] However, far from abandoning the classical compositions of the traditional couplets, they included these in most of their recordings, especially in the album *Bizkaiko kopla zaharrak* (*Traditional Bizkaian Couplets*) (Elkarlanean 1999), a great deal of which is composed of the traditional couplets of Bizkaia compiled by Xabier Amuriza.

Despite following this trend at the beginning of his musical career, Kepa Junkera focused, above all, on musical research. Thus, in the lyrics of his compositions he would give special prominence to the old couplets.

As a result, Tapia eta Leturia and Kepa Junkera created and transmitted a new musical and textual model characterized by hybrid compositions that combine musical innovation with texts that, from a semantic point of view, often account for previous socio-cultural contexts. The "Triki-pop" groups of the 1990s such as Urgabe followed in the footsteps of their predecessors, but the groups of the new millennium like Gose or Koban were mostly inclined towards creating their own lyrics.

Today, newly created groups where female members predominate continue the upward trend of creating their own texts. These texts show an eagerness to transform previous discourses from a gender perspective. The female members of the group Huntza, Uxue Amonarriz Zubiondo, and Josune Arakistain Salas point out that the group has deliberately commissioned female *bertsolaris* of their generation to write some of their lyrics. However, it is worth highlighting the masculine authorship of the song "Harro gaude" (2016), with which the group launched their career in the music industry. According to Josune, they are also aware of the erotic and macho elements of the old couplets:

> gero daude ere oso oso letra matxistak ere,. . iruitze zait esaten zituztela gauzak oso oso argi . . . kamuflatzen baina oso ebidente . . . ta igual bere garaian . . . horrek potentziala zakala baina . . . nik uste gaur egunea allauta hoiek, . . . kriston saltua ikustet letra hoietatik eta gaur egun. . . . Gaur egun holako letra bat idaztezu eta dazkazu kritikak alde denetatikan . . . bai iruitze zait oso oso saltoa daola lengo kantetako letratatik gaur egunea. Ta, hori, bai azkenian gizartearen eboluzioaren islada dia kantak. Nik uste, garbi garbiyak. Ta gizartian nola ikusiya daon emakumia ikusten da oso oso garbi trikitixako kantetan. Oso garbi ikustea.
>
> [there are also some very macho lyrics. It seems to me that in the past they said things very clearly . . . disguising them, but in a very obvious way . . . and that might have worked at the time, but . . . I think that at the present time those, . . . I observe a huge leap between those lyrics and the ones written now. . . . If you wrote those kind of lyrics now you would be heavily criticised from all sides. . . . I think there is a massive difference between the lyrics of the songs before and the present. And that, in short, the songs are the reflection of the evolution of society. In my opinion, they are very clear. In the songs of the trikitixa, it is very clear how women are seen in society. It is very clear.]
>
> (Arakistain 2018)[29]

For her part, the musician who uses the pseudonym Sorkunde confesses the discomfort produced by the numerous compositions that refer to the *neska zahar* or "spinster" and claims to have refused to sing lyrics that conveyed the idea that the so-called spinsters "are a disgrace". For example, Sorkunde remembers refusing to sing couplets created at the end of the first decade of the twenty-first century, in which a husband questions his wife about the number of men with whom she has supposedly shared her bed. According to Sorkunde, the professor who instructed her in the interpretation of *soinu txikia* thought that she was overreacting:[30]

> trikitixa mundu honetan ematen du hori ez dela planteagarria, . . . eta nik ez nuen jo, emanaldi hori ez nuen jo . . . ez dut joko esaten nuen baina talde berri batek trikitixarekin kanta berria nola egiten duen planteatzen nuen eta, noski, pegatzen du hori sartzea eta ez dugu birplanteatzen, konserbadore, oso konserbadorea da, ez? jantzietan askotan, baserritarrez jantzita joan behar gara trikitixa jotzera, zergatik? Ez? . . . ematen du trikitixa kaja batean sartu dugula eta hor kontserbatu nahi dugula baina folklorea batzuk esaten dute dela kultura hil aurreko azken pausoa, ez? . . . eta orduan nik trikitixa ikusten dut edo hortik ateratzen da o, bai, oso polita izango da sagardo egunetan joko dugu baserritarrez jantzita baina ez dugu aurrera egingo horrekin . . . eta orain niretzat trikitixa hori baino harago dihoa . . . instrumentu

bat gehiago da eta uste dut komeni zaiola, ez? hortik ateratzea pixkat zeren bestela hor geldituko da.

[in this world of the trikitixa, it seems that this issue could not be raised . . . and I did not go there, I did not bring it up in that performance. . . . I said that I would not perform it, but what I raised was how a new group of trikitixa makes a new song, and, of course, it sounds good to put that in and we don't rethink it. It's conservative, very conservative, isn't it? To play trikitixa we usually have to go dressed in baserritarra (traditional peasant outfits) clothes, why? no? It seems that we have put the trikitixa in a box, and we want to keep it there, but some people say that folklore is the step that precedes the death of culture, right? Then as I look at the trikitixa it either comes out of there, or, . . . yes, it's very nice to play dressed as baserritarras in the cider season, but we are not going to move forward like that . . . and in my opinion the trikitixa transcends that . . . it is another instrument, and I think it is in its interest to break away from that a bit, otherwise it will stay there [in folklore], won't it?]

(*Sorkunde* 2018)[31]

According to Sorkunde, artists and groups such as her admired Kepa Junkera and Korrontzi are creating a new style of trikitixa from a musical point of view, but she observes that, in general, they tend to fuse the new rhythms with the old couplets. From a gender perspective, Sorkunde questions the semantic and socio-cultural content of this type of literary text and reflects on the need to rethink this question in the world of Trikitixa:

trikitixa kantak egiteak bertso zaharrak eta gainera bertso zahar horien antigoaleko ideia eta tradizioekin sartzea dakarkigu? . . . da mundu bat, ez dakit, . . . goazen ere birplanteatzera egiten duguna, nola egiten den eta aurrera egitera honekin ere, ez? . . . oso tradizionala egiten zait [trikitixa] gauza horietan: nola sortzen den, zein letra sartzen diren

[does making trikitixa songs mean using old bertsos, and revisiting the old ideas and traditions of those bertsos? . . . it's a world, I don't know, . . . we must also rethink what we do, how we do it and move forward with it, musn't we? . . . in those areas [trikitixa] seems very traditional to me: how it's created, what lyrics are used.]

(*Sorkunde* 2018)[32]

Xabier Amuriza[33] reflects on the current usefulness, the content, and the transmission of the old couplets from a gender perspective, in response to the questionnaire posed in this research project.[34]

For the Bizkain *bertsolari*, writer, musician, and researcher, cultural heritage always serves a purpose and, even more so, in cases such as the Basque language, Euskera, where written references are not as abundant or as ancient as in other languages. In his opinion, the old couplets constitute a musical-textual genre of oral literature with unique and interesting characteristics, which constitute a musical genre and whose fascinating elements continue to bring delight, including to *trikitilaris* women. For

Xabier Amuriza, the old couplets are a heritage that portray a context in a more or less appropriate way:

> Kopla zaharretan (hain zaharrak ere ez direnak, bestalde), ondare bat daukagu, batez ere, erromeria eta eske giroan sortzen joan dena. Horietan, dudarik gabe, testuinguru bat agertzen da, batzuetan oso era dotorean adierazia, eta batzuetan, nahiko modu zakarrean ere bai.
>
> [in the old couplets (which are not so old), we have a heritage that has mostly come about during the romería and the tradition of going from farm house to farm house playing music and asking for food, drink and money in return. Undoubtedly, in those couplets, a context is portrayed, sometimes expressed in a very elegant way and sometimes also in a coarse way.]
>
> (Amuriza 2019)

Through different examples, the multifaceted Bizkain exposes and reflects on the satirical, offensive, and non-offensive nature of some couplets. Through an example of eight couplets, Amuriza demonstrates that both men and women are objects of the art-genre of satire, but he adds that it certainly affects women more. The *bertsolari* also suggests creating new couplets as an alternative to those couplets that may now be considered in bad taste:

> Ikusten denez, adibideetan gizonezko zein andrezko hartzen da burlagarritzat. Horretan, koplaritzak alde bietara banatzen du satira, nahiz eta, seguruenik (egiaztatu beharko litzateke), andrezkoen figura maizago satirizatua izan den. Kopla satiriko horiek anonimoak dira. Eta pertsonen prototipo bati eskainiak. Satira pertsonalak balira, irain bat izan litezke, eta sortu zirenean, beharbada, halaxe izango ziren.
>
> Gaur egun iraingarri izan al daitezke? Agian, gustu onekoak ez. Baina sor daitezke berriak, gustu txarrekoak izan beharrik ez dutenak. Satirak arte-jenero bat izaten jarraitzen du, ondo eginez gero, indartsua, eta gaizki eginez gero, deitoragarria. Egilearen dohainen baitan dago kalitatea....
>
> [As can be seen, in the examples both men and women are taken as objects of mockery. The verses deal out satire on both sides, although, surely (it would have to be confirmed) the figure of the woman has been satirized with more frequency. These satirical couplets are anonymous and dedicated to a type of person. If they were personal satires, they could be an insult, and perhaps that is what they were when they were created....
>
> Can they be taken as offensive today? You could say that they are not in good taste. But new ones can be created, which don't necessarily have to be in bad taste. Satire continues to be an art form if done right, but deplorable if done wrong. Quality stems from the qualities of the creator.]
>
> (Amuriza 2019)

In the case of the couplets that he classifies as offensive, Xabier Amuriza confirms the existence of offensive couplets with respect to the figure of

The Representation of Women

the woman and suspects that there are fewer couplets of this nature about the figure of the man. In this case, the Bizkain *bertsolari* states the advantage of removing them from the contemporary iterations of the genre:

> Zer egin kopla mota hauekin gaur egun? Ezabatu? Ez dut uste egokiena denik. Soilik, ez erabili. Daudela hortxe eta kito. Hitzaldi baterako balio dezakete, zerbait adierazteko adibide gisan. Ezabatzeak ez du zentzurik, zeren alaba "saltzea" egia izan zen garai batean. Eta, zoritxarrez, egia izaten jarraitzen du munduko eremu [sic] askotan. Eta gaurko gure gizartean ere, nahiz ez literalki, esanahi figuratuan, arlo horretan jarraitzen dute "salmentak" presio eta mehatxu bidez, bortxa ekonomikoak barne. Baina onartu behar da nonahi eta nolahai kantatzekoak ez direla [sic].
>
> [What should we do with this type of couplets today? Eliminate them? I don't think that's appropriate. Just don't use them. Let them stay where they are and that's it. They can be used for a conference, as an example to express something. Eliminating them doesn't make sense, because "selling" a daughter was true at one time. And, unfortunately, in many parts of the world it is still true. Also in today's society, although not literally, figuratively, through pressures and threats, including economic violence, in that kind of environment there still are "sales". But we have to agree that they are not suitable to be sung anywhere or in any way.]
>
> (Amuriza 2019)

As for the couplets that transmit non-offensive content, Xabier Amuriza states that in literature and, in particular, in couplets, flirtatious remarks have been directed mainly at women and, in his opinion, this can lead to the adoption of misguided attitudes. On this matter, Amuriza gives as the example of a couplet that alludes to the fear shown by mothers of daughters towards obscene boys. He asserts that couplets can be the elegant expression of a worrying reality, which emphasizes the true validity of the content:

> Bi elementu edo irudiren gainean (logika bisuala), pentsamendu edo gogoeta bat doa, ama baten figurari eskainia. Nik ez dut kopla horretan inolako irain-kutsurik ikusten. Bai, ordea, errealitate kezkagarri baten adierazpen dotore bat, edukiaren bigentzia eternoa nabarmentzen duena. Zeren: 1) Alaba "galantak", objektiboki, edo beste gabe, "alabak" (ama batentzat berea beti da galanta) badira eta beti izango dira. 2) Beldurrez edo kezkaz bizi diren amak badira eta beti izango dira. 3) Eta, zoritxarrez, mutil "zantarrak" ere badira eta beti izango dira.
>
> [In two elements or images (visual logic) there's a train of thought dedicated to the figure of a mother. In that couplet I do not observe anything offensive. On the other hand, it is an elegant expression of a worrying reality that highlights the eternal validity of the content. Because: 1) Objectively, there are "elegant" daughters, or just "daughters" (for a mother, hers is always beautiful) and there always will be. 2) There are and always will be mothers who live with fear or worry. 3) And, unfortunately, there are, and there always will be, "obscene" boys.]
>
> (Amuriza 2019)

Therefore, Xabier Amuriza is of the opinion that the heritage should remain unchanged. He believes that if couplets are considered to be inappropriate, other quality couplets should be written, but without altering couplets that are old or are written by a well-known author. The *bertsolari* points out that we have the right not to sing them and considers that the modification of lyrics impairs them. As far as their use is concerned, he is in favor of leaving aside those pieces of heritage considered as offensive and choosing non-offensive couplets of recognized artistic quality. Thus, Amuriza states the need to develop a way forward for the art form, based on the aesthetic of an acceptable heritage:

> 3) Sen artistikoa daukanak sorkuntza berrietan asmatu behar du. Ondare onargarri baten estetikatik, etorkizunerako arte-bide bat garatu. Estetika bera ere joan daiteke aldatuz eta garatuz, baina beti ere, eredu zentraletik erabat aldendu gabe, hori beste eredu batera pasatzea bailitzateke"
>
> [3) People with an artistic instinct have to get new creations right. From the aesthetics of an acceptable heritage, develop a way forward for the art form. The aesthetics themselves can also change and develop, but always without moving away from the central model, because otherwise that would be to move on to a different model.]
>
> (Amuriza 2019)

Finally, regardless of the quality of the content, Xabier Amuriza emphasizes the interest of this literary-musical genre as long as it remains a source of enjoyment:

> koplaritzaren funts interesgarria jenero literario-musikala da. Horrek balio du orain eta beti, jeneroa gustatzen den bitartean. Edukia izan daiteke ona, neutroa edo txarra, egileen dohainen arabera. Baina hori arte guztietan eta jenero artistiko guztietan gertatzen dena da.
>
> [what is interesting about the couplets is the musical-literary genre behind them. That's what counts now and always as long as you like the genre. The content can be good, neutral or bad, depending on the qualities of the artist. But that's what happens in all art-forms and artistic genres.]
>
> (Amuriza 2019)

CONCLUSIONS

The final results of this research project give an account of the historical evolution that has taken place in the way that gender roles are represented in the trikitixa, particularly those that women inhabit. The examples given show the diversity of these forms of representation, both in the past and the present day; they also portray the tensions.

The tensions observed indicate complex processes of perpetuation, transformation, and creation that, on the one hand, perpetuate previous stereotypes and, on the other, allow the construction of new discourses in each and, therefore, throughout social, economic, political, and cultural contexts. The woman continues to be the focal point of praise. There is a marked tendency to extol (and, where appropriate, criticize) the physical characteristics of women. Nevertheless, the cases presented by Xabier Amuriza also make visible the (presumably few) cases of negative criticism of the male figure. In addition, physical stereotyping persists. The changes of physical standards over time should also be borne in mind for interpretative purposes. However, there are also cases in which the woman is portrayed to be especially valued as a person, by bands that could be considered as more conventional. To the contrary, some examples from "Triki pop" groups of the 1990s highlight the tension between the perpetuation of previous roles and an emerging worldview.

The musical career of some male and female artists also makes this discursive transformation visible. It is worth highlighting the active role that certain male artists play in achieving parity. Although on the one hand they seem to perpetuate the idea of romantic love, their proactive attitude, composing and transmitting a remarkable amount of songs that constitute a clamor for the affective-sexual freedom of women and against gender violence, is particularly important in the case of current male groups that are followed by a mostly adolescent audience. As for the representation of gender in political symbols and political activism, there is an increasing tendency to represent women as active subjects.

From a linguistic point of view, the use of neutral language and the tendency to name both men and women when speaking in general terms prevails. Likewise, Gose's song "Naizena izateko" (2012) warns us of the need to create new gender-independent linguistic modals that allow the semantic representation of transgender people (among others). Esne Beltza's song "Ni" (2018) also draws attention to this issue. Artists such as Ines Osinaga (Gose) highlight the importance not only of the semantic part of the texts but also of the choice of musical expression and the attitude that is represented when transmitting the messages.

In this respect, it is worth highlighting the musical, behavioral, and discursive transformation driven by millennial and post-millennial bands partially or mostly composed of female members. Since the middle of the new century, they have been transforming and creating new discourses based on the questioning of traditional gender roles and the way that the representation of women has been perpetuated. In order to do so, they use direct language, irony, and body language. It could be argued that, through this artistic instinct to which Xabier Amuriza referred (but which in this case the post-millennial bands consider to come from an unacceptable semantic heritage), they are creating their own discourse that constitutes an alternative to the offensive compositions alluded to by the Bizkain *bertsolari*.

In the light of historical changes and the results and reflections presented throughout this work, we consider that the emerging discourse, derived and impelled by the current dynamic feminist movement, constitutes another important link in musical expression and, therefore, in a Basque society in constant transformation also from a gender perspective. In our opinion, the audio heritage that has been passed down can contribute positively, from a gender perspective, to this discursive transformation. Thus, we consider that a critical understanding of what we have been not only allows us to trace our history of mentalities and/or emotions but also its understanding enables us to transform and, therefore, create discourses that contribute to the formation of an egalitarian society.

NOTES

1. This publication is part of the projects US 17/10 (UPV-EHU) and FFI2017-84342-P (MINECO) that are being carried out by the research group IT 1047–16.
2. In the present work we will take into account Pio Perez's definition. According to anthropologists, the definition of the concept "Basque culture" is complex due to the political and ideological differences that can determine this definition, as well as the language that can "qualify, determine and in some cases complicate" it (Perez 2013: 16). As pointed out by Pio Perez, in order to obtain the greater consensus, in the present work the "Basque culture" is defined as a Basque cultural space (2013: 16). In this way, Pio Perez understands "Basque culture" as a differentiated cultural space (Gobierno Vasco 2004: 17) that is between two strong cultural spaces: Spanish in the south and French in the north; that it is not alien to the influences of other cultures (Perez 2013: 16) and that corresponds to a geographical, linguistic, and emotional space (Perez 2013: 17).
3. Euskal Herriko Trikitixa Elkartea. Available at: http://trikitixa.eus/?page_id=44 [Accessed May 27, 2019].
4. Ibid.
5. The other three are the *Fandango*, *Arin-Arin*, and *Porrusalda*. For more information about these dances, see: Ansorena, J.I. (2011). "La creación del baile suelto vasco", *Euskonews* 594. Available at: www.euskonews.eus/0594zbk/gaia59401es.html.
6. Euskal Herriko Trikitixa Elkartea. Available at: http://trikitixa.eus/?page_id=44 [Accessed 27 May 2019].

    To find out about the current characteristics of the diatonic accordion in the Basque Country, see: Berguices, A. (2016). *Organologia popular y sociabilidad: el baile de La Casilla Abando-Bilbao y la expansión del acordeón en Bizkaia (1880–1923)*, [Vitoria-Gasteiz]: Departamento de Historia Contemporánea- Historia Garaikidea saila, UPV-EHU. Available at: https://addi.ehu.es/handle/10810/19955
7. Manuel Lekuona analyzes the characteristics of the *Kopla* (couplet) in his entry entitled "Kopla zaarrak" in the *Auñamendi Eusko Entziklopedia*. Available at: http://aunamendi.eusko-ikaskuntza.eus/en/kopla-zaarrak/ar-54816/.
8. To find out about the characteristics of the *bertso*, see the Bertsozale Elkartea website. Available at: www.bertsozale.eus/en/bertsolaritza/what-is-a-bertso?set_language=en. For more information about the art of *bertsolarismo*, see the audiovisual "What is bertsolaritza?" Available at: www.youtube.com/watch?v=6Lk0FhAN3S8.

9. According to the Auñamendi Eusko Entziklopedia, the *irrintzi* "is a loud, shrill and prolonged shout, of a single breath, that the shepherds like to make echo in the valleys and that the Basques in general gladly release to express joy". Available at: http://aunamendi.eusko-ikaskuntza.eus/en/irrintzi/ar-76986/. It is also used nowadays to study the voice box: www.youtube.com/watch?v=gcSaW6JUnUc [Accessed May 27, 2019]. To find out more about the *irrintzi*, see the documentary "Erraiak" by the anthropologist Ekain Martínez de Lizarduy Stürtze. Teaser available at: https://arteman.eus/es/erraiak/.
10. For more information about this instrument, see: Barrenetxea, J.M. (1998) "La alboka", *Euskonews* 10. Available at: www.euskonews.eus/0010zbk/gaia1009es.html.
11. A defined style, repertoire, and way of performing the trikitixa, created through collaboration between different tambourine and diatonic accordion players.
12. Urteaga, M. Interviewed in the Basque language Euskera by Lasa, G. (30 April 2019).
13. Ibid.
14. For more information about published work, see the Euskal Herriko Trikitixa Elkartea website. Available at: http://trikitixa.eus/trikimailua/sailak/diskografia.htm.
15. ERESBIL – Basque Archive of Music. 1941. *Jota vasca*. [Photograph]. https://www.eresbil.eus/opac/abnetcl.exe/O7041/ID41ae6e39/NT1 Accessed 03 December 2019.
16. Urteaga, M. Interviewed in the Basque language Euskera by Lasa, G. (30 April 2019).
17. Bands commissioned these artists to write new lyrics.
18. The Euskal Herriko Trikitixa Elkartea association promoted the initiative "Trikitixa eta generoa" (Trikitixa and gender) to identify women who had performed the tambourine and the diatonic accordion; more were relegated to the private sphere. Among others, the tambourine players Maurizia Aldeiturriaga and Primi Erostarbe gained a position of great importance in the public sphere. At the age of 13, Miren Etxaniz Marizkurrena became the first woman to play the diatonic accordion in the 1980 Trikitixa Championship: "Campeonato Absoluto de Trikitixa". In 1986, Elene Sustatxa and Elena Bezanilla formed the duo Tarratada. This project, known as "Triki punk", lasted until 1988 and took the Trikitixa to an area that had until then been alien to this musical expression: the alternative scene. The songs they performed live had an important feminist style and are available at: https://mierdadebizkaia.bandcamp.com/album/tarratada-lasarte-87-01-28.
19. The journalist Kepa Matxain coined this term to name the musical groups created after the permanent ceasefire declared by the Basque nationalist group ETA on January 10, 2011.
20. For the purpose of this article, "band" includes both individual artists and groups of two or more people.
21. Lasa, G. (2019). *Table 1: Semantic Content of Musical Texts by Category*. Unpublished table.
22. Esteban, M.L. (2019). Email sent to Gurutze Lasa Zuzuarregui, October 10.

23. Ibid.
24. Osinaga, I. Interviewed in the Basque language Euskera by Lasa, G. (June 28, 2019).
25. Esteban, M.L. (2019). Email sent to Gurutze Lasa Zuzuarregui, October 10.
26. Tapia, J. Interviewed in the Basque language Euskera by Lasa, G. (November 11, 2018).
27. Ibid.
28. Ibid.
29. Amonarriz, U. and Arakistain, J. Interviewed in the Basque language Euskera by Lasa, G. (October 1, 2018).
30. In order to preserve the identity of the source, the author of the study has transcribed the content of the testimony into standard Euskera.
31. *Sorkunde*. Interviewed in the Basque language Euskera by Lasa, G. (September 19. 2018).
32. Ibid.
33. Xabier Amuriza has carried out an exhaustive compilation of old coplas and has collaborated with Tapia eta Leturia and Kepa Junkera for their publication, among others, in the albums "Bizkaiko kopla zaharrak" (Elkarlanean, 1999) and "Beti bizi" (Hiri Records, 2010).
34. Amuriza, X. (2019). Email sent to Gurutze Lasa Zuzuarregui, June 11.

FURTHER READING

Euskal Herriko Trikitixa Elkartea. *Inpernuko poza trikitixaren jardunaldiak* (Zarautz: Euskal Herriko Trikitixa Elkartea, 2008) collects communications on the history, characteristics, and genres of the ancient coplas of the experts Xabier Amuriza, Kalzakorta, Jabier, and Junkera.

Joxemari Iriondo, K. *Trikitixaren historia txiki bat (A short history of Trikitixa)* (Pontevedra: Fol Música, 2014). This is a CD/book that contains the history of the Trikitixa, the musical trajectory of Kepa Junkera, a list of the most relevant tambourine players of history (chapter 4). and a biography of Maurizia Aldeiturriaga (chapter 7).

Paia, X. and Paia, F. *Aupa Maurizia!* ([Bilbo]: Bizkaiko Trikitixa Elkartea, 2008). This is a detailed biography of the tambourine player Maurizia Aldeiturriaga.

López, E. *Neskatxa maite: 25 mujeres que la música vasca no debería olvidar (25 women that basque music shouldn't forget)* ([Vitoria-Gasteiz]: Aianai Kultur Elkartea; [Mungia]: Baga Biga, 2015). This contains interviews conducted with Elene Sustatxa and Elena Bezanilla.

REFERENCES

Agirre, J. (14/05/1983) "Trikitixa: gure gurasoen doinu martxosoa", *Egin* 1–2. Available at: http://trikitixa.eus/trikimailua/hemeroteka.htm.

Aguirre, R. (1992) *Trikitixa*, Billabona: Martin Musika Etxea.

Amuriza, X. (2010) *Beti Bizi*. Xabier Amuriza, Leioa Kantika Korala, Kepa Junkera. [CD] (Hiri Records [HR-006]).

Berguices, A. (2019) "Euskal panctereterea: patriarkatuaren urradura. La panderetera vasca: una brecha en el patriarcado", in J. Barrio (ed.) *Ezkutuko ondarea: emakumeen errealitateak Bizkaian = Patrimonio oculto: realidades de las mujeres en Bizkaia*, Bilbao: Bizkaiko Batzar Nagusiak = Juntas Generales de Bizkaia. Available at: www.enkarterrimuseoa.eus/EnkarterrietakoMuseoa/ficherospublicaciones/archivos/Libro-Mujeres.pdf?hash=6c057f0d7cb1fc3dd36ccaaae0fd6541.

Del Valle, T. (2002) "1. Marco teórico y metodología", in T. del Valle (coord.) *Modelos emergentes en los sistemas y las relaciones de género*, Madrid: Narcea.

Esteban, M.L. (2009) "Identidades de género, feminismo, sexualidad y amor: Los cuerpos como agentes = Gender Identities, Feminism, Sexuality and Love: Bodies as Agents", *Política y Sociedad* 46 1/2 27–41. Available at: https://revistas.ucm.es/index.php/POSO/article/view/22978.

——— (2011) *Crítica del pensamiento amoroso*, Barcelona: Ediciones Bellaterra.

Gobierno Vasco. (2004) *Plan vasco de la cultura*, Vitoria-Gasteiz: Servicio Central de Publicaciones de Gobierno Vasco. Available at: www.euskara.euskadi.net/r59738/es/contenidos/informacion/argitalpenak/es_6092/adjuntos/plan_vasco_cultura_c.pdf.

Harana, M.A. (1986) "Trikitixa, 'infernuko auspoa': ehun urte eta gaztetzen", *Argia* 1116 27–38.

Herrera, C. (2018) *Mujeres que ya no sufren por amor: transformando el mito romántico*, Madrid: Catarata.

Maltzeta aita-alabak. (1981) *Maltzeta aita-alabak*. [cassette] (Soinutek [IZ-139 IZ]).

Manchado, M. (2011) "Caminos escondidos bajo techos de cristal: perspectivas feministas en torno a la discriminación de las mujeres en la creación musical", in R. Iniesta (ed.) *Mujer, versus música: itinerancias, incertidumbres y lunas*, Valencia: Rivera Mota.

Martin, J.A. (2013) *El rock de las noticias: la actualidad y sus canciones: "de la tradición anglosajona al caso vasco"*, Leioa: Universidad del País Vasco/EHU, Departamento de Periodismo II (Unpublished doctoral thesis provided by the author).

Matxain, K. (2017) "Bizi luzea Post Pum-ari!", *Argia* 2554 44–47. Available at: www.argia.eus/astekaria/docs/2554/azala/2554_argia.pdf.

Perez, P. (2013) *La asignatura de cultura vasca en el grado de antropología* (Unpublished paper provided by the author).

Ramos, P. (2003) *Feminismo y música: Introducción crítica*, Madrid: Narcea.

Tapia eta Leturia. (1999) *Bizkaiko kopla zaharrak*. Tapia eta Leturia, Xabier Amuriza (compiler). [CD] (Elkarlanean KD-540).

Viñuela, L. (2003) *La perspectiva de género y la música popular: dos nuevos retos para la musicología*, Oviedo: Ediciones KRK.

5

"Hey boy, hey girl, superstar DJ, here we go..."

Exploring the Experience of Female and Non-Binary DJs in the UK Music Scene

Rebekka Kill

For Tami Gedir (2016), global alternative dance music culture is not as non-discriminatory or non-patriarchal as it may initially seem. She also highlights "how troubling gender discrimination is present even in 'alternative' settings where participants are more conscious of political and social issues, including gender". Using an informal interview methodology, this chapter will explore these issues with women who have been "playing out" in nightclub contexts over the last 25 years. Gavanas and Reitsamer describe how across "historical, local and cultural contexts, DJ cultures have been, and continue to be, overwhelmingly male dominated" (Gavanas and Reitsamer 2013: 51). Their research focused on in-depth interviews collected between 2005 and 2011 with predominantly white, middle-class, female "well-established" DJs between the ages of 17 and 45. They ask, "what does this overwhelming male dominance say about DJ culture and the conditions for women DJs?" (Gavanas and Reitsamer 2013: 52). The key themes that they explore in under-representation of female DJs are that:

1. The majority of the researchers in this field are male.
2. The way that music history and musicology has been constructed, regardless of genre, is described as being framed "by a mutually constituting relationship between technology and masculinity" (Gavanas and Reitsamer 2013: 54).
3. They also state that collecting records is "a social practice associated with masculinity" (Gavanas and Reitsamer 2013: 57), and the record store is a 'coded' male space.

Gavanas and Reitsamer (2013) go on to describe how female DJs are seen as:

> Participating as anomalies in contexts where they are into seen as the normative figure of authority, the capacities of female DJs are viewed with suspicion; there is a significant burden of doubt as to whether female DJs possess the required competencies and capabilities to measure up to the job (Puwar 2004: 59; Wahl et al. 2011).
>
> (Gavanas and Reitsamer 2013: 57)

Many music scholars have described the music industry as male dominated, and in particular there are very few women in positions of power. This is also true in DJing:

> Researchers who have examined the gendered divisions within the dance music industry thus far have noted that most DJs and producers are men (Fikentscher 2000; Reynolds 1999); few women are found in other powerful industry positions such as label owners or club managers.
>
> (Farrugia and Swiss 2008: 83)

Writing in the 1980s, Angela McRobbie wrote that the vast majority of the key decision-making jobs across the music industry were dominated by an "old boys network" or "boys club," leaving women to other jobs, with less power and status. She states:

> It is still much easier for girls to develop skills in those fields which are less contested by men than it is in those already occupied by them. Selling clothes, stage-managing at concerts, handing out publicity leaflets, or simply looking the part, are spheres in which a female presence somehow seems natural (1994: 145).
>
> (Farrugia and Swiss 2008: 83)

So, when working in the music industry, women occupy these other roles and the men in these more powerful contexts do little to alter the status quo. For Kruse, "nothing about the social and economic organization of alternative music necessarily seeks to subvert the white, patriarchal structures of the mainstream music establishment" (Kruse 1993: 40).

There are also a number of barriers to learning how to DJ. How do women learn the skills required? Farrugia and Swiss state that the women they interviewed described learning how to DJ as a solitary process. They say, "any training or instruction women received rarely went beyond a brief 'how-to' session, usually from a male friend" (Farrugia and Swiss 2008: 90). Gavanas and Reitsamer also make observations about the scarcity of female DJs. They claim that this

> originates partly in the gendered social construction of technology and partly in the informal character of working environments and social networks in electronic dance music cultures, which are dominated by a combination of gendered images that present men as music technicians. and women as sexualised "objects".
>
> (Gavanas and Reitsamer 2013: 73)

They also thoroughly explore the claim that this issue isn't restricted to dance music, citing examples from scholars of punk, hip-hop, and indie rock (Reddington 2003; Rose 1994; Guervara 1996; Pough 2004; Kruse, 1993; Leonard 2006).

In order to explore these issues in a little more depth, I have divided this chapter into three sections:

1. Becoming
2. Doing it
3. The Next Generation

Initially I will explore how and why women, and non-binary people, become DJs. I've worked as a DJ for more than 20 years, so in the first instance I will explore my own experience. How did I get into music? How did I end up DJing? How have I framed this exploration? During 2019 I conducted detailed semi-structured interviews with four female and one non-binary DJ. I will also outline how they became DJs, looking for similarities between their stories and my story. The second part of this chapter, *Doing It*, outlines the lived experiences of woman and non-binary DJs; I have selected the participants to represent a range of DJ contexts, and they describe some very different experiences of working and of gender difference and misogyny. They also all had very different views on their particular experience. In the final section, *The Next Generation*, I ask each participant what they think about the current lack of younger female and non-binary DJs and what they believe the barriers are to young people who might want to be DJs. I also ask what is being done to address this issue and how we might redress gender imbalance in the future.

BECOMING

I was always interested in music. I watched *Top of the Pops* every Thursday with my sister. I listened to the charts on Radio One every Sunday, poised over the pause button on my cassette player, ready to block the annoying talky bits. My parents had a reasonable record collection; I found Bowie, Pistols, and Buzzcocks albums, and I started buying my own records, mostly 7-inch singles, with my pocket money from Woolworths at around 12 years old. My first album, aged 13, was *Boys Don't Cry*, by The Cure. By this age I also had some of my own kit: a small record player with cheap, tinny sounding inbuilt speakers, a transistor radio, a tape recorder and a box for my records. I gravitated towards friends who also liked music. I was at an all-girls' school, and although single-sex education is not something I chose personally, I wonder if my obsession with music flourished more freely because I was out of view of the boys. My childhood was full of live music: blues gigs, hippy festivals, squat parties. I started going to see live music from around age 14: my first gig was Prince at Wembley Arena; I had been listening to the *Purple Rain* album obsessively and even managed to get into the late-night showing of the *Purple Rain* film. Deep down I wanted to be a rock star, but four years of guitar lessons and the subsequent misogynistic put-downs from muso lads from the boys' school had put paid to that.

By the age of 17 I was regularly going to gigs, mostly punk, and what we called "American post-punk", sometimes referred to as sub-pop. But even saying that irks the teenage music snob in me; Sub Pop was a record label, not a genre. And then rave happened. It was 1988, and I didn't know anything about dance music. The only club in my provincial town was dreadful, full of older people and "squaddies", and played chart hits by Stock Aitken and Waterman. Rave was a revelation. We would all pile into a car and drive up from Kent to the M25, listening to pirate radio the whole way. Then we'd be told an approximate location and head for it. You could hear the bass from miles away. Five thousand people in a big top, in a field, all dancing. I mean all dancing . . . there were a few lasers, a few podium dancers, but otherwise very little that was visual. Just the people and the music.

By the time I was into my twenties I had quite a few records, and people would ask me to bring them to parties. So, I guess my first experience of "DJing" was sitting next to a record player, i.e. one deck, and putting one record on after another. My first experience of playing on decks was aged 26; I was heavily pregnant, and I could barely reach the decks due to the size of my bump. And my first paid DJing gig was in my thirties, when my DJing career bloomed. It was never a full-time job, but at one point it was around 25% of my income. It was loads of fun, a real confidence boost, a great social life – I earned some money and I spent most of it on records.

I've never been a very technical DJ. I'm a music geek, a collector, and it's about the archive for me. I play one record after another, roughly on beat, so that the crowd don't fall over. At one point I did do a DJing course and learned some software, some scratching, and some other things, but that course just made me realize the kind of DJ I didn't want to be. Some DJs are both; that is to say they are both archivists and technically amazing. In my case, playing the tracks that people want hear, want to dance to, and want to sing along with is what motivates me. But often DJs gravitate towards one end of the spectrum of the highly technical to the highly archival.

In both my visual and performance art practice and my academic practice, I have drawn on my experiences as a DJ and working in clubs and festivals. This includes making art works about and in these spaces, writing and delivering curriculum around DJing and clubbing, and publishing academic texts. In my performance work called *24/7* I played 7-inch records, in approximate alphabetical order, for 24 hours in a public space in Leeds. In many ways this was the ultimate DJ as archivist statement. As I played the records, I laid them out on the floor of the art space, celebrating the visuality and the graphical nature of the sleeve designs. It was durational, monumental, and epic; I felt like a music geek hero. But in this work it was the visitors' narratives that were important. They said, "I had my first kiss to this song", "I remember dancing to this when I was at uni", and this then prompted my *Socialist Jukebox* performance work, where collecting stories was the main aim.

However, if I think about *24/7* in a more analytical and auto-ethnographic way, I have to ask: what exactly was I doing, and why was I doing it? Why was it important for me to stand there, playing records, making a music geek statement and playing the endurance hero for 24 hours?

I now understand that it was, at least in part, a statement about being a female DJ. I was making a point about the size of my record collection, about my knowledge, and about my approach to music and collecting. And I was also making a point about resilience.

After *24/7*, Socialist Jukebox, and several written outputs I wrote a performance work called '*Facebook is like Disco, and Twitter is like Punk*'. This was a performance that was formatted like a TED talk. Except I didn't talk – I used the decks to play records, while animated text and images told a story. This performance has been incredibly popular and has been performed numerous times in the UK and overseas (there is also one version available on YouTube). This was also my most autobiographical work to date.

The following are some extracts of the text from this performance:

> In this "talk"
> I'm not going to talk
> . . . at all.
> This is a story of obsession
> and of music.
> I used to live a double life.
> I had two careers that I deliberately kept separate.
> I was worried that, if I came clean. . .
> it would be a disaster both personally and professionally.
> By day I was
> . . . an art lecturer
> By night . . .
> a nightclub DJ.
> *whispers* nightclub DJs don't talk btw
> Eventually, I became frustrated with the duality of my life, after a decade I was tired of burning the candle at both ends and I decided it was time to confess all . . .
> I began to make art about music.
> In 2007 I did a performance work that involved playing 7-inch records for 24 hours
> in 'approximate' alphabetical order.
> I started at 9am on a Saturday morning with The 5th Dimension Aquarius/Let the Sunshine In
> I ended with . . .
> Yazoo Don't Go, Yeah Yeah Yeahs, Rockers to Swallow, The Young Knives The Decision, Yazz Stand Up For Your Love Rights and The Zutons Oh Stacey.
> . . . at 9am on a Sunday morning.
> Over 500 7-inch singles in 24 hours.
> I am obsessed with music
> totally obsessed.
> It all started in 1977.
> In 1977 I was old enough to be allowed to stay up to watch Top of the Pops.

In 1978 my mum went to university, at Portsmouth Poly, and over the next three years I attended all of the end-of-term all-day discos.

So, from '78-'81, three times a year, I danced for 12 hours in the students' union, I watched Top of the Pops religiously and . . .

I listened to my tiny transistor Frosties radio (I'd found it in a charity shop) and this began my obsession.

1981 was a turning point.

I had enough pocket money, and enough freedom,

to go to Woolworths and buy records.

I started collecting.

So, what's all this got to do with social media?

Well, at the beginning of this formative period, in Portsmouth Poly Students' Union, I developed a love for both punk and disco . . .

. . . it was almost like there was a punk-me and a disco-me, and this schizophrenia has continued throughout my life.

I couldn't decide if I wanted to be Donna Summer

or Siouxsie

. . . deep down I really wanted to be Debbie Harry.

I love disco and I love punk,

and post-punk was a kind of happy resolution.

I go on to theorise punk and disco . . .

I'll think a little more about punk and disco first.

Richard Dyer wrote his paper 'In Defense of Disco' in 1979.

He argued against the characterisation of disco as 'capitalist music'

. . . by saying that all music is inherently capitalist.

And he also spoke in defence of the 'ambivalently, ambiguously, contradictorily – positive qualities of disco'.

For Dyer, the three key characteristics of disco are:

eroticism romanticism materialism

Central to this is a desire to escape the mundanity of life, the culture of work, of the office, of the boring job.

'Disco is part of the wider to and fro between work and leisure, alienation and escape, boredom and enjoyment that we are so accustomed to (and which *Saturday Night Fever* plugs into so effectively)'.

'Disco can't change the world or make the revolution. No art can do that, and it's pointless to expect it to.'

A lot has been written about Punk.

Simon Frith's article Post-punk Blues, published in *Marxism Today*, is nostalgic about the 'heyday of political pop' in the late 70s.

Frith states that

'pop music has failed, then, to realise the political fantasies that were piled on punk'.

'Punk failed to change the way popular music worked because it is, in capitalist practice impossible to construct an alternative . . .'.

He goes on, 'The tragedy of punk was not that it 'failed' to change pop but that so many people thought that it could'.

> In his essay *Listening to Punk*, written in 1985, David Laing stresses that an important legacy of punk was the introduction into lyrics of vernacular language.
> He states that, 'It was up to punk rock to introduce 'fuck' and the rest wholesale to popular music'.
> Laing finishes with the construct of the 'punk listener'. The punk listener has two key qualities . . .

1. The expectation of challenging listening i.e. potential for shock, acceptance of avant-garde elements.
2. The punk listener's enjoyment of other listeners' discomfort and trauma.

So, in this performance work, DJing, my childhood, social media, my academic life are given a soundtrack. I'm in the work, it's always a live performance, and I'm presenting myself as a female DJ, an archivist, a punk and disco lover, and a music geek, and I'm also framing it in a highly analytical way.

DOING IT

During my interviews with DJs, I asked the participants to describe how they got into DJing, how they see themselves, and how it intertwines with their sense of self, their careers, or their politics.

The DJs were carefully selected because I thought they might have a unique take on these questions. Hatty Lovehearts is someone who has had the experience of international touring. With Hedkandi, she was DJing across the world, living the rockstar lifestyle; she was a brand ambassador and at one point the face of a L'Oreal collaboration. She was also the only woman on their international roster, touring the world in her early twenties. Alice Bailey got into DJing through being on pirate radio, and throughout her career she has worked in radio, both as a DJ and as a journalist and also in clubs. Lucy Lockett is a DJ whose archival and technical skill is legendary. She has worked in clubs, with a focus on LGBTQ spaces for two decades. She also works for Equaliser, an organization that runs DJ workshops for female, transgender, and non-binary people and puts on parties with these DJs. DJ Miss Melodie is a DJ and producer; she set up Miss Melodie's DJ Academy to encourage more women and girls to DJ. She provides one-to-one tuition, and opportunities to DJ for girls as young as 9 years old. Sayang is a non-binary DJ. They came through the Equaliser program, and they run a QTIPOC Vogue house, in addition to producing and DJing. Sayang is also the youngest and least experienced DJ that I interviewed, only having DJed for around two years.

During the interviews, I identified some significant similarities in the participants' routes into DJing. The first similarity, which is probably the obvious one, is an early obsession with music. All of the DJs had a figure in their early life who encouraged or nurtured this interest – parents, siblings, i.e. usually someone older who helped them navigate the coded male space of the record store. All of them collected music and were fairly serious collectors by their late teens. One of the questions I asked all of the DJs was if their interest in music was different, or more intense, than other girls in their peer groups. There were mixed answers to this line of questioning. My own experience was that as a teenager, I spent time with a group of friends who were also into music, both dance music and rock music, and that so long as I didn't enter the masculine space of performance my interest, and that of my female friends, was encouraged. We also supported each other, playing records for each other and discussing bands. Other participant DJs talked about similar experiences, but also some spoke about being the outsider, or odd one out, or becoming one of the lads.

The next crucial moment in a DJ's career is often in their late teens. Two of the DJs I spoke to chose their university city based, at least in part, on the club scene, although at this point neither were playing out as a DJ. Also, a notable similarity is that both of them went out to a club, within a week of arrival, alone, or with a group of strangers. There was a sense of urgency in these stories, a need to find a particular community, as quickly as possible.

The step into DJing publicly was different for all of the DJs I spoke to, ranging from getting a gig, by saying she was a DJ, and then having to learn very quickly in advance of the gig, to being shown the basics and then practicing intently as a bedroom DJ for a long period of time. But there were also some common themes in this part of the interviews. The most significant commonality was finding an appropriate context and promotor. In all but one case these were spaces that were already employing female DJs, and in the other case, the clubnight was specifically looking for a female DJ. There was a clear sense that some spaces were unapproachable unless you were a young, white heterosexual male. The other issue was access to equipment to practice on, and a commitment to buying very expensive equipment early on.

All of the DJs I spoke to have had very different DJing careers. They have worked at a range of different levels and in a range of different contexts. Two of them had DJing or DJing/producing and teaching DJing as their main job, and two more described DJing as a significant second job. The DJs had worked in bars, nightclubs, festivals, radio, and at parties. They had worked in international, national, regional, and local contexts. They had been both employed and self-employed, had produced their own music and promoted their own nights, as well as being booked by promoters or venue owners. They had worked at all income levels from earning tens of thousands to a hundred or less per night. At one end of the spectrum, the DJ made enough to buy a house, whereas at the other the DJing and the equipment costs probably outweigh any income. All of them had experienced different intensity and workload across their careers. Income, frequency, type, scope, and context of DJ may have dramatically changed from one month to the next. Some of these shifts were related to childbirth and having young children, and other times the shifts occurred due to needing stability, or wanting to work more social hours, or to have a social life or a partner. It is extremely difficult to be working several nights a week, travelling for work, and have a successful relationship or family life. Although this can also be true of men working as DJs, pregnancy, breastfeeding, and support for this is much more complex for women. I did a gig when I was seven months pregnant, and I struggled to reach the decks around my bump. I also really struggled with tiredness and working late at night when my children were very young.

When it came to questions about the DJs' experiences of misogyny or sexual harassment in the workplace, all of the DJs responded differently. Hatty Lovehearts described how she, as the only female DJ in her first club DJ roles, and on the Hedkandi international circuit, created a hyper-femme version of herself. She has always had cropped hair, and she would wear a real hair wig and corsetry. She used her gender to her advantage to get work as what she admits might be the "token" female DJ. Hatty also said that she never experienced sexism from her fellow DJs – sometimes from sound technicians, and often abuse from clubbers, but never her colleagues.

Lucy Lockett, on the other hand, talks about numerous difficult moments, particularly around changeover. I thought this was a really useful moment to discuss. Let's say one DJ is working 10pm until 11:30pm, and then another will take over. The DJ box is a very small dark space, and in the ten minutes or so around changeover both DJs will be in that space. Walking in, you need to know which deck is on which channel and if there are any technical issues with the decks – there usually are. It's also polite to say hello, introduce yourself, and the outgoing DJ would usually say which song they would finish on and where they will leave the sleeve, so that the incoming DJ has a few minutes to think about what they will start with and will be able to carefully put the exiting DJ's record away. All of this is done in a space that is often unlit and always tiny. It's a complicated dance. Lucy describes this moment as one that is often fraught with misogynistic micro-aggressions. Sometimes it's rudeness, or not talking at all. Sometimes it's more patronizing. Sometimes it's about not making space. My own experience of changeover when DJing has been very varied. I have made some amazing friends and colleagues because of this interaction, but I have also been ignored, patronized, and had a number of pretty unpleasant moments.

It was useful during the interviews to think about different types of engagements at work in the DJ space to explore these issues, starting with:

1. Other DJs

   For around half the group of interviewees, other DJs were not an issue – they described supportive and nurturing relationships and friendships. However, for some of the DJs the story was very different. They described being ignored, patronized, and there are also a significant number of stories, from the DJs I interviewed, that relate to being mistaken for the girlfriend of the DJ, a makeup artist (with a box of singles) or a fan or clubber, who had gained access to the DJ box and needed to be removed by security staff. In all these cases, the staff reacting like this were described as being taken aback that these women were actually DJs.

> [G]endering is so normalised that it is often invisible to participants. One such example is the understanding of DJs as male by default. In many dance music communities, it is a common experience for women to have difficulty convincing club security that they are there for DJing work, not to simply "jump the queue" as a clubber. Even once inside clubs, many women are presumed to be in attendance only to assist or support their partners (who are presumed to be male).
> (Gadir 2017: 61)

2. The audience
   Gadir suggests that,

   > claims that some dance floors are free of discrimination while not acknowledging that other dance floors are not, are comparable with postfeminist ideas that gender inequality is a problem of the past. Although the two political positions are contrasting, both utopian and postfeminist perspectives of dance music cultures ultimately avoid and deny the hostility and violence that takes place because of gender – behind DJ booths, on dance floors and in-between gigs.
   > (Gadir 2017: 65)

   Almost all of the DJs I spoke to had experienced verbal abuse from clubgoers. Although this may be common across the sector, it was the nature of these comments that was disturbing. For example, "When is your boyfriend coming back on? He was better than you", "How many c**ks did you have to suck to get this gig?", and so on. We also talked about customers assuming that the DJ is available, and the vulnerability of being behind the decks on your own, as a woman.

3. Venue staff

   Sound technicians, or venue staff with a responsibility for equipment, were flagged as a group who may need some training or development in issues of equality. This isn't just about verbal abuse or patronizing behavior – here were numerous examples of these staff leaning across equipment, during a set, and attempting to adjust levels.

   One young woman said,

   I was setting up my equipment and I disappeared for a minute. The band had taken my power strip and started plugging their stuff into it. The sound guy took it and gave it to the band, but I was going on before the band. [It] threw off my whole vibe. I'm a firm believer in making friends with the sound guy 'cause they make you or break you. Having the power strip stolen was a huge symbol of disrespect to

me 'cause if I was a guy, and I hate saying this, I hate bringing in this if-I-was-a-guy statement, but it's true. If I was a guy setting up all my gear and I went to go to the bathroom and came back I doubt he would have stolen my power strip you know.
(Farrugia and Swiss 2008: 79)

4. Promoters

Promoters are a very complex group. They can be the best thing about a DJing career: wonderful, supportive, nurturing, flexible, and creative. Promoters can also be the worst part of the DJing experience: rude, arrogant, and sexist. One notable commonality amongst the DJs I interviewed was that they had worked with a disproportionate number of female promoters. It is a fairly small sample to draw any conclusion from, but I'm tempted to say that women book women, and men often don't. Several of the DJs referred to an "old boys club" of promoters and DJs, who rarely booked anyone outside of their specific, very male, community.

Cockburn, Rose, and Baker believe that the confidence of women grows when men are removed from the scene (Cockburn 1985; Rose 1994; Baker 2008). In other words, women find the presence of other women more motivating than the presence of men
(Farrugia and Swiss 2008: 91–92).

This means that the women who book women create a minority community in which women and non-binary DJs thrive, whereas the rest of the industry seems not to have made progress in terms of diversity in hiring. This is in part due to the informal, often underground, nature of these opportunities. However, there is significant evidence of the need for training, and awareness raising, across all aspects of the industry.

THE NEXT GENERATION

The final set of questions in my interviews were around the issue that there are still so few women DJs. I asked the DJs what they thought the barriers were now, and also what they knew about any projects or activities that were happening that attempted to redress this issue.

In the section of this chapter on Becoming, I explored my own experience and also that of the DJs I interviewed. Key issues in this were firstly being interested, or maybe even obsessed with music, acquiring knowledge about music, usually with someone older supporting this knowledge acquisition. Several of them talked about being a music geek or nerdy about music. Other key issues in becoming a DJ were finding the right community, context and environment, access to equipment, and overcoming anxiety. Being nerdy, or being perceived as a nerd, has been described as a potential barrier to girls and young women getting interested in DJing.

> In many music scenes – as Simon Frith (1981), Sarah Thornton (1996), and others have found in their own research – people gain access to "insider" knowledge and become "in the know" by networking, hanging out in record stores, clubs, cyberspace, and studios. But as we mention above, the social spaces and conversations that may encourage women to produce are often off limits to them.
>
> (Farrugia and Swiss 2008: 91)

Gadir also challenges what she calls "claims about the lack of interest of women in the obsessive pursuit of recorded music"; she states that this attitude is pervasive across genres. This has been "identified by Will Straw as a marker of masculinity – reinforced by understandings of record collecting as a mode of expertise" (1997: 4–5), and in Sarah Thornton's terminology, of subcultural capital (1996: 60–61). This "mastery" excludes women due to a process of male "homosociality" – realms of socialization where men are the sole bearers and sharers of their "nerdish" obsessions" (Straw in 1997: Gadir 2017: 59).

When I asked the DJs what they thought were the barriers to teenage girls getting involved in music and aspiring to be DJs, they all had similar insights but with difference emphases. Lucy Lockett described the process of beatmatching (aligning the beat of two records in order to seamlessly move from one to the next) as intrinsically patriarchal. All of the DJs talked about role models, communities of practice, the cost of equipment and records, and confidence issues. For most of the DJs the focus was on what young women can do to get into DJing, and the projects that were available to support them in this endeavor. The DJs that I interviewed were involved in several projects that aimed to redress the gender balance in DJing. For example, Miss Melodie's DJ academy was established to help encourage more females to become DJs.

Her website states that Miss Melodie set up the academy after,

> Having struggled to promote herself in a time when DJing was a much more male dominated business, something like the one-to-one tuition offered here at the academy would have been very beneficial. (www.melodiesdjacademy)

The academy now runs courses for all genders and includes working with people referred as part of an occupational therapy program. They also run a very successful junior DJs group for children from the age of nine. One of the essential things for Miss Melodie is that all of the tuition packages include an opportunity to perform. She has set up a club night specifically to showcase her students, and also arranges opportunities at festivals and events. She sees this as an essential element missing from many other forms of support and training. In other words, Miss Melodie's academy addresses the need for a safe space for new DJs, with a supportive promoter and audience.

Equaliser is an organization that two of my DJs have worked with. They run DJing workshops and events for "cis-women, trans women, non-binary and trans people". They encourage collaboration, and they share music events and are building a community to encourage and advise each other.

Their Facebook group says:

> Equaliser is a Leeds-based DJ collective that aims to nurture and promote the talent of cis-women, trans women, non-binary and trans people. They host monthly, free-to-access DJ workshops for non-cis men, helping to highlight new talent and offering access to an intimate, safer space to practice and learn. They also put on parties for everyone, celebrating the talent and artistry of DJs and performers who are underrepresented in the music industry. All profits from their parties go into funding the workshops.

Equaliser create a space where subcultural capital and music knowledge can be shared; where mentors and role models can be found. This space is also one where gendered ideas about being nerdy about knowledge acquisition or technology are challenged and debated. And like Miss Melodie's academy, performance opportunities are also part of the package.

These two projects, and many other similar ones across the country, are focusing on the 'becoming' issue. The issues around sexism and misogyny in the workplace are far more complex to address. In my interviews, there was one notable point of difference in our discussion of the issue of promoters and bookers. Sayang, the youngest of the group and non-binary, clearly placed the responsibility for change with promoters and bookers.

In her *Dancecult* article, 'Forty-Seven DJs, Four Women: Meritocracy, Talent and Postfeminist Politics', Tami Gadir describes how in 2016, only four of forty-seven DJs booked for Musikkfest, a festival in Oslo, Norway, were women. She goes on to describe how employment law or other measures are often not in place in these contexts and how in clubs "a prevalent postfeminist sensibility' is unchallenged and leads to the "persistence of discrimination". This echoes my own, and other DJs', experiences where either behavior went unchallenged, or when we or other colleagues highlighted discriminatory behavior club owners were often surprised that we did.

The Keychange program is an international program designed to highlight the under-representation of women in the music industry, both on stage and behind the scenes. They ask music festivals to pledge a 50/50 gender balance on their line-ups by 2022. The program calls for targeted investment to address under-representation, research, and education, but most importantly for initiatives aimed at improving working conditions and the lack of senior role models. These need to address 'recruitment, remuneration, career development and sexual harassment policies in a male dominated workforce' (Keychange website). Vanessa Reed is the CEO of the PRS Foundation and a major investor in the Keychange program. Reed stated that the program was interested in "solutions not statistics", and believes there's "no excuse" for not working towards equality (Cafola 2018).

In spite of mixed experiences with other (male) DJs, promoters, clubgoers and venue staff, all of the DJs including myself have found the experience of DJing liberating, confidence-building, and above all a huge amount of fun. All of the conversations that we had were predominantly focused on the positives of working as a DJ, and most of the DJs notably had a genuine and generous commitment to developing other young or inexperienced DJs. Another common theme was that the DJs had a sense of disappointment that the instances of misogyny were still a feature of working as a female, or non-binary, DJ. They had all hoped that by 2019 these behaviors would be a thing of the past. However, I was struck by the growing numbers of development opportunities and networks that it is possible to get involved with as a new or young DJ; I certainly would've got involved with these initiatives had they been available to me 25 years ago. There are far more opportunities to start out in a safe space as an aspiring DJ now. Also, there are people, like Reed, working at the level of policy, and people working at grass roots to redress the gender imbalance in the music industry. This is going to take time, but with organizations like Equaliser striving to "create open conversations, challenge and teach perceptions of gender within music, with an aim to create safer music spaces that everyone can be equal in", there is hope for the future. Equaliser's strap line is a good one: "Party for everyone, party for equality".

REFERENCES

Baker, S. (2008) From Snuggling and Snogging to Sampling and Scratching: Girls' Nonparticipation in Community-Based Music Activities. *Youth and Society* 39 (3): 316–339.

Cafola, A. (2018) *New Manifesto Details Steps the Music Industry Can Take for Gender Equality*. www.dazeddigital.com/music/article/42294/1/music-industry-gender-inequality-manifesto-keychange-europe. Accessed 21 May 2019.

Cockburn, C. (1985) *Machinery of Dominance: Women, Men, and Technical Know-how*. Dover, NH: Pluto.

Farrugia, R. and Swiss, T. (2008) *Producing Producers: Women and Electronic/Dance Music*, Current Musicology, No. 86, 79–99. New York: Columbia University.

Fikentscher, K. (2000) *"You Better Work!" Underground Dance Music in New York City*. Hanover, NH and London: Wesleyan University Press.

Frith, S. (1981) *Sound Effects*. New York: Pantheon.

Gadir, T. (2016) Resistance or Reiteration: Rethinking Gender in DJ Cultures. *Contemporary Music Review* 35 (1): 115–129.

Gadir, T. (2017) Forty-Seven DJs, Four Women: Meritocracy, Talent and Postfeminist Politics, Dancecult. *Journal of Electronic Dance Music Culture* 9 (1): 50–72.

Gavanas, A. and Reitsamer, R. (2013) DJ Technologies, Social Networks and Gendered Trajectories in European DJ Cultures. In *DJ Culture in the Mix: Power, Technology and Social Change in Electronic Dance Music*, edited by Bernardo Alexander Attias, Anna Gavanas, and Hillegonda C. Rietveld, 51–78. New York: Bloomsbury.

Guervara, N. (1996) Women Writin' Rappin' Breakin'. In *Droppin' Science: Critical Essays on Rap Music and Hip Hop Culture*, edited by William Eric Perkins, 49–62. Philadelphia, PA: Temple University Press.

Keychange website https://keychange.eu/. Accessed 20 May 2019.

Kruse, H. (1993) Subcultural Identity in Alternative Music. *Popular Music* 12 (1): 33–41.

Leonard, M. (2006) *Gender in the Music Industry*. Aldershot, UK and Burlington, VT: Ashgate Publishing.

McRobbie, A. (1994) *Postmodern and Popular Culture*. London and New York: Routledge.

Melodie's DJ Academy website. www.melodiesdjacademy.co.uk/. Accessed 18 May 2019.

Puwar, N. (2004) Space Invaders. Race, Gender and Bodies Out of Place. Oxford: Berg

Pough, G. (2004) *Check It While I Wreck It: Black Womanhood, Hip Hop Culture, and the Public Sphere*. Boston: Northeastern University Press.

Reddington, H. (2003) "Lady" Punks in Bands: A Subculturette? In *The Post Subcultures Reader*, edited by David Muggleton and Scott Weinzierl, 239–251. New York: Berg.

Reynolds, S. (1999) *Generation Ecstasy: Into the World of Techno and Rave Culture*. London and New York: Routledge.

Rose, T. (1994) *Black Noise: Rap Music and Black Culture in Contemporary America*. Hanover, NH and London: Wesleyan University Press.

Straw, W. (1997) Sizing Up Record Collections: Gender and Connoisseurship in Rock Music Culture. In *Sexing the Groove: Popular Music and Gender*, edited by Sheila Whiteley, 3–16. London and New York: Routledge.

Thornton, S. (1996) *Club Cultures: Music, Media, and Subcultural Capital*. Middletown, CT: Wesleyan University Press.

Wahl, A, Holgersson, C., Hook, P., & Linghag, S. (2011) *Det ordnar sig*. Teorier om Organisation och kon. Lund: Studentlitteratur.

6

She Plays the Pipe

Galician Female Bagpipers in the Production of Local Tradition and Gender Identity

Javier Campos Calvo-Sotelo

INTRODUCTION

In the extreme northwest of the Iberian Peninsula, Galicia has remained for centuries isolated from its neighboring communities, due to its rough terrain and underdeveloped communications systems. As a result, progress and modernity arrived there later, allowing the survival of interesting cultural practices, including some anachronistic remnants and demographic constrictions, although not necessarily deterrents of the contemporary emergence of female pipers studied in this chapter. In fact, the gender marker was reified in unique ways around the instrument assumed as soul par excellence of the Galician community and outstanding androcentric (that is, male-dominated) fetish, which is the bagpipe.

However, within the Iberian sphere, Galician society is not essentially different with regards to the secular discrimination of women. Moreover, in some respects the machismo found there surpasses that of other regions. For example, in the *muiñeira* (miller dance, the most popular Galician dance) the woman must keep her eyes down at all times, while the male dancer looks at her frontally: "she receives submissive the masculine homage, with the sight set on his feet" (Castroviejo 1983 [1970]: 579. Spanish original, hereafter SO[1]). According to Crivillé this forced visual reverence is "almost unique in the varied typology of Spanish dances" (1988: 255. SO). Clear evidence of discrimination, directly concerning bagpipes, can be read from the nineteenth-century Galician dictionary by Cuveiro; the entry *Gaiteiro* (male piper) is simply defined as: "Piper. He who plays the bagpipe". Instead, *Gaiteira* (female piper) means: "Frivolous and slutty woman" (1876: 144. SO). Galician folklore is unequivocal: bagpipes belong to men, and females very much appreciate the masculinity of the instrument. For instance, in the popular song "La gaita de Cristovo" (The pipe of Cristovo), he is kicked out of the house by his wife because he sold the bagpipe. A phallic metaphor is explicit in the following song: "A muller do gaiteiro/Muller de moita fortuna/Ela toca dúas gaitas/Outras non tocan ninguna" (The piper's wife/Is very lucky/She plays two pipes/Others play none. Losada 2002: 44. Galician original, hereafter GO).

Women devoted to popular music were rather infra-cultural figures in Galicia, largely confined to begging; when American journalist Ruth Anderson toured the provinces of Pontevedra and Coruña in 1924–1926, she met several scenes of mendicancy. One of them consisted of two women playing at a rural fair in Lalín (Pontevedra): "a blind violinist and a seeing tambourine player, who went about collecting pennies" (Anderson 1939: 109–110); other sources describe a similar marginality (Arias 1980). There are scarce testimonies of women playing bagpipes. It is likely that they did play at times, but this almost never took place in public. In Scotland "[g]irl pipers, though they certainly existed, were relegated to the background" (Collinson 1975: 193), and the circumstances were similar in Ireland (Harper and McSherry 2015). The Galician figure of Aurea Rodríguez (1897-?) has received some attention; she founded a pipe band in Orense with her orphan brothers a century ago, but despite this and a few more isolated precedents, women were detached from bagpipes for centuries. The masculinity attributed to the instrument embodied a kind of moral order, even a religious dogma, and any deviation was simply unimaginable (Campos 2007, 2015). In Galicia:

> Women's musical practices were focused on singing accompanied by small percussion (usually the tambourine) or linked to dance. A woman was an ornament for popular music, lacking remuneration. By contrast, men reigned supreme over pipes. . . . He was the musician, the "professional", and was accordingly paid.
> (Barreiro 2012: 143. GO)

On the other hand, Nacho González (bagpipe teacher) points out that concerning the skills required to play pipes, gender is irrelevant; "even young girls can do it just as an adult man" (Personal Communication, hereafter PC, 2018[2]). According to Hipólito Cabezas (influential piper and teacher), "a girl can play the bagpipe for more minutes than an adult male if she masters the technique of the instrument. Piping is not a matter of strength" (Graña 2013. SO).

Despite the weight of history, the "Celtic decade" of the 1990s (Campos 2017) witnessed a sudden emergence of young female pipers in Galicia, who challenged the male-gendered realm of the instrument. The most outstanding were Cristina Pato, Mercedes Peón, and Susana Seivane. A new generation is currently taking over their initial outbreak, with significant activities that revitalize Galician cultural capital. Forerunners of all of them were a 1960s pipe band called Meniñas de Saudade (Nostalgic Girls), from Ribadeo (Lugo), whose socio-cultural role was quite different due to the historical context. The phenomenon here studied is rather uncommon, as the figure of the female piper is not frequent in the world. Susana Seivane stressed in an interview how it was "outside Galicia" where people were surprised to see a female piper (Rejas 2005. SO). This fact might be explained considering Galician background culture. Firstly, the Galician matriarchy: an internal hierarchy that made it normal to see a woman pulling the oxcart, driving a tractor, organizing the family businesses, and ultimately playing bagpipes. This matriarchy was not due to the supposed femininity of the Celtic ancestors, or similar fantasies, but to the scourge of Galician significant male emigration.

> The high levels of male migration from Galicia produced a remarkable demographic imbalance in many areas. In some jurisdictions, there were as few as eighty men for every one hundred women. . . . Thus, Gallegan society was characterized by high rates of female-headed households.
> (Poska 2000: 314)

Local nationalism decisively interfered with the social reality. Miguélez-Carballeira establishes how the myth of a "feminine" Galicia (sweet, nostalgic, and traditional) was discarded at the beginning of the twentieth century as inefficient in the construction of an active and pressing Galicianism: "the early texts of Galician political nationalism reacted against such metaphors by means of a heightened masculinist discourse bent on recasting national insurgence as a question of virility" (2012: 367; see Palacios 2009). The female "inherent inferiority" pervaded the twentieth century; when Maruxa Miguéns (wife of Xosé Romero, founder of the Galician traditional band Os Rosales) started rehearsing with her husband's band, the male members said: "what a shame she is not a man to come with us" (Couto 2017: 57. GO). Years after she declared that in the early 1960s "a woman was deprived of everything. . . . It was totally inconceivable that a woman played in a men's band" (Busto 2005: 65. Portuguese original). In the Casa da Gaita Museum (House of bagpipe. Viascón, Pontevedra), some instruments of the exhibition are classified as typically played by women; none of them are melodic, only percussion, as if to suggest low female intellectual, rather than physical, status (author's visit, July 31, 2018). Therefore, the pipers' revolution of the 1990s probably involved an assault not so much on masculine strength, but on talent and intelligence, and as characterized by a strong determination:

> [In the last decades] bagpipes are the most played instrument by female musicians, above others such as the piano or the violin, which had enjoyed a much wider preference among the institutions of music education in Galicia.
> (Barreiro 2012: 143. GO)

Secondly, and likewise related to male migration, a little-known fact is that there have been many female bagpipers in Galicia in the last century and up to the present. They constitute an essential support for the feast, the maintenance and sustainability of Galician musical heritage, and the production of spaces of positive social interaction: in the patron celebration, summer events, carnivals, weddings, and the like. These girls and children, wearing the regional dress, are totally apart from the music industry, and normally ignored by the people around; they play traditional pieces with moderate technique, and will barely obtain any profit from their performance. They articulate a conventional-touristic image of Galicia – according perhaps to the interests of a local Council – but their contribution to the celebration must not be disregarded; in many cases their formation and activity come from individual initiatives, not from institutional and/or commercial strategies (Figure 6.1).

Thirdly, with the Spanish transition to democracy and its cultural tide (after 1975), there was a rediscovery of Galician traditional music by artists and bands like Milladoiro, Emilio Cao, Fuxan os Ventos, and others, which triggered a conscience of local-ness as the triumph of the aesthetic sovereignty of Galicia (Campos 2009). The pipers of the 1990s thus incorporated a burden of overcoming symbolically the past, helped by a strong revival that spread worldwide since the mid-twentieth century. Finally, despite Galician long-established backwardness, the feminist movement also left its mark, and a woman holding a bagpipe began to suggest refinement and cosmopolitanism (the reverse of older times), in addition to the pure bodily attraction (on female creativity, gender stereotypes, and domination models, see Eisler, Donnelly and Montuori 2016).

To summarize, in the late twentieth century a synergistic combination of facts paved the way for the success of female pipers in the old and remote Galicia. Pato stated in 2005 that:

> I have never felt gender discrimination. . . . In my record label [Fonofolk] in 1999, the slogan was something like "Cristina Pato, the first

Figure 6.1 Pipers in Cambados (Pontevedra), celebrating the Albariño Wine Feast. August 6, 2016

Source: Photo J. Campos.

woman to release a bagpipe album in Spain" [*Tolemia*]; as far as I know, that headline opened for me many doors to the media, and maybe my success is due to that phrase. . . . Honestly, my case has been a clear example of "positive discrimination".

(PC 2005. SO)[3]

Seivane concurred when asked if to be a woman hindered the access to solo piping: "not at all. Nowadays there are many women playing the bagpipe. . . . It is already a normalized fact" (Mouriño and Sánchez 1999: 23. GO). Galician cultural micro-cosmos, at the turn of the century, was experiencing via bagpipes a process of redefinition of the female figure, the birth of a local ecofeminism, and a remarkable gender leveling that went far beyond the strict limits of music. Three pipers stood out within the process, but they had a noteworthy precedent.

MENIÑAS DE SAUDADE

Meniñas de Saudade were a pipe band founded in the northern village of Ribadeo (Lugo) in 1961 on the initiative of the cultural promoters Amando and Carlos Suárez Couto; their teacher and manager was the piper Primitivo Díaz. Initially they were eight girls, and eventually ten or more, forming a main section of pipes accompanied by snare drum and tambourine or square drum. María Acuña, journalist and researcher, is the daughter of Teresa Rodríguez, piper of Saudade. According to María, the 1960s were a turning point in the history of Galicia, a time of sociopolitical effervescence in which the vindication of women's role in music relied upon a group of pipers, Saudade, the first band entirely formed by women in Galicia and Spain, although there were precedents abroad (PC 2018–19).[4]

From a contextual perspective, nonetheless, the band possibly responded to the Francoist patriarchal imperative, playing in demure style, and confined to an ornamental function (see Campos 2009; Hernández 2014).[5] Moreover, their organization and decision making belonged to men exclusively. To some extent they could fall within the category of 'sexual exoticism' (Giuliani 2016), meaning gender otherness exploited as entertainment (playing pipes in this case). Some marginalizing mechanisms were influential; for example, Teresa Rodríguez remembered how joining the band altered positively her adolescence, because at that age girls were deprived of continuing to study: "we had to devote ourselves to learn sewing, to be dressmakers" (Cuba 2012. SO). A particular trait of Saudade is that they always played in strict unison, avoiding the complexities of polyphonic and rhythmic combinations (PC 2018–2019). Their monophony perhaps reflected the then widely accepted belief that women were unable to understand anything that required mental effort (more information and some pictures of the band are available on http://bit.do/eS8YH, and mostly on Acuña's 2013 online paper: http://bit.do/eTefc).

Despite the submission connotations, Meniñas de Saudade achieved an impressive success all over Galicia and beyond, displacing women from the hearth to the stage, something unthinkable in the past. They demonstrated that the audience wanted females, that pipes were attractive (and not sacrilegious) in feminine hands; and that young girls could play them as well as men. However, their activity barely surpassed the level of live performances; Meniñas de Saudade were filmed for the documentary *Así es Galicia* (This is Galicia), directed by Santos Núñez in 1964; and for the film *España Insólita* (Amazing Spain) directed by Javier Aguirre in 1965. The band also had fleeting appearances in NO-DO (a Spanish state-produced series of newsreels from 1943 to 1981). In 2010 Acuña completed the documentary *Saudade, Retrato en Si Bemol* (Saudade, portrait in B flat), on the history of her mother's band (Acuña 2013; PC 2018–2019). Nonetheless, the girls of Saudade never recorded an album,[6] spoke on the radio and TV, or verbally addressed the public. Their legacy is thus rather silent and hard to grasp. In this facet, there is a strong contrast with the next generation, who unexpectedly broke through gender constraints in the 1990s.

THE THREE PIPERS: MERCEDES PEÓN, CRISTINA PATO, AND SUSANA SEIVANE

Although before and after them there have been other important female bagpipers in Galicia, Cristina Pato, Mercedes Peón, and Susana Seivane represent somehow a different category, a step above the rest, both because of the quality and extensiveness of their output and because of their iconic capacity to act as catalysts of the female uprising. They have also exerted an important role in the renewal of Galician musical traditions and nation-building. In times of cultural globalization, permanent crisis, and feminist struggle, they became the best ambassadors of the country worldwide.[7]

No one doubts that they are musical forces. All three have won quite a few awards and deserved international recognition. But far from being a homogeneous trio, among them there are notable differences. With respect to their public image, they showed a spicy sensuality in record covers, promotional imagery, and live performances.[8] Their narrative involved the transgression of the Galician woman canon, subject to her father/husband/children, aiming above all to marry, and dressing only in black once the husband had died. However, despite the active and oppositional femininity exhibited by Seivane and Pato, probably the most disruptive in the breach of heteronormativity and production of a new femaleness was Mercedes Peón; around the year 2000 she shaved her head (Sinéad O'Connor style) and further addressed gender issues, occasionally in radical terms (see next). Musically there are also remarkable divergences among them. Let us consider their figures separately.

Cristina Pato Lorenzo remembers that her first contact with music was at home, where the whole family used to sing around the accordion of her father; she would never forget that social condition of music, "shared IN community" (uppercase original; PC 2019). Pato's debut record was *Tolemia* (Madness, 1999), and it was a very successful one.[9] The front cover showed a young and smiley Pato with the bagpipe, in cheerful colors (see Padrón and Sánchez 2013). Her ulterior green hair dyeing separated her from mainstream stereotypes; despite her dynamic and varied career, bagpipes are always near (Figure 6.2).

Periodically she has released solo albums following an organized process:

> [Lately] I usually work with concepts (*Migrations* [2013], *Rustica* [2015], *Latina* [2015]). I start with the idea of the story I'd like to tell and from there I develop the content and profile of the recordings.... I usually order works or arrangements from composers.... Once I have everything ready, I start production: musicians, music, image, design, videos, public image, record label and promotion. Since I moved to NY in 2005, I control and direct the process from start to end.
> (PC 2019)

Her participation in other recordings includes a notable guest appearance in the Yo-Yo Ma & Friends 2008 album *Songs of Joy and Peace*, with a Galician carol ("Panxoliña"), which she performed with the eminent cellist (fragments of the rehearsal are available on http://bit.do/eTdV6).

Figure 6.2 Cristina Pato, the glocal Galicia

Source: Photo courtesy Xan Padrón, 2018.

The record received a Grammy Award the following year, in the category Best Classical Crossover Album.

Pato soon became a renowned pianist no less than piper. In fact, she decided to abandon her life as a bagpiper in Galicia, around the beginning of the twenty-first century, to devote her career to the piano in the USA. There she works actively in the composition and performance of avant-garde pieces (as a bagpiper and as a pianist), collaborating especially with the Russian-Argentine composer Osvaldo Golijov. She has played with the Chicago Symphony, Hartford Symphony, Baltimore Symphony, and New York Philharmonic orchestras, with hybrid repertoires (Pato 2018). About her setting up in America, she emphasizes the symbolic significance of bagpipes:

> My own experience in the field of music in the USA . . . was based on the simple strategy of accomplishing as many activities and alliances as possible with the means I had around me. No one needs a bagpiper on the other side of the ocean, but the qualities stemming from the metaphor of the instrument itself (pastoral origin, a catalyst for the survival of cultural identity) and its practice (centered in the community and maintained by it) are universal values.
> (Pato 2018: 60. GO)

In an interview in *La Voz de Galicia*, Pato explained: "[the bagpipe] is 30% of what I do. Had it not been for it, I wouldn't be here, it opened the doors. It's just that there's much more than Cristina bagpiper" (Méndez 2018: 35. GO). Her initiative The Gaita and Orchestra Commissioning Project is a professional frame for avant-garde compositions with bagpipes (see http://bit.do/eS8Zq). However, and a bit surprisingly, Pato's strongest vocation is teaching, and the reason for going on with the instrument:

> Teaching is the strongest call I've had in recent years, with passion and vocation. But the bagpipes are the means that open the doors to the teacher, and so I continue my career as a soloist.
> (Pato 2018: 62. GO)

Pato is the educational advisor of Yo-Yo Ma's Silk Road Ensemble, affiliated with Harvard University. Over the years she has formed her own stable band (the Cristina Pato Jazz Quartet) and the Galician Trio (with accordion and percussion), organizing personally their tours and recordings, spanning almost the whole world. Additionally, Pato has created the music festival Galician Connection (2012–2014), which was a fascinating experience for her, and has received ample online attention. Her intellectual career also deserves attention: among other titles she holds a doctorate in Musical Arts from Rutgers University in 2008. In June 2019 Pato was appointed Chair in Spanish Culture and Civilization at the King Juan Carlos I Center (University of New York).

The complete list of Pato's discography, videography, publications, organization of events, educational activities, and others can hardly be summarized here. Only on the subject of musical recordings, it includes six solo albums, nine in collaboration, and participation in 28 more (see her personal website, http://bit.do/eS8Z2; under her name it reads: "Bagpiper, pianist, educator, writer, producer, composer . . . independent artist!"). Overall her productions are characterized by a "glocal" and successful bias, where Galicia and bagpipes still play an important role, but within a spirited and ambitious transnational frame. The notion of "glocal" implies hybrid cultural productions resulting from a dynamic interaction between the local and the pressure of modern globalization; and therefore "a potential new paradigm that is both post-national and post-peripheral" (Colmeiro 2009: 217). Cristina Pato is an outstanding representative of the exogamy dimension of current Galician-ness, as a dialectical tension between the local homeland and the global infinity. In her own words:

> Personally, I am particularly proud of my work with Yo-Yo Ma and the Silkroad, and of having contributed to the projection of the Galician bagpipe as a global instrument; through my work with them in classical music, with my American jazz quartet, with my Galician trio in world music, and as a soloist with orchestras like the Chicago Symphony or The New York Philharmonic.
>
> (PC 2019)

Mercedes Peón Mosteiro self-defines as: "folklorist, researcher, choreographer, bagpipe and tambourine player, vocalist, composer, conductor, presenter, producer, and businesswoman" (PC 2019). She is considered to be a remarkable singer besides her pipe career, and has additionally carried out a relevant labor as compiler of the oral Galician legacy in extensive fieldworks.[10] Peón was for eight years presenter of the show *Luar* (Moonlight) at the TVGA (Galician TV), where she introduced to the audience many unknown Galician musicians, singers, and even just peasants, intensely advocating local repertoires as alive and sustainable heritage. She has published a few academic articles stemming from her research (e.g., Peón 2004).

At the beginning of her musical life, Peón did not like bagpipes: "I loved the tambourine and the telluric trance that my body experienced playing or dancing to the sound of these impressive ladies [tambourine players]" (PC 2019). It was due to the fascinating influence of José, a piper from Paradela (Coruña), that she felt attracted to the instrument: "It was love at first sight. . . . I suddenly found an instrument that really was part of the authenticity of the musician; coherent, free, unregulated, intuitive and wild" (PC 2019). However, coinciding in this with Pato and unlike Seivane, in her recent works bagpipes have lost weight: "the bagpipe is a brushstroke at some moments in my last albums. I think until the third one it was an indisputable part of most of the tracks, but in the last 10 years it's no longer" (PC 2019).

Born in Coruña, currently Peón lives in the small inland village of Oza dos Ríos (Coruña). She describes her career as follows:

> My formation is linked to the collective creation of micro-habitats in Galicia through fieldwork for 20 years, until being able to understand

those musical codes that reach the twentieth century. On the other hand, working as a music producer stemmed from not wanting to depend on external assessments; I am totally self-taught. My facet of composer is linked to improvisation, political and social positioning and the search for new systems consistent with each emotional moment translated into sound spaces.

(PC 2019)

Unlike Pato and Seivane, she has typically addressed gender issues in her public discourse, e.g. discussing feminism (http://bit.do/eS82i). On July 1, 2006, she opened a concert in Santiago de Compostela proclaiming the Gay Pride. Currently she affirms that: "discrimination exists because we live in a patriarchy in a sexist, racist, colonialist system" (PC 2019).

Her TV programs in *Luar* became a fundamental part of her trajectory. They enabled Peón to bring to the fore songs and dances of the innermost Galicia, which was rewarding for her:

> I was there eight years bringing village people. They taught us collective creations and Galicia's true musical language. . . . I made 278 programs; it was one of the best things I did in my life. . . . This resignified the villages, their speech, their customs.
>
> (PC 2019)

Interestingly, Peón discovered a great musical variety within Galicia, which most likely influenced her future works:

> Each place, each village, had its own music resources that were constantly evolving. Complex melodies, quarter tones . . . diverse nomenclatures . . . different ways of grasping the instruments. Continuous inventions, compositions and improvisation.
>
> (PC 2019)

Mercedes Peón has released five solo albums since her 2000 debut with *Isué* (that is, in Galician slang) – totally based on her compilation research labor, the same as many of her live performances (Figure 6.3; see her personal website, http://bit.do/eTeit).

Her motivations to compose involve critical assessments:

> I always had an urgency to denounce through my own freedom. Freedom to speak of the situation of the language in Galicia (where only 10% of boys and girls speak Galician), of the patriarchal system that surrounds the nation states.
>
> (PC 2019)

Concerning her working method, Peón states that currently she does not necessarily plan her recordings in advance, but when she accumulates enough pieces, she goes into the studio with a view to creating a

Figure 6.3 Mercedes Peón, performing Galicia

Source: Photo courtesy Mercedes Peón.

concept album. The material procedure is via Nuendo, a digital program with which she designs the songs. In the process, Peón funds the recording costs herself and when the record is released, she cedes the operating license and receives the copyright benefits. She has also spoken in many radio stations and acted regularly at the TVGA (not only in *Luar*). In a normal year she performs around 25–40 concerts, playing and singing with different ensembles. She has performed all over the world and given lectures at the universities of Sorbonne in Paris, Wales in Cardiff, Porto and Lisbon in Portugal, and others (PC 2019). Peón has created several film soundtracks and theater music (like the music for *SóLODOS* play, by Maruxa Salas), as well as musical choreographies (e.g., the music of *O kiosco das almas perdidas*, Roberto Oliván's show for the Galician Choreographic Center. Information from http://bit.do/eTere).

Musically Peón has never abandoned the local roots, but her evolution towards complex trans-genres is clear. The official video of her song "Déixaas" (meaning "Leave them alone", the first track of the 2018 album of the same name) displays a gloomy shipyard in Ferrol (Coruña) where three women perform a mixture of hard beats, vaguely Galician melodies, and the background of heavy machinery sounds, with a somewhat hypnotic result for the audience (http://bit.do/eS82Y). Like all her vocal music, it is sung in Galician. Commenting on the ideological contents of the record, she outlined four elements related to Galicia as inspiration source: folklore, feminism, language identity, and the industrial crisis (PC 2019).

In summary, Mercedes Peón is a woman committed to her country, its customary practices and its people, to the rural and working Galicia, and a zealous guardian of the country's intangible cultural heritage (ICH). She represents tradition renewed with cross-cultural repertoires within the fundamental affirmation of Galician otherness. In the thorny intersection between gender and nationalism, she reconciles both axes in a synthesis that becomes central to her production. Her placement is never far (even geographically) from the motherland, advocating women's civil rights and sexual freedom.

Susana Seivane Hoyo is the most purely consecrated piper of the three. Her life is the instrument and its strict defense from a determined notion of Galician authenticity. She keeps the family tradition of bagpipes fabrication at the workshop her grandfather founded in Cambre (Coruña). Occasionally she plays the piano and the accordion and sings regularly, but above all she is an outstanding piper. A detailed 2010 documentary-interview devoted to her, with many musical interpolations, was broadcast by the TVGA program *Alalá*, no. 146, on May 6, 2010 (http://bit.do/eS83x). In it Seivane describes her world, family, interests . . . and bagpipes as the leitmotiv of her life, declaring: "the bagpipe is my passion, my whole life. . . . I was born with two arms, two legs and a bagpipe. . . . It's the soundtrack of my existence" (see also Cronshaw 2001; and Seivane's personal website: http://bit.do/eTgNA).

Seivane has released five solo albums. The first one, entitled *Susana Seivane* (1999), was produced by Rodrigo Romaní, front man and founding member of Milladoiro, the most important Galician neo-folk band of those years. Her second album, *Alma de Buxo* (2001), produced by Seivane herself, included the collaboration of Uxía Senlle, Kepa Junquera, and her musical advisor Rodrigo Romaní. In this record Seivane introduces drums, electric bass, and some compositions of her own. Nevertheless, *Alma de Buxo* is first and foremost a tribute to bagpipes, starting with the title, which means "soul of boxwood", alluding to the Galician tree species used to fabricate the pipe model assumed as a paradigm of Galician musical authenticity (even though most of the "Galician" pipes are actually fabricated with foreign woods, same as is the case in Scotland or Ireland with their flagship pipes).[11] She performs live mostly in Galicia, but over the years she has played worldwide.

Seivane was a jury member of the bagpipe contest *Vai de Gaita* (About bagpipes, TVGA), where in December 13, 2013, she had a strong argument with a participant because he played a bagpipe that she considered alien to Galician tradition. The debate followed in the next days on the Internet, with many web surfers outraged for one reason or another. However, Seivane's defense of local values has not led her to reject global culture; she participated in a publicity campaign of Microsoft in Galicia in 2011, playing with her bagpipe the Galician national anthem at the foot of the cathedral of Santiago (subtitled in English on http://bit.do/eUk5M). Seivane provided the music for the TV advert of the Windows 7 operating system in Galician language, under the slogan *Vive en Galego* (Live in Galician).[12] Additionally, she has presented some of her records at the

FNAC store of Coruña,[13] publicly advocating for both FNAC and music copyrights. These tensions between Galician historical traditionalism (to the extreme of repudiating a pipe as sacrilegious in a TV contest) and the immersion in a global market of the latest generation (that of powerful multinationals) converge within a progressively glocalized Galicia in some of its social actors (Colmeiro 2009; Hooper 2011; Romero 2012. See the comments above on C. Pato). According to Romero: "Seivane represents the best example of *fusion* of tradition and modernity that Microsoft wants to communicate" (2012: 29, italics original).

THE 2002 CONCERT IN CORUÑA

A staged collaboration of the three women was expected by many people after their arrival on the scene. Finally, it took place in 2002, becoming an outstanding milestone in the Galician transition to the narratives of modernity, as well as a sort of empathetic icon for the future. The triumph of the female piper – as it might be called – was publicly celebrated on August 22, 2002, at the María Pita square of Coruña, where a significant concert of Pato, Seivane, and Peón, accompanied by the Orquesta Sinfónica of Galicia conducted by Víctor Pablo Pérez, took place. It represented somehow the burst of a renewed Galician cultural capital, both in the female-gendered assault on the historical hyper-masculinity of the instrument, as well as in the aesthetical syncretism of the fusion between pipes and a classical orchestra. Nine thousand spectators attended the event, entirely filling the capacity of the square. Interestingly, most of the pieces performed were original by the three women, with arrangements for orchestral accompaniment. By Peón: "Marabilla," "Serea," "E xera," and "De seu." By Seivane: "Marcha procesional dos Mato," "Vai de polcas," "A farándula," and "Sabeliña." And by Pato: "Noite de lúa," "Africa," and "En o Sagrado en Vigo" (information from different 2002 media. All the songs had Galician titles). Pato's memories of that concert, 17 years later, emphasize the emotional environment present there:

> In hindsight, what I remember was the atmosphere that was breathed in the square. The expectation for something unique that was happening and the illusion of putting the bagpipe in front of the orchestra at a historical moment for the instrument.
> (PC 2019)

This is Peón's narration of the experience:

> I remember most the orchestral performance of my piece "Marabilla" [Wonder]; it was one of the most amazing moments in my life because, considering that I lacked formal education, I never had the opportunity . . . to have so talented musicians at my disposal, with immense affection, as well as the conductor Victor Pablo. . . . Also being with Cristina and Susana was a pleasure. . . . I left off the stage with a huge hug to all these people and my admired piper companions.
> (PC 2019)

The public, media, and musicians themselves praised the event, highlighting its unusual nature as a cathartic gathering. A deep social and aesthetic metaphor was implicit at that moment: women were dethroning men in the use of the bagpipe (a millenary symbol of manhood), same as the bagpipe itself was dethroning the violin (the king of cultured music for centuries) as a soloist in front of a classical orchestra.

A NEW GENERATION: LETICIA, DUNIA, AND MARINA

To complete this study, it was necessary to devote attention to recent pipers. The following commentary derives from personal interviews carried out face to face in August 2018 with three young girls, who together provide a true picture of their generation.

Leticia Oubiña, 20 years old, represents a very important dimension of Galician popular music and piping in particular, which could be called the purely hedonistic. She plays because she likes it, ignoring ideological nuances and ethnic or gender commitments. Far from anomalous, her positioning is shared by many current pipers (female and male indistinctly), who approach piping with similar purposes. For example, she states that the cloth of her instrument is pink to deviate from mainstream black and red (worn by many pipes), not due to any gender issue (PC 2018).

Leticia started playing percussion and singing from an early age. At eight years old, she began with the bagpipe. Shortly after, she incorporated to the school of González in Cambados (Pontevedra), where she still is. At the same time, she is finishing a Nursing degree at the University of Santiago and knows that she will not make her living with bagpipes. Her band is Xironsa, which won the pipe bands contest of Galicia in 2015. Despite the success, their media repercussion is null, apart from a brief appearance in a local TV program. They have not recorded any albums, nor participated in other recordings; by mid-2019 there were no available videos of them on the Internet. With Xironsa she plays assiduously at patronal feasts dressed in the Galician costume. The band charges about EUR 350 for each participation, which only covers the expenses; there are never benefits-sharing. Interestingly, Leticia's main complaint concerns the instrument itself. Asked if she ever felt discriminated for being a girl playing a conventionally masculine instrument, she answered that not at all: "the worst thing is the negative image of the bagpipe among Galicians, not being a woman; the bagpipe is despised" (PC 2018). These almost anonymous pipers unveil a different side of piping in Galicia, far from the attention achieved by other performers.

The personality and mentality of Dunia Alvarez (34 years old) are virtually the opposite of Leticia's, although they coincide in several revealing points. To start with, Dunia asked to have our PC in Galician, as a means of vindicating language. She holds a degree in Audiovisual Media (University of Vigo; her bachelor thesis will soon be published) and works as a computer technician. Her beginnings in music were early: at the age of four she learned Galician dance, but she did not start playing bagpipes until she was 30, although she always liked the instrument.

Dunia is an extremely active woman. She represents a new generation of Galician females in music who take a multifaceted approach to music-making (Dunia is a tambourine player, bagpiper, dancer, singer, compiler, teacher, editor, writer, musicologist). Importantly, these women are selfless with the causes they esteem, evincing a strong Galician awareness that raises "the maintenance and transmission to new generations of Galician culture and Galician language" as sovereign goals that must be fought for daily (PC 2018). Not ignoring gender issues (that also affect her), Galicia is the emotional epicenter of Dunia's work: its land, culture, language, and otherness. In Dunia nationalism and feminism thus coalesce, invigorating each other, in a leftist and rebellious compendium in search for an audience to convince. In fact, on stage she frequently indoctrinates the public with ideological content. Dunia is openly nationalist and hopes one day the independence of Galicia will take place, "because it is a different country, with its own language". Concerning gender, she believes that in Galicia there is still "a lot of patriarchal chauvinism, although shrinking" (PC 2018).

In 2016 she founded an only-females band called Cinco en Zocas (Five in clogs; to enhance the rustic wooden shoes old peasants used to wear). They sing and play traditional percussion (rarely bagpipes), with intense beats and frequent harmonizations in parallel thirds. Cinco en Zocas have achieved some popularity, and a few videos of them are available (e.g. http://bit.do/eS84b, Dunia is second from the left; and http://bit.do/eS84B, Dunia in the middle[14]). Cinco en Zocas also compile Galician traditional music; they do not publish the pieces collected but use them to enrich their own repertoire. They receive no grant or help for the task; their motivation is to preserve the Galician ICH: "so that the Galician tradition is not lost; to transmit it to the young people and to keep it alive". In the same way, they impart occasional summer courses, also at schools, normally for free (PC 2018).

Concerning the practical problems of current piping, Dunia corroborated Leticia's assessments: "pipe bands are poorly paid". She complained about the disproportionate distribution made by public entities of funds for festive events. For example, the Council of O Grove (Pontevedra) spent EUR 170,000 on the management of the 2018 Festa do Marisco (seafood festival). Paradanda (a local band conducted by González) "marched in the streets, encouraged dancing, with several stops, etc., for four hours in full summer sun," and received EUR 500, which were destined to the group expenses; the musicians received nothing. On the other hand, "any pop band, singer or DJ with a minimal reputation, will cash-in from ten to twenty thousand euros in the same event". Dance orchestras charge around EUR 5,000 per performance. If a pipe band demands more money, the council member who negotiates the contract "will not listen to them, and ultimately get angry; next year the band shall not be called" (PC 2018).

Compared to the *hedonist* Leticia and the *Galician* Dunia, 25-year-old Marina Fernández represents the *female* piper, as gender is the core of her personal and musical narrative. In 2014, she and some friends founded the band Maraghotas (formed by two pipes, snare drum and bass drum), and would not allow any men to join them.[15] In their performances, they "educate" the audience about the importance of women in society and for the traditional repertoire. Marina also affirms Galician-ness, but moderately, as her musical-ideological positioning is mostly concerned with gender. When performing, she hears from time to time: "being girls, you don't do it that bad, but better if you go out in a bikini". And if she enjoys "playing the bagpipes" (maliciously).[16] "That is, the instrument is still strongly associated with the male phallus". She believes that the bagpipe is dominated by men in Galicia (PC 2018). In fact, Maraghotas originated as a reaction to the perceived maleness of the instrument, at the Ponteareas pipe contest (Pontevedra; an obligatory rite of passage for every Galician piper). In the contest "almost all musicians are male, as well as the entire jury," which exasperated Marina. Nevertheless, she admits that being female is fashionable now, and female bands "currently have an advantage because of the pro-feminine social awareness;" in fact, "the combination of girls + pipes sell well," and Maraghotas had played recently in O Rosal (Pontevedra), Pardiñas (Lugo), Foz (Lugo), and were going to perform in Santiago (PC 2018).

Figure 6.4 shows the band in ripped jeans and with no Galician-traditional semiotics in the scene, but the instruments and the music certainly are. It is the new female interpretation of tradition, performed in their own style, and relying on determined aesthetic codes.[17]

Marina holds a degree in English Philology and is currently completing her bagpipe studies at the Conservatory of Vigo (Pontevedra).[18] She also earned a master's degree in Theatre and teaches bagpipe in two cultural associations, but the benefits are limited. Marina entirely concurs with Leticia and Dunia that "the valuation of the bagpipe by Galicians is really low". Maraghotas charge between 600 and 800 euros per performance, a bit

Figure 6.4 Maraghotas, performing femaleness
Source: Photo courtesy Maraghotas, 2018.

more than Cinco en Zocas, "but it leads nowhere", they cannot make their living on that basis (PC 2018). In his 2018 PC González agreed, stating that bagpipes are popular in Galicia but assumed as low culture, not really appreciated, being impossible to earn a living just with the instrument.[19]

Interestingly, Maraghotas introduce regularly avant-garde subtleties in their tradition-based pieces, which doubtlessly stem from their ample musical formation. They transform vocal pieces into instrumental ones and arrange tonal melodies inserting strong dissonances (of major seventh) and atypical progressions. Their instruments are "authentically" Galician, disdaining the massive and controversial novelties of the last decades. Maraghotas are intending to record a CD, but the negotiations are far from solved. They have recorded the repertoire of Los Pacheco, a quartet from Mondoñedo (Lugo) a century ago; it is part of the PhD thesis of Xiana (piper of Maraghotas) but not yet published (PC 2018). However, some video clips featuring Maraghotas are available on the Net; like the "Maneo de Caión" (http://bit.do/eS85A), where the avant-garde orientation is applied to a very traditional piece; Marina is the brunette piper. Part of their intervention at the 2018 Festival of Pardiñas can be seen on http://bit.do/eS85T.

With respect to Seivane, Pato, and Peón, in their respective 2018 PCs Leticia stated that "they were a model of female vindication"; for Dunia they had "a lot of projection", but she did not experience them as "a model to imitate"; and Marina believes that they were "fundamental as symbols", although she is not on the same musical wavelength. Going ahead with the generational comparison, it seems relevant that none of the recent pipers has recorded an album, to some degree because of their limited repertoires so far, but mostly due to the lack of interest from record labels and general audience; their public dimension is restricted to live performances, a few files on the Internet, and presence in social networking sites. However, compared to Peón, Seivane, and Pato, it is unclear whether pipers like Marina, Dunia, and Leticia are not gifted or musically so inferior; maybe the surprise effect has passed, and today a female piper does not generate massive attention, with the correspondent consequences for the music industry.

CONCLUSIONS

The main content of this study has relied upon the contributions of the pipers. They were and are the protagonists of the female revolution that has shaken the realm of Galician bagpipe practice since the 1990s. The social perception of their labor was positive in general, with virtually no adverse gender segregation. A simple conclusion could be that females on stage sell but, in the cases considered, the social background and the strictly musical contents go far beyond this. In this sense the musical quality of their output should be underlined, to avoid an approach limited only to sociological criteria.

From the three stages considered, Meniñas de Saudade opened a door to reverse the passiveness and low status imposed to Galician women in the past. Nevertheless, it was not possible for a true female-musical revolution at that time. With the three bagpipers, however, a real change took place in the local scene, triggered by a quite different socio-cultural frame and collective mentality. The very fact of their individuality is relevant in itself: Pato, Peón, and Seivane did not definitely integrate in a band; rather, they formed their own ensembles, giving their names to them. Their different musical trajectories, ideological projection, and business involvements also prove a remarkable sense of independence, leadership potential, and creative capacity. They were successful and outstanding musicians as well as representatives of a new Galician feminine profile which laid the foundations for the future. In fact, they were hardly paralleled among the female personalities of Galician culture during those years. In their wake, Leticia, Dunia, Marina, and other pipers are developing interesting combinations of musical tradition and hybridization, with gender identity as a cross-cultural marker that pervades their production to different extents. They also unveil a less successful side of current piping, perhaps constituting a form of subculture in comparison to the high-status environment of the instrument. Certainly the relationship with, and repercussions within, the music industry have been quite different, from the 'silent' Meniñas in the 1960s to the explosion of the 1990s. Currently the number of recordings, concert tours, and media interest has declined, as the figure of the female piper is already installed in Galician society and fails to attract much attention.

Geographically there has been a shift from the rural rooting to the city; i.e., from Meniñas (Ribadeo) to Maraghotas (Vigo). However, nowadays the phenomenon of new traditional music economies and rural cultural recovery is aiding the return of certain repertoires. Concerning the distinct faces of identity involved, Galician nationalism has been assumed in a number of ways by the pipers. For example, Peón and Dunia convert it into the emotional and ideological core of their lives and activity, while Pato and Leticia are more distanced from this positioning. This may constitute a helpful locus to increase synergistically the alterity marked by femaleness, but also an obstacle due to the possible conflict of interests. Within this complex dialectic frame, glocality can thrive as a rather de-nationalized combination of local elements and the global culture/economy.

The profile of the average female piper has evolved considerably in the last decades in Galicia. Today these women access the university and undertake standardized studies of bagpipes. Frequently they will become involved in the defense of identity and gender issues, even at the expense of their time and economic interests. The metamorphosis has been radical, from a pleasant ornament (Saudade, without ignoring their merits) to engaged women with remarkable critical capacity and well-defined objectives. These pipers develop a struggle and resistance discourse intertwined with the purely musical, thus adding an ideological factor to the contemporary interpretation of Galicia, and a referential framework of realization to femininity.

ACKNOWLEDGMENTS

I am deeply grateful to the Galician female pipers who have generously contributed to this chapter, with relevant personal testimonies, opinions, photographs, and assistance in accessing other sources. They have been the real basis for this study: Dunia Alvarez, Marina Fernández, Leticia Oubiña, Cristina Pato, and Mercedes Peón. María Acuña was the main source for the study of her mother's band, Meniñas de Saudade. At the beginning of the research, the contributions by Marco Antonio Costa (cultural promoter based in Cambados, Pontevedra) and bagpipe teacher Nacho González were important in order to focus the fieldwork and contact several pipers.

NOTES

1. In this paper all the translations into English are by the author.
2. The specific language of each PC is indicated in the list 'Personal communications cited'.
3. In this chapter, the working method has mostly consisted of nine personal communications with the protagonist pipers and three more related people. They are cited in quotation marks or in different indentation, with the common abbreviation PC (personal communication) and the year.
4. The Dagenham Girl Pipers, from London, were very popular; they started in 1930 and recorded at least two albums with Capitol Records in the 1950s.
5. Francoism is the period during which the military government of Francisco Franco ruled Spain, from the end of the Spanish Civil War (1939) until his death (1975).
6. Here, and also in regard to the other Galician female pipers and bands discussed in this paper, the lack of material recordings is an outcome of various circumstances. For example, Galicia has historically been a poor region, where records were a luxury reserved for the wealthy few. Record labels have also been conservative in attitude, and more so when audiences for such music are likely to be scarce.
7. Also in the 1990s, it started the controversial Real Banda de Gaitas (Royal pipe band), based in Orense and integrated both by males and females. Women wore uniforms that covered the full body, so that the face was barely seen; and the colors and dress style were equal to the masculine ones. This band did not exploit the feminine image, but subordinated it to a collective and anonymous performance, rather gender-less and dominated by discipline. They received strong financial and media support, recording regularly and touring the world (Campos 2019).
8. For example, in the record covers of Seivane's *Alma de Buxo* and Pato's *Xilento* and *Latina*.
9. Released shortly before Seivane's first record, it was not the first record ever by a female piper, as has been wrongly stated. For instance, Kathryn Tickell released as a pipe soloist *On Kielder Side* in 1984 (Saydisc).
10. An illustration of her singing skills, followed by a remarkable pipe performance, in the beginning of her participation in *Luar*, can be seen on http://bit.do/eS88G.
11. Marina Fernández (see next) clarified that: "boxwood is a protected species, and therefore very scarce. In addition, its wood is not very dense in comparison with the grenadilla's, having a softer timbre and less power; over time it tends to twist. Nowadays a boxwood pipe is almost a whim" (PC 2018).

12. Advert produced by Microsoft Ibérica and Xunta de Galicia (Galician Autonomous Government) in 2011; broadcasted by TVGA.
13. E.g., on October 20, 2018, she presented there *FA*, her last record, followed by a short concert.
14. In this video, performed at the Radio Pontevedra station, Cinco en Zocas accompany voice with shells and pans, to underscore the musical-traditional roots of those objects.
15. The name "Maraghotas" comes from a Galician coast fish. It involves an untranslatable word game that transforms the phonetics of the end of the word into "jotas", a very popular and lively musical genre. It also vindicates the Galician traditional pronunciation, quite different from the official one.
16. In Spanish and Galician, to "play" (an instrument) is "*tocar*", which means both to play and to touch.
17. The Galician costume became a cultural marker of integration with Francoist folklorism. Maraghotas wear jeans with that critical intention, but all of them with the same style. So, there is still a sense of discipline, although self-imposed and not from an external authority.
18. At this official center, the degree in Bagpipes is actually long: four years of Infant level; six years of Professional level; and four more years of Superior level (Marina Fernández, PC 2018).
19. These grievances about the contempt for bagpipes may seem contradictory with the aforementioned status of the instrument as the soul of Galicia and an irreplaceable fetish. Both social perceptions are quite real, in the same way as the Galician population is diverse. The cultural division that surrounds the reception of the instrument forms part of a deep cleavage installed in Galician society since two centuries ago.

REFERENCES

Acuña, M. (2013) "Grupo Saudade. Gaiteiras de Ribadeo, pioneiras da música folclórica galega liderada por mulleres," http://bit.do/eS8Vf. Accessed June 9, 2018.

Anderson, R. M. (1939) *Gallegan Provinces of Spain: Pontevedra, La Coruña*, New York: The Hispanic Society of America.

Arias, J. (1980) *Viajeros por Galicia*, Sada: Edicións do Castro.

Barreiro, M. (2012) "O imaxinario feminino na música popular galega," in M. Capelán et al. (eds.), *Os Soños da Memoria*, Pontevedra: Deputación Provincial de Pontevedra, pp. 133–146.

Busto, M. (2005) "Gaiteros com senhorita," *Etno-Folk* 1: 61–66.

Campos, J. (2007) *Fiesta, identidad y contracultura. Contribuciones al estudio histórico de la gaita en Galicia*, Pontevedra: Diputación Provincial de Pontevedra.

———. (2009) *La Música Popular Gallega en los Años de la Transición Política (1975–1982). Reificaciones Expresivas del Paradigma Identitario* [PhD Diss.], Madrid: UCM Eprints. http://eprints.ucm.es/8801/.

———. (2015) "Around the Origins of Bagpipes: Relevant Hypotheses and Evidences," *Greek and Roman Musical Studies* 3: 18–52. Doi: 10.1163/22129758-12341026.

———. (2017) "*I Celti, la Prima Europa*: The Role of Celtic Myth and Celtic Music in the Construction of European Identity," *Popular Music and Society* 40 (4): 369–389. Doi: 10.1080/03007766.2015.1121642.

———. (2019) "In the Name of Ossian: Celtic Galicia and the 'brothers from the north'," *Social Identities* 25 (6): 828–842. Doi: 10.1080/13504630.2018.1564267.

Castroviejo, J. M. (1983 [1970]) *Galicia. Guía Espiritual de una Tierra*, Madrid: Espasa Calpe.

Collinson, F. (1975) *The Bagpipe: The History of a Musical Instrument*, London and Boston: Routledge and Kegan Paul.

Colmeiro, J. (2009) "Peripheral Visions, Global Positions: Remapping Galician Culture," *The Bulletin of Hispanic Studies* 86 (2): 213–230.

Couto, G. (2017) "Entrevista a Maruxa Miguéns e Xosé Romero: 'Os Rosales'," *Aturuxo* 16: 56–61.

Crivillé, J. (1988) *Historia de la Música Española. 7 – El Folklore Musical*, Madrid: Alianza Música.

Cronshaw, A. (2001) "Celtic Iberia," in K. Mathieson (ed.), *Celtic Music*, San Francisco: Backbeat (Third Ear), pp. 140–175.

Cuba, A. F. (2012) "El grupo de gaitas Saudade fue el primero en España solo de chicas," *La Voz de Galicia*, January 22. http://bit.do/eS872. Accessed May 20, 2019.

Cuveiro, J. (1876) *Diccionario gallego: el más completo en términos y acepciones . . .*, Barcelona: N. Ramirez.

Eisler, R.; Donnelly, G.; Montuori, A. (2016) "Creativity, Society, and Gender: Contextualizing and Redefining Creativity," *Interdisciplinary Journal of Partnership Studies* 3 (2): 1–33. https://doi.org/10.24926/ijps.v3i2.130.

Giuliani, G. (2016) "Gender, Race and the Colonial Archive. Sexualized Exoticism and Gendered Racism in Contemporary Italy," *Italian Studies* 71 (4, Cultural Studies): 550–567. Doi: 10.1080/00751634.2016.1222767.

Graña, A. (2013) "Estaría bien poder enseñar también zanfoña y acordeón," *Faro de Vigo*, October 4. http://bit.do/eS9Mu. Accessed May 16, 2019.

Harper, C.; McSherry, J. (2015) *The Wheels of the World. 300 Years of Irish Uilleann Pipers*, London: Jawbone Press.

Hernández, C. (2014) *La agrupación de danza de Sección Femenina de A Coruña: dimensiones políticas, sociológicas, folklorísticas y vivenciales* [PhD Diss.], Vigo: Universidad de Vigo.

Hooper, K. (2011) *Writing Galicia into the World: New Cartographies, New Poetics*, Liverpool: Liverpool University Press.

Losada, O. (2002). *El sentir de un pueblo. Alma de gaita*, Noia: Toxosoutos.

Méndez, M. (2018) "Cristina Pato. Compositora, pianista e gaiteira," *La Voz de Galicia*, July 31, p. 35.

Miguélez-Carballeira, H. (2012) "From Sentimentality to Masculine Excess in Galician National Discourse: Approaching Ricardo Carvalho Calero's Literary History," *Men and Masculinities* 15 (4): 367–387. Doi: 10.1177/1097184X12455398.

Mouriño, N.; Sánchez, F. (1999) "Susana Seivane," *Na Gaits* 7: 23–24.

Padrón, X.; Sánchez, P. (eds.) (2013) *Cristina Pato. Galicia no Fol*, Trasalba: Xunta de Galicia & Fundación Otero Pedrayo.

Palacios, M. (2009) "Within and Beyond the Nation. Contemporary Irish and Galician Women Poets," *European Journal of English Studies* 13 (2): 193–206. Doi: 10.1080/13825570902907235.

Pato, C. (2018) "Alén das fronteiras e dende a diversidade. Unha experiencia personal," *Grial* 56 (219): 58–63.

Peón, M. (2004) "Introducción conceptual ó noso folclore e baile tradicional," in C. Pujales (ed.), *O Son da Memoria. Actas do I, II e III Encontro*, Santiago de Compostela: Consello da Cultura Galega & Arquivo Sonoro de Galicia, pp. 151–157.

Poska, A. M. (2000) "Gender, Property, and Retirement Strategies in Early Modern Northwestern Spain," *Journal of Family History* 25 (3): 313–325. Doi: 10.1177/036319900002500303.

Rejas, C. (2005) "Entrevista a Susana Seivane," *Fusión*. http://bit.do/eTajk. Accessed February 13, 2005.

Romero, E. R. (2012) *Contemporary Galician Culture in a Global Context: Movable Identities*, Lanham: Lexington Books.

WEBOGRAPHY

Cited websites (all of them accessed in May 2019):

- I recommend the personal websites of Pato, Seivane, and Peón. They are available in English and include ample information and links to other sites:
  - Personal website of M. Peón: http://bit.do/eTeit
  - Personal website of C. Pato: http://bit.do/eS8Z2
  - Personal website of S. Seivane: http://bit.do/eTgNA

- Meniñas de Saudade:
  - http://bit.do/eS8YH
  - http://bit.do/eTefc

- Mercedes Peón:
  - http://bit.do/eS82i
  - http://bit.do/eTere
  - http://bit.do/eS82Y

- Cristina Pato:
  - http://bit.do/eTefc
  - http://bit.do/eS8Zq

- Susana Seivane:
  - http://bit.do/eS83x
  - http://bit.do/eS83P
  - http://bit.do/eUk5M

- Cinco en Zocas:
  - http://bit.do/eS84b

- http://bit.do/eS84B
- http://bit.do/eS84Z

- Maraghotas:
  - http://bit.do/eS85A
  - http://bit.do/eS85T

Further online information:

- Galician Statistics Institute: http://bit.do/eTAL4 offers interesting gender figures.
- On Galician relevant women: http://culturagalega.gal/album/
- Website of Seivane's workshop: http://bit.do/eTgJq
- C. Pato's discography and other publications: http://bit.do/eTegs
- M. Peón interviewed on different issues (in Galician): http://bit.do/eTvK2

PERSONAL COMMUNICATIONS CITED

The contributors knew that other pipers were participating in this research but ignored their responses. These were the communications, in chronological order:

- Pato, Cristina. July 2005. Written questionnaire in Spanish (resumed from a previous collaboration).
- Costa, Marco Antonio. May–August 2018 (Cambados, Pontevedra). Email and personal conversation in Spanish.
- Acuña, María. May–June 2018, and May 2019. Email and telephone contact in Spanish.
- González, Nacho. August 2, 2018 (Cambados, Pontevedra). Personal conversation in Spanish.
- Oubiña, Leticia. August 10, 2018 (Cambados, Pontevedra). Personal conversation in Spanish.
- Alvarez, Dunia. August 11, 2018 (O Grove, Pontevedra). Personal conversation in Galician.
- Fernández, Marina. August 17, 2018 (Vigo). Personal conversation in Spanish.
- Pato, Cristina. January–March 2019. Written questionnaire in Spanish.
- Peón, Mercedes. January–April 2019. Written questionnaire in Galician.

7

Rare Bird

Prince, Gender, and Music Production

Kirsty Fairclough

Long before his passing in 2016, Prince's artistic boldness and refusal to be typecast within generic, cultural, or political boundaries had helped establish his legendary status. Although his frequent changes of direction unsettled critics and some audiences, he maintained a high degree of commercial success within and beyond his first decade as a recording artist. In the immediate aftermath of Prince's death, the recognition of his work and the extent of his posthumous presence were clearly both critical and commercial. This ability to confound critics and make music for himself rather than pandering to momentary musical fads made him at once impossible to understand as well as deeply intriguing.

Prince was lauded critically and commercially for his live shows. Anyone who saw a Prince live show was in no doubt that they were in the presence of musical greatness. He was a performer who channeled talent and performance prowess as well as racial and gender subversion in often kinetic ways in the spaces he occupied. Prince often combined an overt expression of sexuality in his performances, an often perceived flamboyance in his costuming, and moreover, he challenged notions of hegemonic masculinity, especially Black masculinity perpetuated within American society and by his male contemporaries. Prince's styling has left a subversive mark upon popular culture, one that expands expressions of gender and eroticism for both musical performers and the consumers of his image and music.

As Nancy Holland has argued in her paper, "Prince: Postmodern Icon":

> His music provides the basis for deconstructing the obvious hierarchies of race, gender, and sexuality, but also those of the sacred and the profane, the writer/composer/producer and the performer, and even Self and the Other. He touched the hearts and lives of a far wider audience than any theorist and helped create a new cultural climate without any apparent awareness of academic postmodernism. This opens the possibility for a deconstruction on the meta-level of another traditional hierarchy, the one between reason and intuition, or put differently, between words and music.
>
> (Holland, 2017)

Unusually, given the overtly regressive aspects of the music industry at the time Prince emerged and in subsequent years, he achieved huge global commercial and critical success throughout his lifetime, despite his rejection of conventional notions of masculinity present in the music industry throughout his career. One of the primary reasons for this was his unique use of stage presentation, including costuming and other aesthetic markers, such as his dance style, which subverted traditional understandings of masculinity. Prince emerged as what appeared to be a fully fledged artist in the 1970s and rapidly achieved global recognition through the late 1980s until his death.

Prince is one of the few American artists whose career as an established, respected act spans the conservatism of Ronald Reagan in the 1980s through to the unexpected ascent and progressive perspectives of Barack Obama in the 2000s. Prince lived long enough to witness the materialization of presidential political polar opposites, representing ideals which he either spoke against or supported. One can only theorize as to the likely nature of his social commentary if he had also witnessed the dawn of the anachronistic and prejudice-tainted Trump administration.

One of the most consistent elements of Prince's entire career was that he championed female creatives. Throughout the decades he supported a series of acts including Vanity6, the girl group formed in the early 1980s; Jill Jones, the soul-based former backing singer, whose 1987 album was written and produced by Prince; and Wendy & Lisa, Wendy Melvoin and Lisa Coleman, who played with Prince as part of the Revolution during his particularly prolific period in the first half of the 1980s. Up until his death in 2016, his last significant collaborators were 3RDEYEGIRL, the three-piece all-female band who worked on the *PLECTRUMELECTRUM* album and a number of his live shows. He also gave hit songs to Sinead O'Connor (*Nothing Compares 2 U*), Martika (*Love Thy Will Be Done*), and Sheena Easton (*Sugar Walls*). He also supported female session musicians, most notably Sheila E, who played percussion with Prince from 1984 to 1989 and again from 2010 to 2011.

Not only did he consistently support female musicians and artists, but he also embraced a female perspective in the studio. Sound engineer Susan Rogers began working with Prince in 1983, a time when female engineers were an anomaly in a male-dominated arena. Rogers' studio work with Prince spanned some of his most successful and commercial output during the 1980s, and she was afforded both responsibility and power in the studio at this time, which the few female engineers in a contemporary context were rarely given.

Issues of gender representation within the production workplace did not feature in Prince's musical landscape. His approach to supporting female engineers in a traditionally masculinized space was a rare viewpoint in a largely male-dominated and driven industry. Rogers held a unique position in Prince's musical legacy, and his apparent subversion of the long-held concept of the studio as the domain of the alpha male was one that has permeated much popular discourse on his legacy. Prince's fluid performances of gender as a performer seeped into the studio, creating an environment that fostered inclusivity and equality in a traditionally masculine arena.

Susan Rogers has become known in Prince fan and critic circles for her years working as Prince's staff engineer in Minneapolis from 1983–1987, a period in which she not only encountered Prince's own unique approach to his work but also created his now-infamous music vault. Rogers is a particularly important figure, not only for her work with Prince, but as one of a handful of female sound engineers working in a male-dominated arena. She began her career as an audio technician working at Audio Industries Corporation in Los Angeles, where she trained as a maintenance technician, and studied recording technology outside of her working life. In 1980, Rogers went to work at Graham Nash's Rudy Records again as a technician, and this led to occasional assistant engineer positions. In 1983 she began work with Prince as his staff engineer until 1988 when Paisley Park opened, and she left to work with other musicians including the Jackson family and David Byrne.

In addition to capturing countless recordings of Prince's material, including his own recording of *Nothing Compares 2 U*, which he recorded alone with only Rogers at the desk, she was also present for his wildly prolific period of recording artist projects including The Time, Apollonia 6, the Family, Sheila E, Mazarati, and Madhouse. Rogers was aware that she was an anomaly in a male-dominated industry, and her move to work with Prince made her one of the rare examples of global success in the field.

In gendered spaces and presentational contexts, Prince achieved huge global commercial and critical success throughout his lifetime, despite his rejection of conventional notions of masculinity on a number of levels that were present in the music industry throughout his career.

Hegemonic notions of masculinity in the 1970s, the decade in which Prince began his musical career with his debut album in 1978, were largely conservative and centered on characteristics such as domination and often an oppression of women as exemplified in popular culture. Prince spent much of his career undermining these understandings of masculinity by his approach to supporting female artists and creatives and via his mode of expression. In the now iconic 1980 live performance on *American Bandstand* where he sang *I Wanna Be Your Lover* and *Why You Wanna Treat Me So Bad*, with blown-out hair and a skintight costume, Prince performed in a sexually suggestive manner and made no apologies for it. Later in the performance, he gyrated against one of his male bandmates, turning his back to the audience and flipping his hair as he plucked the strings on the shaft of his guitar. His was not a version of masculinity that subscribed to dominant norms in any aspect of his career.

Prince's career officially began in 1978 with the release of *For You*. The album showcases Prince's abilities to play a host of instruments, including guitar, bass, drums, keyboard, and more, and to produce, compose, and arrange an album independently. Despite the modest commercial success of *For You*, the producer, Warner Bros, was interested in the long-term development of their artists, thus Prince continued to hone his musical style and image that had developed from his years prior to his signing with the mainstream label. Prince as we recognize him today fully appeared on the scene in 1979 with the release of his second album, *Prince*. This album marked the beginning of his legacy in global popular culture. His eclecticism in musical taste is a way to understand Prince's desire to be an artist undefined by musical genre and also by sociological categorizations such as race and gender. Prince's sense of non-conformity was clearly evident. As Sarah Niblock and Stan Hawkins state, this position has ultimately "enabled him to avoid categorisation", and fostered his interest in gender deconstructive performances.

> Coupling his disavowal of societal norms in songs with the stage antics he delivered during the live promotion of his second album, Prince showcased not only his vast capabilities as a musician, but also his ability to captivate an audience with his expressive performance style that displayed both conventional femininity and masculinity.
> (Hawkins and Niblock, 2011)

The explicit eroticism of Prince's performance and costuming also played into his defiance of convention and categorization. Specifically, he used choreography to aid his rebellious aesthetic. Prince's choreography was thrilling: he traversed a stage both with grace and style and combined it with unambiguous sexuality. He essentially created his own artistic presentation and began to reimagine the possibilities of gender and performance in music and popular culture more broadly.

Prince renounced familiar notions of gender by undermining the concept of hegemonic masculinity. The term is understood as the perpetuation of practices that allow the dominance of men to continue without any boundaries and has been used to identify the behaviors and practices of men that subjugate femininity and subordinate other forms of masculinities. Prince did not subscribe to these ideals and instead invited women into all spaces in his musical universe in order to support them, but also because he recognized that men would routinely challenge him for dominance, whereas he believed women would not.

Rogers explains,

> He liked working with women. One of the reasons he liked it was we would not challenge him for that alpha dominant position. The man just didn't have the bandwidth to deal with that; with anyone competing with him. Women were less inclined to compete for dominance in a professional situation.
> (Stamp)

Prince worked at a pace that most individuals could not replicate. Rogers could, and she possessed the musical knowledge that Prince recognized as similar to his own.

> I had the skillset he needed and then some. I had the stamina to stay up all night. And then two other things: I'd been a Prince fan since the very beginning and had all his music. And I knew his frame of reference – I listened to the same music he did. So he could reference any R&B or soul band, and I knew those references.
>
> (Stamp)

Prince was known to work extremely quickly and worked fastest in the studio. Often Rogers would record him with an entire band, but those times were generally in rehearsal. She was able to maintain the prolific pace that he demanded. If he had written a new song and they were also in tour rehearsal, Prince and Rogers would work out all parts of the full arrangement in the rehearsal. They would record the basic track, with each band member contributing his or her original part, and then take the tape and complete the track in the home studio in Minneapolis. Throughout all of the interviews that Rogers has given, it is evident that she was afforded a voice in the studio, and this was not only rare, but indicative of Prince's approach to working. He respected those with the required skills and knowledge regardless of gender.

Since his passing, Prince has been eulogized as a stellar performer, a musical innovator, but less so as a champion of women artists and creatives. These three factors intersect on a point of gender subversion. Prince defined his persona as something that many could not comprehend, a position which he strived for throughout his career. He challenged and succeeded in breaking through the constructed boundaries of race, gender, and the many ways in which they intersect. He provided his audience with in-between spaces and worked toward dissolving accepted and oppressive notions of identity. His support of Rogers and other women who worked with him have provided a view of a musical icon who not only had a deep and nuanced knowledge of musical language, but also was fully cognizant of the push and pull of the complexities of the industry; his refusal to buy into dominant tropes and industry practices made him a true innovator.

REFERENCES

Hawkins, Stan. & Niblock, Sarah. (2011). *Prince: The making of a pop music phenomenon.* Farnham: Ashgate.

Holland, Nancy. (2017). Prince: Postmodern Icon. *Journal of African American Studies.* 21. 1–17. Doi: 10.1007/s12111-017-9363-7.

Part Two
Women in the Studio

8

Slamming the Door to the Recording Studio – Or Leaving It Ajar?

Henrik Marstal

A male producer's recollections regarding negotiations of gender in the studio work of female recording artist Where Did Nora Go (DK).

INTRODUCTORY REMARKS

And the day came
When being curled up inside
Was no longer enough

And the day came
When staying secure
It lost its frantic grip

Can you feel it
Can you see it

And the day came
When the longing to share what is sacred
Grew inside

Grew vaster than the fear
Of being left out, all alone, cast aside

I see a window open
I let the bravery run free
I see a window open
First part of 'And the Day Came'
(Where Did Nora Go, 2013)

This chapter investigates the creative process in the early studio productions of the Danish alternative pop artist Where Did Nora Go (a pen name for the singer, cellist, and songwriter Astrid Nora Lössl). These productions include the EP *Away, Away, Away* (private release, 2012) as well as the albums *Where Did Nora Go* (Für Records, 2013) and *Shimmer* (G Records, 2014), the two latter being released and promoted in the territories of Denmark, Germany, Austria, and Switzerland. The investigation concerns the creative process of making these recordings by analyzing the dynamics of the process and decision-making seen in a gender perspective.

The purpose of the chapter is to map out some of the pitfalls of gender relations in the recording studio, which on a daily basis relate to notions of gendered expectations and assumptions between people, but which at the same time can be understood as a result of historical and ideological male dominance strategies in the creation of popular music. According to Swedish historian Yvonne Hirdman's polemical reflections on the logics of the genus system, the labor division between men and women could and should not at any rate be mixed together (Hirdman 1988: 51), since that would challenge the power balance of the sexes and – for instance – threaten the notion that any handling of technology in the studio is a man's job. According to Hirdman (ibid.), the genus system is made to prevent women from being in power, and in the traditional studio work this seems indeed to have been the case. In the recording studio, the notion has the potential to be a pitfall in itself, since the reliance on long-established norms overrules critical reflection on the ongoing power balance between the sexes. Even for otherwise well-balanced and critical founded males this seems to be the case, since the pitfalls can be so subtle and tacitly accepted by everyone involved that they are hard even to notify and challenge. I therefore agree with Paula Wolfe's notion that "the historical predominance of men who have practiced music production has resulted in a gendering of both the technical expertise and artistic creativity associated with the profession" (Wolfe 2020: 60).

It is indeed true that we lack knowledge on how studio productions and gender issues relate to each other. Stories and evidences need to be unfolded much more extensively, if everyone involved in the decision-making of studio recordings – producers, engineers, musicians, A&Rs and alike – should be better informed about how gender issues of various kinds can pervade the atmosphere and creativity of the studio in inappropriate ways.[1]

To provide a little bit of evidence on this topic, I will examine how notions and negotiations of gender were constructed and conducted during the process of the three aforementioned recordings between Lössl and her two producers, Kasper Rasmussen and myself. This will be done by reconsidering the recording and mixing processes, which took place in Rasmussen's small, vibrant studio called Mikroskopet on the outskirts of the hipster area of Vesterbro, Copenhagen. I will do this by combining academic consideration and practical, as well as artistic, insight in trying to recollect the pros and cons of the process. Moreover, a number of considerations by Lössl herself are incorporated, thus bringing Alistair Williams' remark in his book *Constructing Musicology* into mind: "Like music, musicology does not just reflect what happens elsewhere; it offers ways of inhabiting and shaping the world" (Williams 2001: 140). The chapter is informed by feminist theory complexes, which are used to clarify the gender-related conditions of the collaboration, as well as the involved constructions of power and decision-making.

THE MALE ISSUE OF STUDIO RECORDINGS

And the day came
When feeling estranged
Didn't seem so startling anymore

And the day came
When clarity dawned
And the illusion of distance, being separated, faded

Can you feel it
Can you see it

And the day came
When fear no longer would scare
But be part of the quest

And the day came
When freeing ourselves
Was no longer an intimidating fantasy

I see a window open
I let the bravery run free
I see a window open
      Second and last part of 'And the Day
      Came' (Where Did Nora Go, 2013)[2]

The advent of studio recording seems indeed to be a male issue, even in the early twenty-first century. If that is to change, it might be necessary to reclaim the notion that music-making activities of all kinds are always gendered, since more or less stereotyped gendered relations and conceptions always interact with each other, between the people involved. This is not only the case when female musicians or, much more seldom, female engineers or producers are involved. Male-to-male relations in the studio are always gendered as well, since, as Sara Cohen notes, (heteronormative) gender relations are "constructed through relations and distinctions between men and women, men and other men, women and other women" (Cohen 2011: 230). One could add that this is also the case between all kinds of men, women, and other genders, no matter how their feminine and masculine energies are distributed and used in creative relations. This is the case not least in the history of recorded music, since music is a discourse through which gender is constantly coded (Williams 2001: 69) and negotiated.

Historically, the recording studio has always been an arena in favor of male genderness, since musicians, engineers, and producers have – with a number of notable exceptions (Djupvik 2017: 118) – traditionally been men (Ibid.: 117, 129). Women have usually been kept away from decision-making when in the recording studio, Yoko Ono's infamous appearance in the control room during the making of The Beatles' eponymous album from 1968 (Harry 2000: 108–109) – usually referred to as *The White Album* (MacDonald 1997: 286) – being one notable exception to the rule.

Since the studio is dedicated to the performance of music in relation to recording technologies, and since technology as such according to Lucy Green "has long been gendered masculine because of men's development and uses of it" (Green 1997; quoted after Kearney 2017: 121), performances of gender seem to be inscribed (see Citron 1993: 11) in every action concerning the recording processes. This process is usually understood as "a culture of masculinity" (Leonard 2007: 67) – masculinity defined, I would add, as "the total cultural, social and political expression of maleness" (Halberstam 1998: 1, quoted after Djupvik 2017: 117). I thus take my starting point in the notion that gendered aspects are always an interacting component in the relations between agents in a recording studio – not only when the producers, engineers, and perhaps all musicians are male and the artist female, but in every conceivable relation.

Stereotyped, gendered preferences do play a role since there is no such thing as 'gender neutrality' in musical-creative relations. And since the fabric of music production and engineering have been historically taken care of almost exclusively by males, there are reasons to presume that modes of listening, methods of production, approaches to creativity, mechanisms of priorities and so on are framed by a general hegemonic masculinity (Connell 2005/1987; Connell & Messerschmidt 2005), which seems to dominate popular music discourse in what Will Straw calls a "masculine gendering of cultural habits" (Straw 1997: 5). Since there are countless ways of approaching the art of music production, there is of course not a certain 'male' way of doing things. But it makes sense, I would argue, to maintain the assumption that hegemonic masculinity has had its say concerning what is acknowledged as defining acceptable studio production approaches and what is not.

Lucy Green notes that especially female vocalists are not a threat to "the patriarchal status quo" of the recording studio male practices, since "unlike instrumentalists, they do not rely on technology to produce musical sound" (Green 1997; quoted after Kearney 2017: 121). Although female vocalists not unusually accompany themselves on instruments such as the cello (as in Lössl's case) and more often the piano or the acoustic guitar, the technology involved is often defined by being acoustic, not electronic, that is: femalized, not masculinized, so to speak. Moreover, as Ian Biddle claims, "[t]he phallogocentrism of Western rational epistemology marks music as a particularly volatile yet profoundly effective (and affective) cultural resource in the imagining, policing, and managing of discourses on gender" (Biddle 2003: 208). Similarly, Sara Cohen has pointed out in a seemingly general precondition that rock and pop (and other popular music genres as well) are "closely associated with gender – with patterns or conventions of male and female behaviour and with ideas about how men and women should or shouldn't behave" (Cohen 2011: 226). Finally, she states that:

> [w]omen have been associated with a marginal, decorative or less creative role within rock culture, hence the popular stereotypes of glamorous women who act as backing singers for male groups or feature on their videos and other merchandise, and girls as adoring fans who scream at male performers.
>
> (Ibid.: 232)

These considerations all allude to the point that bringing a female singer, songwriter, *and* cellist to the historically male-dominated studio with two males as producers – as was the case with Where Did Nora Go – might involve the risk of creating resilient gender obedience or pure gender trouble. This is the case even though, according to George Dvorsky and James Hughes (2008: 18), "a post-genderist future is made possible not just from breakthroughs in science but through the 'decline of patriarchy' resulting from our 'slowly dismantling the heritage of patriarchal power, culture and thought' " (ibid.) (quoted after Wolfe 2020: 10).

ABOUT THE CONTEXT OF THE RECORDINGS

So you wanted me to be your doll
All pretty and quiet
I nod and I smile
Never out of line
But it's not me, it's not me

I could do it so well
I'd be the sweetest thing
I'd be willing
Oh, this is me, but it is not me

I could do it so well
I'd be the sweetest
More than willing
Oh, this is me, but it's not me
         First part of 'Your Doll,
         Your Maid, Your Toy'
(Where Did Nora Go, 2012)

In 2011, I was approached by Lössl, whom I did not know at the time. She had come across *Thieves Like Us* (2011), the EP debut by the Danish singer and songwriter Penny Police (née Marie Fjeldsted), which I – like her later debut album *The Broken, the Beggar, the Thief* (2012) – had produced together with Kasper Rasmussen. We had also played all the instrumental parts on the EP apart from the ones Fjeldsted had done herself.

Lössl wanted to make a radical departure from her earlier work as a singer and songwriter, and she considered Rasmussen's and my collaborative production work as an opportunity to make that happen – and invited us to do an EP with her. Lössl added that she also wanted to work especially with me because I had recently written and contributed to a few articles in the daily newspaper *Information* about the role of women in popular music, and the necessity of providing more space for female presence in Danish music (see Marstal 2010; Tholl 2010).

Prior to our first meeting, Lössl had sent me some demos with her own material. She later informed me that she had written more than a hundred songs primarily in a modern soul idiom, which we decided not to use, though, in favor of completely new material. She also pointed out that she needed some input considering the artistic direction of the new project. I was personally attracted to the idea of working with her not only as a producer, but also as an artistic collaborator and co-writer. Since she had played the cello earlier on, and since I played the cello myself and used the instrument on a regular basis in the recording studio, I suggested that we did a number of jam sessions together in her home, playing together on cellos only. And so we did, both of us placed on an old beautiful carpet in her living room, which in a sense became magical to us, since the communal music-making was framed by it. This endeavor resulted in a number of instrumental song sketches, which we both liked. Based on live home recordings of these sketches, Lössl then wrote melodies for them. During this process, she also made a number of songs on her own, using a similar artistic concept. After some more sketches recorded on our mobile phones, the project began to take shape. Lössl considered it to be a solo project with me as her prime collaborative partner, and Rasmussen and I as her producers. She told me that she wanted the lyrical themes to be about gendered matters, among these the burden of being a woman forced to live in an everyday world where male dominance and patriarchal norms rule. At one point, she allowed me to contribute with lyrics for one of the songs on the eponymous debut album, 'Sing, Ye Birds' about a troubled relationship, including a possible rape of a woman and a possible revenge at the possible rapist.

The lyrics of the repertoire of Where Did Nora Go deal not only with themes of marginalization, low self-esteem, doubts and frustration, though, but also of longing, liberation and joy, bringing forth a feministic-philanthropic mindset combined with considerations about human existence as such. The project name referred to the final scene in Norwegian playwright and poet Henrik Ibsen's world-famous, proto-feministic play *Et Dukkehjem* [*A Doll's House*] (1879).[3] Here, the young wife Nora Helmer leaves her world-weary husband Thorvald Helmer and their three children after a complicated series of events in favor of a life of her own command. She does this in order to escape the feeling of being constantly belittled or *bedolled* by her husband, who lacks the skills or the will to acknowledge her as she really 'is'. The play literally ends with the sound of the front door slamming as a result of her leaving the house. Since then, the phrase "where did Nora go when she left?" has been a perennial question in Norwegian and Danish everyday thought (it is even the title of a play from

1968, *Hvor gik Nora hen, da hun gik ud?*, by Danish playwright Ernst Bruun Olsen). While introducing me to this pen name, Lössl informed me that she actually was named after Ibsen's Nora (her middle name), which undoubtedly provided the whole project with a substantial authentication. By using her own middle name in the project name, she made it clear that it should be regarded as a strong feminist statement and project all together. In this way, she intended to also react against male dominance in the music business and elsewhere.[4] Even though the two organization theorists Barbara Orser and Catherine Elliott's work on Feminine Capital (Orser and Elliott 2015; see also Geoffroy-Schwinden 2019) was not known at the time, Lössl's artistic vision could easily be said to be the result of a similar capital, since it was closely related to the necessity to react to the personal, societal, and musical experiences of being a woman in a man's world.

Taking its departure from Ibsen's character, Lössl's artistic project Where Did Nora Go is concerned with the limitations of the 21st century woman within a patriarchal discourse, encouraging the listener alone to consider how social constructions of gender in popular music as well as in every other aspect of society are reproduced. Musically, the project is melancholic and thoughtful, held in a *dream pop* idiom and using the cello as the main instrument. The songs adhere primarily to a melodic pop song formula, but differ from this formula to a large extent due to the use of the cello at the expense of the typical pop instrumentas such as synths, guitars, bass/synth bass, and drums/beats. Moreover, the music has a strong emotional content due to Lössl's intense yet smooth singing style and the overall *necessity* of the project fueled by the longing for fulfillment, respect, and acknowledgement.

Lössl's artistic vision of cello-based songs with lyrics about modern womanhood determined the whole process and thus prevented her from being what Alistair Williams determines as "a passive feminine other" (Williams 2001: 89). Before the project was established, Lössl had originally rejected the cello for artistic purposes, finding it "too voluminous, female-like and soft".[5] She changed her mind though, finally becoming attracted to the sound of the cello in connection with her own mature singing voice, thus acknowledging the cello as a vehicle for the gender-specific discourse of the project with its round shapes resembling a woman's body and its unamplified, 'soft' sound concept related to notions of feminization. This identification with the instrument continued to play a crucial role throughout the entire period covered in this chapter, also in visual representations such as music videos (see, for instance, Magnussen 2012).[6]

PRECONDITIONS AND MOTIVATIONS

So you wanted me to be your maid
Wash and clean and feed you every day
Keep my mouth shut
Unless when needed for a blow
But it's not me, it's not me

I could do it so well
I'd be the sweetest thing
I'd be willing
This is me, but it is not me

And, oh, I could do it so well
I'd be the sweetest
More than willing
Oh, this is me, but it's not me
    Second part of 'Your Doll, Your
       Maid, Your Toy' (Where Did
           Nora Go, 2012)

My own learned experience as a collaborator and producer has primarily evolved around the presence of female solo artists and all-female bands based in Denmark. I made my first appearance as a producer in 2008 with *Lille filantrop* [*Little Philanthropist*], the debut album by the renowned singer and songwriter Annika Aakjær. Moreover, my own artistic merits as a musician and songwriter have involved close relationships with female collaborators in, respectively, the duo marstal:lidell (with singer, co-producer, and co-writer Anna Lidell (DK)), the duo White Flag Society (with singer and co-writer Jullie Hjetland (DK/NO)), and recently the Danish-Norwegian trio The Baby Zebras (with singer and co-writer Live Foyn Friis as well as musician and producer Mikael Elkjær). Without having any intention of it, I have happened to *like* working with women rather than men after quite a number of years in the music business involved in male-only bands backed by the services of male-only live crews, male-only studio engineers, male-only producers, as well as male-only business partners.

Many of my productions have been done as a teamwork together with Kasper Rasmussen, though, a long-time fellow musician and friend with whom I have felt a strong human and artistic kinship ever since we first worked together during a studio recording in 2006. Our relationship has been based on mutual musical and personal respect, and as well as a certain ongoing *bromance* between us, driven by a informed sense of humor, a mutual preference for ambient soundscapes as well as loud rock noises, and not least a shared interest in the artistic process as much as the end product. Our relationship has always benefited from the fact that he is a drummer and I am a bass player, enabling us to provide backing rhythm tracks to the productions ourselves on our primary instruments, while at the same time being multi-instrumentalists, both of us playing guitar, keyboards, and other instruments – and as already mentioned: in my case the cello as well.

Our collaboration has contained male-gendered aspects not only with regards to our sex, but also genre. The framework of the long-established indie rock discourse, which more or less has defined our musical partnership, does hold traditional core values of masculinity, since the hegemony of rock is rooted in the male-gendered genre privilege which it represents, reproduces, distributes, and benefits from regardless of intention (Marstal 2016). On the other hand, we are both quite 'metrosexual' individuals with a rather 'feminine' appearance. For my own part, during the collaboration with Lössl I became an activist male feminist (or pro-feminist, according to which definition one prefers), focusing on gender issues in popular music discourse and in society as such, resulting in a large number of articles in daily newspapers and even a book (Marstal 2015). Just like Paula Wolfe, I therefore made an "interrogation of patriarchal norms of control associated with music production" (Wolfe 2020: 12) in order to challenge the norms of studio work. It is the question, though, whether these conditions had any impact concerning *The Bro Code* between Rasmussen and I, since the bonds which constituted it seemed to be stronger than any appearance or ideological stance were able to alter.

In the years that followed the recordings of the EP and the two albums with Where Did Nora Go, I felt a need to reflect upon my own gendered positionings and biases not only as a producer or musician, but also as a person. My interest in feminist thinking had created a necessity in me to learn more about how I could *do my thing*, so to speak, without reproducing too many of the male gendered stereotypes of the music business, which I so easily and willingly had adopted and internalized a long time ago.

In retrospect, this also affected my partnership with Rasmussen: I had to reflect upon the gender-explicit alliance which we sometimes, whenever convenient, made for ourselves more or less *against* Lössl. This notion was perhaps fueled by the fact that the three of us were almost exclusively the only people present in the studio during the recordings, which amounted to several weeks or even months. Here are some of my *post festum* considerations: was the creative decision-making as producers based upon blind faith in our own abilities, rather than an ongoing creative dialogue with Lössl? Did we establish too many male-oriented habits in the studio related to language, behaviour, and ethics? And, did we position Lössl as an 'other' not only because she was a woman, but also because she was the only one of us who took responsibility in all extra-musical matters such as providing snacks, making coffee, supplying take-away dinners or even, at times, doing the dishes herself while Rasmussen and I were busy in the control room with our ever so 'important' or 'serious' production duties? In short: were we – without really being aware of it – running the risk of turning the whole project into a male-styled version of Lössl's original female vision in the name of a gender-biased norm? And would she in the end have made much better records with a female recording team instead of us?

The irony soon enough dawned on me, since the gender-focused theme of her lyrics apparently had had trouble being decoded by the very people who were brought in to facilitate the emergence of the songs in the first place. This became especially clear when I read historian and feminist Rebecca Solnit's (in)famous essay 'Men Explain Things to Me' in her book by the same title (Solnit 2014). Because Rasmussen and I had, in a joyous and informal yet slightly patronizing way, numerous times 'explained' to Lössl what her project was really about, it turned out that we had not really been paying enough attention to what she had had to say before we took over the discussion and let our own interpretation of her artistic vision rule instead. This was, I realized, especially the case for me personally. So I felt affected especially by this paragraph in Solnit's essay:

> The battle with Men Who Explain Things has trampled down many women – of my generation, of the up-and-coming generation . . ., not to speak of the countless women who came before me and were not allowed into the laboratory, or the library, or the conversation, or the revolution, or even the category called human.
>
> (Ibid.: 9)

All of these considerations forced me to rethink the whole process of my engagement with Lössl. I really *liked* the artistic outcome of our collaboration, but I was at the same time reminded of the processes of making the music with her, which were not without problems of different kinds. In the most tense moments, I remembered, it was always Lössl who lost her temper, burst into tears, and – this was at least my impression – was the solitary one, while Rasmussen and I always stuck together at the expense of the collectiveness of the project, while relying on the firm assumption that rationality and even calmness are always constructed as a male position. In those situations, we performed the role of being representatives of 'reason' or 'sense' while she happened to be the 'emotional' or counter-intuitive one. Perhaps we would have been better off with the reversed situation – *that* situation might have ensured a much more interesting and perhaps even more reasonable dynamic, instead of the traditional gender-specific one I just described.

It occurred to me that the *bromance* between Rasmussen and I on the one hand actually did contribute to making the atmosphere of the studio recordings joyous, light-hearted, and inspiring, but on the other hand created a sense of exclusion, since the relation between us depended on homosocial (Sedgwick 1985; Farrell 2001) oriented values, in which any female presence by definition threatened to disturb our collective devotion to each other. Mechanisms of gender identity and even gender privilege were undoubtedly at work here, and without any of us being really aware of it, we relied on historic assumptions as well, since homosocial male bonding processes have always enjoyed the status of being a core value within the production of popular music – not least concerning the dynamics of the rock band, as well as the dynamics of studio production.

It also occurred to me that the recording studio is a male-gendered space, which men regard as a completely natural habitat, while women can have considerable concerns doing so, because they do not necessarily feel invited to participate. This has been noticed by Mavis Bayton, who states that "women . . . have less access to space than men. They take up a smaller physical area by the way they sit and use their bodies and there are fewer 'female' spaces" (Bayton 2006: 349). Bayton also quotes Valerie Hey for the observation that even public houses such as pubs are not really public for women, but rather "male 'playgrounds' to which women are 'invited' on special terms" (quoted after Bayton 2006: 351). There are, it seems, "masculine connotations of space-claiming in popular music" (Björck 2011: 55), and claiming space can thus be seen not only as an act of "extrovert self-promotion" or "introvert focus on the musical craft", but also as "an ongoing struggle between empowerment and objectification" (Ibid.). To me, the question of space was important to the relationship between Lössl on the one hand and Rasmussen and I on the other. With the Swedish musicologist Cecilia Björck's dissertation on claiming space for female musicians in mind, I asked myself the question she brings forward: whether space can be only taken or also given (Ibid: 56). Maybe we should have provided much more room for such a relation between us where the latter might have been a more suitable possibility.

LÖSSL SPEAKS: LEAVING THE DOOR AJAR

I am human
Flesh and blood
Human to the bone
I am human
Heart and spirit
Human to the bone

So I wanted to be your toy
All dressed up for fun
Your sweet possession
Not a mind of my own
This is me, but it's not me, I'm not being me

I am human, I am human, I am human
I am human, I am human, I am human
Third and last part of 'Your Doll,
Your Maid, Your Toy' (Where
Did Nora Go, 2012)[7]

To learn more about the whole matter, I decided to approach Lössl even though we had not talked for a while. We met, and I asked her to revisit the project and reflect upon how gender-related issues between her and the production team had been established, negotiated, and – to a certain extent – deconstructed. A few days later, she sent me a number of remarks about our working relationship.[8] From reading her words, I learnt a lot about her thoughts before, during, and after the process. She began by describing her pre-history, stating that her earlier encounters with male producers, engineers, and musicians had been quite devastating for her:

> I had been in the recording studio many times, always with male producers and musicians, both during my years as a student at RMC, the

> popular music conservatory in Copenhagen in the early 2000s, and as a solo artist – always ending up with a result where I could not recognize myself artistically.
>
> I had often worked with charismatic, yet ego-centric male producers, who made me feel more like a puppet following orders than an artist with a right to interfere with the creative process. This had been not only a very costly, but also a very painful experience. Even shameful at times. I had a deep longing of sharing my voice and heart, and I also felt that I had a huge potential, which I could not seem to give birth to, though.

As producers and collaborators, neither Rasmussen nor I were aware of this when we started working with Lössl. We just thought that she shared similar musical aesthetics with us. But the reason for approaching us was of a more serious nature. She continued her reflections:

> I needed to find both brave and sensitive producers who could take on the task of redeeming my potential – helping me emerge as the artist I was, rather than me fitting into their ideas and visions. I mean producers who would *see* me and help me rather than manipulate me. I knew that Henrik was intrigued by working with female artists and also with preserving a sense of vulnerability in music, and this really spoke to me. We finally decided on a collaboration, and the work took off.
>
> As I was searching for guidance, I was of course open to suggestions and had a willingness to be led, yet at the same time trying to maintain a strong sense of self and authenticity. I also attempted to leave aside the insecure, emotional-ridden and pleasing sides of myself and instead attempt a lighter, more confident way of communicating and working in the studio. I even decided on 'desexualizing' myself or at least downplay any kind of erotic capital, since I did not want sexual energy of any kind to be part of the atmosphere in the studio – which I unfortunately had experienced before.

In general, Lössl felt that the studio work which we did together proceeded in an atmosphere of respect and decency, which differed in a positive way from the kind of producers and engineers she had been working with in the past. She also stated that even though notions of gender were of course present during the studio recordings, we were not doing anything wrong, so to speak: discourses of power related to gender were not really present at all, which meant that the recordings felt easy, fun, inspired, and creative for all of us. But even though this was the case, she also had some criticism:

> I really enjoyed working with Henrik and Kasper, and I felt that we succeeded in our attempts to fulfill my potential as an artist and create music on a high artistic level. The albums were both critically acclaimed, and I started to win recognition for my music. But, on several occasions there was tension, especially between Henrik and I. At times, he could be short-tempered or quite a bit distant and withdrawn.

> During these incidents, I would often feel cut off or even belittled. Having been taught to never get angry, I would react by turning the situation inwards, become sad and perhaps even shed a tear, while Henrik did not really comment on my reaction or feel sorry. To me, this could easily been seen as a gender-specific situation, adhering to a traditional male understanding of being the one in charge.

In her notes, Lössl also states that the homosocial bonding between Rasmussen and myself, related to our internal way of speaking and joking, did not really tend to exclude her in any way since she was, as she wrote, invited to join us along the way. But still, he and I were the ones to decide and define how to talk, joke and behave, thus confirming my own afterthoughts about the studio work: that we established and maintained a male discourse of talking related to our informed homosocial bonding. But there were no other options concerning how to talk and speak – she was, I can see at a distance, not really in a position to challenge this firmly established discourse of structure.

Another observation from Lössl had to do with the act of encouragement while recording and arranging in the studio:

> Kasper and Henrik – who also played almost all of the musical parts, which I did not play myself – had developed a habit of not only happily complimenting each other, but also clearly enjoying their own performances when they recorded something they were especially pleased with. To me, this was a characteristic male-gendered way of acting: At least I could never do this myself. I have never met a female musician – yet – who fully indulges in her own work of art or her own musical performances in the same way. But I have often met male musicians who thoroughly enjoy their work and are even proud to say so. The aforementioned habit did not include me, though. My contribution to the music didn't seem to warrant compliments from them all that often. I state this without any sense of bitterness. Nevertheless, it seems that complimenting or even just appreciating performances by female musicians – and this goes for both women and men – is something that requires a conscious effort. Still. Amazing. Also to myself.

The last part of Lössl's observation resonates almost too well with Paula Wolfe's remark with regard to Simon Frith and Angela McRobbie: "[W]hen women perform as vocalists, either in front of a band or as a solo performer, the reception of her skill is often underplayed as it is seen as 'natural' (Frith and McRobbie 1978)" (Wolfe 2020: 13). Regarding Rasmussen's role – since he was not only part of the production team but also in charge of all of the engineering – Lössl had some thought-provoking criticism:

> Kasper had a tendency of being more critical of my recordings on the cello compared to Henrik's recordings on the same instrument, even though – and I am sure Henrik will agree on this – I am just

as good a cellist as he is. Kasper seemed to regard the recording of my contributions as a kind of necessary, but dull work. And, as we know, uncertainty breeds more uncertainty and is reflected back by the surroundings – as is the case with confidence. So I sometimes felt that I was the only one responsible for fulfilling the artistic vision of Where Did Nora Go, which created a sense of loneliness in me. I wonder if this would have been the case, had I been a male musician myself, or had Kasper and Henrik been female producers.

Because the collaboration between Lössl, Rasmussen, and myself included an EP and two albums in the course of around two years, there was a strong sense of artistic development at stake. However, Lössl notes that this development made the second album, *Shimmer*, stand out as different in many respects:

> For me, the success of the EP and the debut album had led to a higher degree of relying on my own skills. Where Henrik until now had co-written a lot of the musical material, I wanted to include more songs on the second album written by myself. Moreover, I developed clearer ideas of how I wanted the music to evolve. This meant that Henrik was less involved in the creation of the music. Moreover, since the days of the EP, he had co-founded a new project, the dreampop duo marstal:lidell, in which he did contributed to all the songwriting himself. For these reasons, perhaps, he seemed to be less enthusiastic about the whole project as a producer.
>
> Another clash of interests occurred, when I proclaimed that I wanted neither bass nor drums to be part of the new album, knowing that Henrik is primarily a bass player and Kasper primarily a drummer. This kind of decision-making on behalf of my own artistic vision had not been possible on the debut album, but I knew that it was what I wanted for now. Kasper initially profoundly opposed the idea, but after heavy debating he accepted it. In all fairness, it should be noted that he and Henrik did not use, though, any homosocial power against the idea or refer to democratic rules by wanting to let the majority decide.
>
> There is another thing which should be mentioned: During the making of the album, I unknowingly became pregnant, a fact which completely turned my world – and my hormones – around. My composure, which I always had been able to control, crumbled on several occasions, and emotions soared. So we had several controversies due to the fact that I had more confidence in speaking my mind, but at the same time did not have the same emotional control as before and therefore tended to be quite sensitive. Needless to say, this development put gender differences in a new light, and there were even occasions were I left the studio crying, feeling unheard, overlooked and belittled.

In combination with Lössl's quite intimate considerations, I will add that the process of recording *Shimmer* ironically resembled the stiffened dynamics between Nora and Helmer: her as the one striving for being acknowledged for who she was and what she did, and me as the one being reluctant to continue engaging in our musical relationship. But, even though Lössl at least at one point, she admits, considered terminating the collaboration, she decided in the end not to slam the door to the studio but rather leave it ajar which made it possible for us to finish the *Shimmer* album as well as a long, creative period of working.

CONCLUDING REMARKS

Oh, sister, dear sister of mine
Let's march down to the sea
And drown all those thoughts that kill us
What they convinced us
Everything false we thought to be true

And we shall walk in the light
Walk in the light
Walk in the light
Leave it behind
Last part of 'Sister of the Dark/
Walk in the Light' (2012)[9]

This chapter has pointed towards the notion that gender(ed) relations between individuals in the recording studio – no matter which gender they adhere to – are crucial to the process of making music together. The chapter has shown that even though the homosociality of male bonding is not necessarily excluding by nature, it can easily run the risk of complicating matters by taking space instead of giving it away. The very case described here, though, shows that afterthoughts and post-reflections of the recording process can provide a better understanding of how gender-related dynamics between individuals work, and how risks of evoking pitfalls, leading to either frustration or disengagement, can be integrated into the discourse of re-imagining the whole recording process.

But the case of Where Did Nora Go also shows that a certain awareness of the gender perspectives of any studio recording can help fulfill the potential of artistic vision and creative wishes. In addition, it shows that all involved – not least myself – still have a lot to learn considering the negotiation of gendered aspects. Bearing this in mind, anyone responsible for or participating in a studio recording process will be able to benefit from more reflection on the gendering process involved. A way of doing this would be to consider very carefully whether the individuals present in the studio identify themselves as either Noras or Helmers, and whether the door to the recording studio will run the risk of being slammed due to unhealthy gender relations. And again, the act of leaving the door ajar instead is an act which acknowledges everyone's right to not just take up space in the recording studio but also to be given space – no matter which gender identities or positions of power are at stake.

The ability to collaborate is essential for every individual entering a recording studio, since good music-making so often is dependent upon the mutual work among the musicians, producers, and engineers involved. But in order to create the best assumptions for a possible collaboration, the empirical results of this study show the importance of regarding each other as equal partners. The more out of gender balance the collaboration is, the greater the risk of someone slamming the door, resulting not only in interrupted collaborations but also feelings of betrayal and being let down in fragile surroundings.

Finally, the study has shown that in recording studio cases where people of different genders work together, it is important to prepare oneself for both the ontological and epistemological tasks at stake. But since the history of recorded music is primarily a male-gendered history, and since male producers and engineers continue to dominate at a numeric level, the tasks should be first and foremost a male issue – and not, as it is usually assumed or even expected, a female issue.

NOTES

1. Unfortunately, and perhaps not too surprisingly, there is not much help to find in literature concerning the gendered aesthetics and practices as well as the histories of the recording studio in popular music discourse. Neither Milner 2009 nor Cook et al. (eds.) 2009 make any mention of the word *gender* in their respective indexes. This is also the case for Cunningham 1996, and while two of the sixteen devoted chapters in Frith and Zagorski-Thomas (eds.) 2012 are either written or co-written by female experts, this anthology does not deal with gender issues at all. This notion calls for even further consideration concerning the topic of gender in music production, even though it has been addressed in a number of studies, including Wolfe 2020.
2. The verses quoted in the sections 'Introductory remarks' and 'The male issue of studio recordings' amount to the lyrics to the song 'And the Day Came' from the album *Where Did Nora Go*, written by Lössl. The lyrics are reproduced with kind permission.
3. Curiously, in Chinese, the word for "feminism" is named after Ibsen's Nora and called *nuola zhuyi*, which translates as "Noraism" (Bailey 2012: 62).
4. Thinking once again of Yoko Ono's somewhat futile appearance in the studio during the recordings of The Beatles' *The White Album*, it seems quite apt that the working title for the album actually was *A Doll's House* (Harry 2000: 108–109).
5. Personal conversation with Lössl, September 2018.
6. Ibid.
7. The verses quoted in the sections 'About the context of the recordings', 'Preconditions and motivations', and 'Lössl speaks: Leaving the door ajar' amount to the lyrics to the song 'Your Doll, Your Maid, Your Toy' from the EP *Away, Away, Away*, written by Lössl. The lyrics are reproduced with kind permission.
8. Email correspondences with Astrid Nora Lössl, March 2–5, 2019. The correspondences are reproduced with kind permission.

9. The verses quoted in the section 'Concluding remarks' amount to the last part of the lyrics to the song 'Sister of the Dark/Walk in the Light' from the album *Where Did Nora Go*, as well as the EP *Away, Away, Away*, written by Lössl. The lyrics are reproduced with kind permission.

REFERENCES

Bailey, Paul J. (2012): *Women and Gender in Twentieth-Century China*. New York: Palgrave Macmillan.

Bayton, Mavis (2006): 'Women Making Music: Some Material Constraints.' Andy Bennett, Barry Shank & Jason Toynbee (eds.): *The Popular Music Studies Reader*. New York: Routledge, pp. 347–354.

Biddle, Ian (2003): 'Of Mice and Dogs: Music, Gender, and Sexuality at the Long Fin de Siecle.' Martin Clayton, Trevor Herbert & Richard Middleton (eds.): *The Cultural Study of Music. A Critical Introduction*. New York: Routledge, pp. 215–226.

Björck, Cecilia (2011): *Claiming Space. Discourses on Gender, Popular Music, and Social Change*. Gothenburg: University of Gothenburg [Art Monitor dissertation No 22].

Citron, Marcia J. (1993): *Gender and the Musical Canon*. Cambridge: Cambridge University Press.

Cohen, Sara (2011): 'Popular Music, Gender and Sexuality.' Simon Frith, Will Straw & John Street (eds.): *The Cambridge Companion to Pop and Rock*. Cambridge: Cambridge University Press, pp. 226–242.

Connell, R.W. (2005/1987): *Masculinities*. Berkeley: University of California Press. Second version.

Connell, R.W. & James W. Messerschmidt (2005): 'Hegemonic Masculinity: Rethinking the Concept.' *Gender & Society* 19/6, pp. 829–859.

Cook, Nicholas, Eric Clarke, Daniel Leech-Wilkinson & John Rink (eds.) (2009): *The Cambridge Companion to Recorded Music*. Cambridge: Cambridge University Press.

Cunningham, Mark (1996): *Good Vibrations. A History of Record Production*. Surrey: Castle Communications.

Djupvik, Marita B. (2017): ' "Working It". Female Masculinity and Missy Elliott.' Stan Hawkins (ed.): *The Routledge Research Companion to Popular Music and Gender*. Abingdon and New York: Routledge, pp. 117–131.

Dvorsky, George & James Hughes (2008): 'Postgenderism: Beyond the Gender Binary.' *IEET Monograph Series* 3, March 2008 (https://ieet.org/archive/IEET-03-PostGender.pdf, accessed 4 November 2019).

Farrell, Michael P. (2001): *Collaborative Circles. Friendship Dynamics and Creative Work*. Chicago: The University of Chicago Press.

Frith, Simon & Angela McRobbie (1978): 'Rock and Sexuality.' Simon Frith & Andrew Goodwin (eds.): *On Record, Rock, Pop and the Written Word*. New York: Pantheon Books, 1989, pp. 371–389.

Frith, Simon & Simon Zagorski-Thomas (eds.) (2012): *The Art of Record Production. An Introductory Reader for a New Academic Field*. Farnham and Burlington, VT: Ashgate.

Geoffroy-Schwinden, Rebecca Dowd (2019): 'Music as Feminine Capital in Napoleonic France: Nancy Macdonald's Musical Upbringing.' *Music and Letters* 100/2, pp. 302–334.

Green, Lucy (1997): *Music, Gender, Education.* Cambridge: Cambridge University Press.

Halberstam, Jack [Judith] (1998): *Female Masculinity.* Durham, NC: Duke University Press.

Harry, Bill (2000): *The Beatles Encyclopedia. Revised and Updated.* London: Virgin Publishing.

Hirdman, Yvonne (1988): 'Genussystemet – reflexioner kring kvinnors sociala underordning' ['The gender system – reflections on the social subordination of women']. *Kvinnovetenskaplig tidskrift* 3, pp. 49–63.

Kearney, Mary Celeste (2017): *Gender and Rock.* New York: Oxford University Press.

Leonard, Marion (2007): *Gender in the Music Industry.* Aldershot: Asghate.

MacDonald, Ian (1997/1994): *Revolution in the Head. The Beatles' Records and the Sixties.* Revised Edition. London: Pimlico/Random House.

Magnussen, Benjamin Maroti (2012): Musicvideo to Where Did Nora Go: 'Please Pleaser' (filmed and edited by Benjamin Maroti Magnussen) (www.youtube.com/watch?v=EpWVzcBKeLg, accessed 20 September 2019).

Marstal, Henrik (2010): 'Køn, fordomme og rock 'n' roll' ['Gender, Prejudices and Rock 'n' Roll']. *Dagbladet Information*, 3 October (https://www.information.dk/henrik-marstals-blog/2010/10/koen-fordomme-rocknroll, accessed 18 January 2020).

Marstal, Henrik (2015): *Breve fra en kønsforræder* [*Letters from a Gender Traitor*]. Copenhagen: Tiderne Skifter.

Marstal, Henrik (2016): 'The Musical Power Triangle of White Male Rock: The Case of Kent (S).' Research paper presented at the conference *Gender and Music: Practices, Performances, Politics.* Örebro University, Örebro, Sweden, 16–18 March 2016.

Milner, Greg (2009): *Perfecting Sound Forever: The Story of Recorded Music.* London: Granta Publications.

Orser, Barbara & Catherine Elliott (2015): *Feminine Capital. Unlocking the Power of Women Entrepreneurs.* Redwood City: Stanford University Press.

Sedgwick, Eve Kosofsky (1985): *Between Men. English Literature and Male Homosocial Desire.* New York: Columbia University Press.

Solnit, Rebecca (2014): 'Men Explain Things to Me.' *Men Explain Things to Me.* Chicago, IL: Haymarket Books, pp. 1–15.

Straw, Will (1997): 'Sizing Up Record Collections. Gender and Connoisseurship in Rock Music Culture.' Sheila Whiteley (ed.): *Sexing the Groove: Popular Music and Gender.* London: Routledge, pp. 3–16.

Tholl, Sofie (2010): 'En rigtig musiker er en mand' ['A Real Musician Is a Male']. *Dagbladet Information*, 24 September (htpps://www.information.dk/kultur/2010/09/rigtig-musiker-mand, accessed 18 January 2020).

Williams, Alistair (2001): *Constructing Musicology.* Aldershot: Ashgate.

Wolfe, Paula (2020): *Women in the Studio. Creativity, Control and Gender in Popular Music Sound Production.* Abingdon and New York: Routledge.

9

Interview With Betty Cantor-Jackson

Sergio Pisfil

Betty Cantor-Jackson became part of the music industry in the late 1960s, working in paradigmatic San Francisco venues like the Avalon Ballroom and The Carousel Ballroom (that would later become The Fillmore West), and subsequently working for concert promoter Chet Helm to create an outpost of the Family Dog in Denver.[1] She soon became part of Alembic, a company conceived to develop sound technology, and worked with one of the most inventive crews in the field of live audio production. Betty first got involved with the Grateful Dead on different studio productions, including the band's third album *Aoxomoxoa* in 1968, pioneering the use of 16-track technology. According to my own research (Pisfil 2019: 43), she was presumably the only woman doing sound on the road in the late 1960s and early 1970s. Betty was interviewed by phone on March 9, 2019, and we mainly discussed the early years of her career.

Sergio Pisfil:

> . . . I would like to make sure everything is clear about your background. I think you started being interested in sound and music before working at the Avalon. You used to book bands for events at your high school, is that correct?

Betty Cantor-Jackson:
That is true. I did do that.

S.P. Tell me more about it. How was that important for your relationship with the music?
B.C.-J. It sort of introduced me to live music, really. Because I was going to hear some local bands, for a dance, at my high school. And that was a band playing live, as opposed to listening to records on the radio. So, it did initiate that for me, having that kind of experience. There weren't that many times actually, only a few. But those did inspire me to go to the Avalon to hear bands whose music I'd heard on the radio.
S.P. It's interesting that a lot of early live sound engineers I've talked to also mention those first experiences at their high schools. A lot of them started getting involved in music production in that specific context.

Were you doing that by yourself or was there a group of people doing that in your high school?

B.C.-J. No, I did check out the bands by myself. In high school there would be a committee of students for dances. Someone handled refreshments, someone else decorations. I picked bands.

S.P. So, you also had to take care of some type of technical issues when you booked the bands, I suppose?

B.C.-J. Not really, nothing technical at that point. It was just "they play well enough, let's have them!"

S.P. I see, so everyone brought their own equipment, then?

B.C.-J. Yeah.

S.P. When do you think was your first connection with anything that had to do with electronics or building things?

B.C.-J. I've always been mechanically minded. When I was a child, I would take apart small appliances that weren't working. I would fiddle with them, put them back together, and they would work. My first experience with sound gear was when I was working at the Avalon. Bob Cohen, one of the owners of the Family Dog (it was Chet Helms and Bob Cohen), would set up the P.A. system; he was into electronics. He had his recorder and I would help him set up microphones, because I found it interesting. It was more technical. And I leaned in that direction, as opposed to what I was doing.

S.P. Yes, because you were doing a little bit of office work at the Avalon as well, I think . . .

B.C.-J. I sold concessions. I took over the management of the concessions, which morphed into running the ballroom night office. I paid the bands, dealt with the contracts, etc. and also paid the ballroom employees. There was a lot of casual labor. It was in the daytime business office that I started doing the bookkeeping, because I like numbers.

S.P. Yes, that's the other part of it. You were really into math and science all the way from the start, right?

B.C.-J. Yes, I was always into math. And it was easy for me.

S.P. Was it at the Avalon that Bob Cohen showed you the ropes of everything that had to do with P.A. systems? You were not handling recordings up until then, I believe.

B.C.-J. Well, he had a two-track on his P.A. output mix. He would record the bands, and I would help set up the microphones. What I understood from him was: "There's a microphone, there's a cable connected to it and there's something that gives it volume. You need to plug it up to a tape recorder and that records what was played". So that was the basics. That simple. No consoles or anything.

S.P. I think Bob Cohen was using Altec equipment, those green devices.

B.C.-J. Yeah, we had Altec Lansing for the P.A., I don't know which one. Maybe a 1567A, for the mixers. I don't remember the kind of tape machine.

S.P. And those Altec mixers didn't have any EQs on it.

B.C.-J. True, you didn't have EQs, you didn't have pan pots. Later I was doing my live recordings with Ampex tube mixers. They were Left, Right, and Center also. No EQs, nothing. All you had was a volume knob per microphone and a master output volume knob. You could go left, right, or center if you wanted to do stereo. That was all you could do! (laughs)

S.P. When you started putting your hands on equipment at the Avalon (or at the Carousel Ballroom, which was a little bit after), were there any women working with you?

B.C.-J. No, there weren't any. I did one set up with Dusty Street, who was doing radio DJ-ing. That was a one-time thing. It was funny because at the time I thought I wanted to be a DJ and she thought she wanted to be a recording engineer. We ended up in the opposite fields.

S.P. But that was it, then, in terms of the technical crew.

B.C.-J. That was it. I don't remember another female doing it.

S.P. What about when the Family Dog moved to Denver? Was the situation there similar, no female workers on the crews?

B.C.-J. Not on the crew, but when I was with the Family Dog I wasn't a recording engineer. I took care of the money in Denver. I didn't really do any recording at that point. When I came back from Denver, that's when I got into recording.

S.P. Right, and that's when you met Bob Matthews?[2]

B.C.-J. Yes, he was working with the Grateful Dead, with equipment. And we started doing recording together.

S.P. At Pacific Recordings?

B.C.-J. Yes, that was down in San Mateo. We did a bunch of demos for bands there.

S.P. So, if I understood correctly, it's with the Grateful Dead that you first started doing studio work?

B.C.-J. Well, I did some location recordings on the second album. *Anthem of the Sun* was a bunch of recordings. I went to a couple and helped to set up, like the Avalon and Kings Beach Bowl. I got to help set up for those recordings, which later became part of *Anthem of the Sun*. I wasn't the recording engineer. I just helped doing the set-up for the recording. Dan Healy was engineering at that time. And then *Aoxomoxoa* happened. Bob [Matthews] and I were doing a lot of demos, and we are using the studio set-up in San Mateo. So we brought the Dead in and that's when *Aoxomoxoa* happened. When we first went in, the studio had an eight-track. By the time we left, it was a 16-track and we also went out the door with a 16-track to do live recording. These were the first 16-track live recordings, and they ended up as *Live Dead* later.

S.P. Was that experience with *Live Dead* your first experience on the road, on a tour?

B.C.-J. I guess that would be my first time doing engineering set-ups for recordings beyond the studio.

S.P. So, let's talk a little bit more about the Grateful Dead, then. I know they are not particularly well known for their business skills.

B.C.-J. (laughs) It's funny because they are well known for not having any. And then they are also well known for having no business plans, but that actually worked for them. They were outside the boundaries of most businesses.

S.P. Absolutely, it did work for them. But I was thinking more about people like you, for instance, who had just arrived to the Dead organization. In terms of having a salary, how did that happen?

B.C.-J. I wasn't paid by the Grateful Dead at that point. We were our own company. We started Alembic.[3] We billed the Grateful Dead; they paid Alembic. The only time that I was actually on Grateful Dead salary was when I was traveling on the road in '77/78. After that I put together our "Front Street" studio on behalf of Jerry.[4] But trying to put everything into proper order in sequence is the hardest part. It's been 40 years or more! (laughs)

S.P. Yes, dates are always tricky, but that makes me think of something else: this is a period (the late 1960s) that is usually portrayed as a very progressive period in terms of politics, creativity, and so on, and the Grateful Dead is perhaps one the most paradigmatic examples of that. Yet [Dennis] McNally's[5] description of the Grateful Dead organization as "sexist" (2002: 215) makes me wonder. Also, I recently watched a documentary about the Grateful Dead (Long Strange Trip 2017), and you can hear someone saying that the organization was generally misogynistic and that there was a misogynistic atmosphere.

B.C.-J. Well, they were misogynistic as hell, as far as that goes. I was allowed to do what I did on the road because I carried the same amount of weight and I lifted the same amount of stuff as any other guy on the crew. I went in before the rest of the band crew and set up my recording gear; then, when the crew came in, I set up Bobby [Matthews]'s stack. When that was done, I helped set up the stage microphones for soundcheck. That's what I did, and I would get the same pay as the other crew members. That's part of why it was sexist. The other side of it was that it was quite unbelievable to most that I had anything to do with the actual technical area or with any decision-making because I was female. I saw I was automatically discounted, and then I had to manifest carefully. If I manifested too aggressively it would cause trouble. The thing is that I didn't want to play any games to get things done. That's how you work when you are a female in a totally male-dominated situation.

S.P. Did you feel that you had to play those games though?

B.C.-J. I didn't have to play it so much with the Grateful Dead, although there were times I had conflicts with my co-workers as to the correct way to record something. "A matter of preference", I responded. With outside folks you just play the game. Which is: you ask them about their equipment, what gear affects what. You ask questions to which you already know the answers. You can't get things done by telling them because that's when people get defensive and shut down the thought process.

S.P. Talking to Dan Healy and Ron Wickersham[6] and reading some of the interviews that Bear [Owsley Stanley] did in the past, I got the impression that the idea of "being in charge" changed each month, each year. Yet there seems to be some implicit hierarchy in the crew. How did you work around the practice of everyone being in charge?

B.C.-J. Well, you have certain areas that you cover, and they were very turf-oriented as far as the equipment. And you have the most, how can I put it, the most bad-ass road crew on the planet; they had their spaces, and you didn't mess with their stuff. Often, I would provide something they needed, and that was one of the reasons I was tolerated as a female on the crew. I would get asked: "why are you carrying all that stuff?" I'd tell them: "because on the next gig you are gonna need something I brought". So, invariably, it happened that way. Doing that and always being there and making sure you are ready to work, absolutely. And that's what you are there for and that's obvious. That's your focus. Therefore, you have earned respect. You have earned the right to say: "Hey! I'm as much part of this crew as anybody else".

S.P. And was it easy to earn their respect, from the Grateful Dead crew?

B.C.-J. (laughs) No! Not at all! I'm the only female that I know of that had that kind of respect from them. And then Candace [Brightman] for what she did. She was a lighting person, and she earned a lot of respect on the road, but that's lighting, that's a different field. I don't know how many females worked there, but I'm sure there was a gender bias in her field. She was like me, she was there to work.

S.P. Was she there already when you arrived to work with the Grateful Dead?

B.C.-J. No, I can't remember when we picked up Ben[7] and Candace, but somewhere along the lines we got them. I'd go out with Alembic and do recording. I wasn't with the Grateful Dead crew at that point. We were the Grateful Dead extended support crew. I mean we started Alembic because we wanted to record the Grateful Dead and we wanted to do it ourselves, our way.

S.P. It's quite interesting the appearance of Candace, because that's another example of a female worker with the Grateful Dead. Were you self-conscious about the specific situation of being female workers?

B.C.-J. Well, you had to be, because we were the only ones out there. We were the only females out there on the road with rock-and-roll bands. You couldn't help it. I don't know how many female lighting people there were at that point. She and Ben, I think, started working together about when Bob and I started working together; and later I did it by myself. Bob was more technical, I was more ears. That's how it worked.

S.P. Now that you mention Bob Matthews again, were you ever confronted by people, back in those days, that maybe looked at you as being just Bob's girlfriend?

B.C.-J. Oh, absolutely. It was definitely that way. I know when it stopped: when Bob and I split, and that was 1972. I ended up taking the main engineering role when we split up.

S.P. But before that, it never made you want to stop working there? I mean, not being valued because of your work, but because you are someone else's shadow?

B.C.-J. Frustrating position at times. I really like learning and experimenting. The Grateful Dead were always open for trying new things. Although not openly acknowledged, I knew what I did made a difference. It was good. And Jerry always called on me and valued my input. So you continue. Keep on keeping on. In other words: "perseverance furthers".

S.P. I believe there were a couple of other female employees. I don't know if they were part of the crew, but I also found out about Melissa Cargil and Rhoney Gissen. Did they work as part of the P.A. crew?

B.C.-J. They did do a lot of soldering, putting together cables and stuff before road tours.

S.P. And going back to your position there in the crew, when did you feel you started to gain their respect? Was it after something specific that you did in a show? Or was this something natural that eventually happened after some time?

B.C.-J. I think it happened naturally just watching me work. I was always trying to make things a little better. I'm really not sure, it sort of happened because we were a family. Members of family gravitated to various areas of interest. We were always pushing the envelope of what was possible with sound. Experimenting with change. Within that framework of change, people seemed to fill the space to which they gravitated. As a family, each found a role.

S.P. It's interesting that most members of the Dead organization considered themselves as being part of a family. Yet, I imagine it must have felt a bit contradictory to feel like a member of the family and, at the same time, you, as a woman, were not getting the same treatment. How did you deal with that contradiction?

B.C.-J. Yes, but it was a family where the females were walking behind the men so to speak. Most of the females in the family were mainly office workers and significant others. I didn't exactly conform. I had to accept I was the only female on the road as crew. There were actually a lot of people on the road at times. Sometimes the band members' children, their personal friends. Some of them were family related, some of them were not, but they were considered extended family. I felt like the Grateful Dead was my first real family. And still is.

S.P. What about your mixing skills? I'm sure you are well aware that the community behind the Grateful Dead, those that proudly carry the label "Deadhead", have a special love for the so-called "Betty boards".[8] These have a special place in their ears and are widely praised. Why do you think that is?

B.C.-J. Well, one thing is that I started with mixers with Left, Right, and Center outputs only, and no EQ. So you have to learn how to mike the stage and mike the instruments. Particularly how to mike the drum set; and make it sound stereo, with limited amount of inputs. Sometimes two microphones, you didn't have a lot. So you learn what kind of mikes to use and where to position them, to get your best sound. For me, what I always tried to do was recreate what's going on onstage. So you hear that stage. And my take for that was to close up. To make you feel you are in the front row, not at the back of the hall.

S.P. The things you learned, you took them from a man-driven world, but you put your own sensitivity into it.

B.C.-J. Oh yes, you first have to do what everybody wants, and then you do what you want to do. Sometimes you need to demonstrate you have ideas that are valuable.

[Once] Jerry [Garcia] was rehearsing to go into a studio in the city to make an album. I recorded the rehearsals, and Ron Tutt, the drummer at the time, liked the drum sound. He suggested that we just record the album at the practice studio. As Ron was Elvis Presley's band leader, Jerry took heed and said [to me], "oh, you could have my 16-track here by tomorrow, right?" I said: "sure, no problem"; which I did, because at that point it was just mixers; two mixers going to the tape machine. I'd put two mics on the drum. . . . And then Jerry let me order the Neve! But with that album, I was put to the test again. *Workingman's Dead* had been my first solo vinyl mastering: the band was on the road, so I went to mastering alone. I didn't like how the normal settings they applied to it sounded. Because I was on my own, and it was going to be on my ass what came out of it, I changed the process to make myself happy. They very much liked the results. That became something I did, I mastered. I would always make the record. But then when I transformed the practice hall into a studio, John Kahn, as part of the band, was skeptical. He didn't trust me yet. So he went out and had another test cut done of the album I had engineered. I was in L.A. doing the same. When we came back into the studio, we played my disc. John put on his other version. He played about 30 seconds and said "take it off" and that was it. John trusted me forever after that. But I had to do that A/B test that I didn't know was happening. I didn't care, I just wanted it to be the best possible. So, if somebody else would have done it better, I'd say, "Oh please, have them do it".

S.P. Even though it was difficult for them to trust you, I get the impression that when you dared to take you own decisions, as you did on that occasion, they did appreciate what you did.

B.C.-J. Yes, the product spoke for itself. When I was the only one there and, like I said, it was on me, then I had to manifest my preferences; I had to make decisions.

S.P. And you took those decisions because you thought you needed to? No one was in charge of making decisions about "who gets to do it?"

B.C.-J. Well, I did that because it had to be done. It had to be mastered. And I didn't like the formula. The formula the industry used didn't work for me. So, I had to come up with my own formula. And it was so well received! I mean, they didn't know they were going to hear that. They didn't know that I had done anything different. I took off the record and I played it, and they went: "What the hell!" (laughs).

S.P. You must have an especial relationship with the music, to be able to be so confident and decide at that time you were going to do it your way.

B.C.-J. I couldn't honestly represent anything but my own choices.

S.P. Were you a musician yourself?

B.C.-J. I do things by ear. I played flute, clarinet when I was a child. But the piano is where you can always find the note.

S.P. Interesting. On a different topic, it has been well documented that women are often invisible in history, in all types of historiographies; do you think it was the same in your case? For instance, regarding the Grateful Dead's history, it took me some time to realize that you had a crucial role. I had to do a lot of research to be able to understand the importance of your role. How can you explain this?

B.C.-J. Well, in terms of invisibility, it's how you move around onstage and nobody even knows you've done that. When you go up and fix a microphone, for instance, it's part of that invisibility. Now when it comes to my mixing, I want to be as invisible. I do not want to be seen. I just want to represent the band, capture the music that's happening. I try to affect the sound as little as possible. I don't use EQ and a lot of processing, because I don't want to interject my own thing. I want it to be purely what it is.

S.P. Let's go to a couple of specific examples. Were you at Altamont?[9]

B.C.-J. Absolutely. I worked the stage, I set the stage.

S.P. Was there a big crew there?

B.C.-J. The recording crew was me and Bob [Matthews]. Bob drove the machine to the site and, when he got there, I had been handling the stage all day.

S.P. I think Dan [Healy] was there as well, right?

B.C.-J. Yes, he was upfront, he was working with the P.A. He wasn't doing the recording at that point.

S.P. And in terms of the recordings, were there any difficulties there? How would you share the role of recording at that specific concert?

B.C.-J. At that specific concert I did the stage, I was the one out risking my life (laughs). I was onstage making sure everything stayed mounted and got recorded. I think I scared Charlie Watts.[10]

S.P. So were there two different sets of microphones? One for the recording and one for Dan Healy?

B.C.-J. No, we split microphones, one output for the P.A. and the other for the recording.

S.P. Interesting, what about the Medicine Ball Caravan?[11]

B.C.-J. Bob and I were fairly interchangeable there, as far as running the stage, although I did a little bit more of that. Part of my expertise was doing that miking thing. So I did more of that and Bob ran more the machine.

S.P. I'm asking about those scenarios, because when you go to festivals and you are doing things on the road, there are other people from other road crews that are also there trying to make their own decisions. The fact that you were the only female setting up microphones onstage, how was that with the other road crews?

B.C.-J. Well, they were surprised! (laughs). "Oh look, there's a girl". And of course they discounted me. It still happens unless they are "Deadheads" and have heard my name.

S.P. You don't think things have changed from those days in the 1960s to now?

B.C.-J. Well, now there are a lot of women running shows all the time, working the stage or working something. A lot of the time they have that same attitude towards other females, and I say "wait a minute" (laughs). So you get that from females, too.

S.P. And now that you cross paths with a lot of women, in terms of music production, can you identify any type of differences between the way women work onstage and the way men work onstage?

B.C.-J. Generally speaking, women are a little more careful and particular about details. They often want to make sure that they did everything in the right order. I think they are a little bit more attentive, because they have to be. It's more or less expected. You have to do more to be equivalent. It's still that way.

S.P. Any possible explanations?

B.C.-J. (laughs) Well, men still think that way, and the basis of society is still male-oriented. It was male-dominated and it's still male-dominated. So therefore, you have to be on it to be considered an equal. You are really careful on how to do things.

S.P. Was this something specifically gender bias, or would you say that any new member of the crew would get that unfair treatment, and would have to learn how to earn the respect of the people in charge?

B.C.-J. There will always be "the new guy". Whether male or female, there's a pecking order. But females have a higher hill to climb.

S.P. Thank you, Betty. This was all very informative.

B.C.-J. Cheers!

CONCLUSION

Besides shedding some light on the cultural context and rudimentary technological circumstances in which live audio production developed in the late 1960s and early 1970s, Betty's testimony is a rare source to identify early attitudes towards female engineers.

Her story confirms the rather unique position of working with a rock band on the road in a completely male-dominated field. On one side, it reveals her determination and highly praised skills to record quality sound; and, on the other, it represents an intimate account of the contradictions and frustrations produced by gender-biased decisions within the Grateful Dead sound crew.

Although it's clear that the working conditions within the band's road crew were regulated by conventions that were unfavorable to women, it is important to highlight that, as ethnomusicologist and live sound historian Nick Reeder states, "the Dead family was a more likely place than others for such conventions to be overturned" (2014: 275). Indeed, the Grateful

Dead did have a substantial role in the counterculture movement and led the way for the transformation of musical and societal alignments that had prevailed until that time. Jerry Garcia's high regard for Betty's technical judgment cannot and should not be used as evidence of the lack of sexual discrimination within the band, but rather, as an example of the complex and multiple ways in which gender-based discrimination can happen in big organizations. Instead of being solely an ensemble of musicians, rock bands – like most labor establishments – have different, sometimes rather independent, social spheres, and discriminatory practices may take place only in specific scenarios.

Although female roadies and live sound engineers are not uncommon nowadays, there is still an obvious imbalance in the percentage of male and female employees in live music production. Moreover, Betty confirms that inequalities do not only remain, but they are sometimes triggered by female employees, too. In response to this situation, an online platform was created in 2013 to encourage women's participation in the field of professional audio and to expand their job opportunities.[12] Working conditions of female live sound engineers are without a doubt different from those in the 1960s but, paraphrasing Peruvian poet Cesar Vallejo, *there is still much left to be done*.

NOTES

1. Created in 1965, the Family Dog was a community-based organization in the San Francisco music scene, and it organized and promoted some of the most influential events in the counterculture movement.
2. Bob Mathews became a recording technician for the Grateful Dead in 1967.
3. Alembic was a company created by Owsley Stanley (aka Bear) to focus on innovation in sound technology. Engineers at Alembic were responsible for the design of what was later called the "Wall of Sound", a new concept in P.A. systems.
4. Jerry Garcia was the lead guitarist, singer, and founding member of the Grateful Dead.
5. Dennis McNelly was the former Grateful Dead publicist and official historian of the band.
6. Ron Wickersham played a key engineering role in Alembic since its inception in 1968.
7. Ben Haller was hired to assist Candace with the lighting set-up during the 1972 Grateful Dead European tour.
8. "Betty Boards" is the colloquial name for the highly regarded soundboard recordings of the Grateful Dead, made by Betty Cantor-Jackson during the 1970s.
9. Altamont was a free music festival held in northern California, in December 1969.
10. Drummer of the Rolling Stones.
11. Conceived to put touring musicians and other counterculure personalities on screen, The Medicine Ball Caravan began in San Francisco and travelled the United States through August 1970.
12. See soundgirls.org.

REFERENCES

Long Strange Trip (2017) Online video, Double E Pictures/Sikelia Productions, U.S., distributed by Amazon Video, directed by Amir Bar-Lev.

McNally, D. (2002) *A Long Strange Trip: The Inside History of the Grateful Dead*. Broadway Book: New York.

Pisfil, S. (2019) *A New History of Rock Music: Documenting Live Sound between 1967 and 1973*. PhD Thesis, University of Edinburgh, UK.

Reeder, N. (2014) *The Co-Evolution of Improvised Rock and Live Sound: The Grateful Dead, Phish, and Jambands*. PhD Thesis, Brown University, USA.

10

Twists in the Tracks

An Interview With Singer, Composer, and Sound Producer Aynee Osborn Joujon-Roche

Liesl King

Aynee Osborn Joujon-Roche is an accomplished singer, composer, and sound editor for film and television who lives and works in Long Beach, California. She is the lead singer and co-founder of Restless Blues Band, the lead singer and co-composer of the albums *There and Back Again* and *Just In Time*, and when you look for her on IMDb, she has 71 credits with the Sound Department beginning in 1998. I first met Aynee as an undergraduate, when we were both studying for a degree in Theatre Studies at the University of Santa Barbara, and then professionally and geographically, we went very different ways – I to the UK, where I pursued a career as an academic, teaching literature, science fiction, and gender theory; and she, as you will hear, to Colorado, LA, Nashville, and then back again to LA, pursuing a career in music and sound post-production for film and television. She and I and a small group of other Californians who met in Santa Barbara in the 1980s have stayed in touch over these many years, and this summer of 2019 I had the privilege of interviewing her in Santa Barbara for this collection on gender in music production. In the interview you are about to read, Part 1 offers a potted history of Aynee's experiences in the music industry; Part 2 explores her experiences as an editor/creator of sound for film and television; and Part 3 expands on her impressions – touched on in Parts 1 and 2 – of the way gender differences can adversely impact women's well-being in the industries of music and post-production, but also the ways in which they can offer fantastic opportunities for collaborations among women and enhanced relationships between men and women, too. As I read through the interview, it strikes me that feminists and gender theorists working across the same historical period have helped to shed light on many of the issues that Aynee raises, and have contributed to the dramatic advancements in gender equality that post-industrial Western societies have undergone in the twentieth century, a dynamic that has (largely) continued to shape gender relations positively in the twenty-first century. Importantly – and the points Aynee raises in the final section, 'On Gender', make this clear – feminists, activists, gender theorists, the LGBT community, AND innumerable women and men who would not officially place themselves in any of the above categories but who simply live their lives with sensitivity and respect for others – have contributed to a radically transformed and transforming reality in this twenty-first century where *men and women of both genders and all sexual identities are becoming more comfortable relating to each other as human beings instead of types*.

As I have suggested, twentieth-century gender theory, which came into its own through second-wave feminism in the 1980s, gathered up individual narratives like Aynee's, garnered statistics and trends, and offered a studied critique of the way in which women, women and men of color, and non-heteronormative men and women were treated within a vast range of professional and domestic cultures that appeared to primarily privilege white, heterosexual men. The point that Aynee makes in the final section of the interview, which is that during the time she was getting started in the music business and coming out as a gay woman in LA, many men still felt it was acceptable to 'grab [her] in the butt', not only foregrounds the sexism that was prevalent in the popular music industry at the time, but also underlines theorist Monique Wittig's point in *The Straight Mind* in 1991 (originally 1978), which was that, to paraphrase, the straight community often simply assumes that everyone else is straight. And in Part 2, her narration of the way in which the female producer she was working with at Sound Deluxe left to start her own business, and invited Aynee to come join her when she was made redundant, brings to mind Adrienne Rich's utopian concept of an all-female, 'lesbian continuum' (Rich 27), in which women of every sexual identity – straight, bisexual, gay – might support one another so that they mutually thrive, and overcome difficulties by looking out for each other whenever they have a chance. And in the final section of the interview, Aynee's discussion of the fantastic, collaborative experiences she has had working with a wide range of men in the music and post-production film/TV industries, when she and they have been able to 'take off the hats of men and women', and be 'in the same tribe, making music', brings to mind gender theorist John Beynon's key point in *Masculinities and Culture* (2002), which is that men too are adversely impacted by hypermasculine environments (in which 'hegemonic masculinity' is culturally produced (3)), and by extension, that they too naturally benefit from collaborative, respectful, non-hierarchical ways of being and making and doing.

As soon as I lay claim to the position that 'we' (and we all must be careful about 'we', as often our sphere of understanding is far more local than we believe it to be) in many arenas within Western culture are making dramatic strides in terms of gender equality, a reader will no doubt disagree profoundly, and point to heinous examples of sexism that she or he has experienced or observed. Of course, My literary specialism is feminist science fiction utopias of the '70s and '80s, and we are a long way from the difference-respecting culture that Marge Piercy imagined in *Woman on the Edge of Time* in 1976. However, I would argue that in many ways, the collective, Western 'we' (and of course this is not a dynamic exclusive to the West, but for the purposes of this essay, I will talk about the geographical areas with which I am familiar) is experiencing and in many ways contributing to a gender-quake that in terms of the long history of *Homo sapiens*' relations, is marking out a profound and dramatically important sea change. Aynee Osborn Joujon-Roche's experiences, observations, and conclusions, in sum, show me a glimpse of what our species can be, and what it ideally will become.

PART 1 – BACKSTORY

Interviewer

Our edited collection, *Gender in Music Production*, is hoping to acknowledge the contributions that female sound producers, engineers, and musicians bring to the music production process within a studio environment. We want to identify and celebrate the positive attributes women bring to the production and the ultimate outcome of a musical artefact, and we want to discuss the attitudes or cultural behaviors that potentially impinge on this process.

Firstly, can you outline your background and tell us something about how you arrived at where you are now?

Aynee Osborn Joujon-Roche

My God – I'm out of college, I was under development with a girl trio, called, are you ready? Trimm. And we were thrust into the studio. We were thrust into roles, where I fit a certain niche of a certain girl 'type'– a white girl, a black girl, me whatever – and was put in an outfit and hair, makeup, and totally doing something that had nothing to do with who I was other than the fact that I could sing. So that was my first sort of exposure to the business, and Berry Gordy and all these people were involved with Motown and it was so, so interesting. And yet, it was so not me. So that was part of it. And then that kind of went into 'okay, what can I do?' I wasn't a songwriter in the sense that I felt proud of at the time. So I could sing; so then I got a piano player. And I did the whole cabaret route. And so I would sing songs that I liked – you know – it's interesting because it's almost full circle, because now I'm singing with a jazz band – and I started out doing jazz stuff like Ella Fitzgerald and Billie Holiday, and Dinah Washington and all these old beautiful, phenomenal American songs. And I got some attention through that, in the form of I made a friend – and he comes into this later – I was singing in the cabarets of Los Angeles, and I met this guy named Paul Rothchild. And Rothchild ended up being this icon of '60s rock, where he produced some Dylan; he recorded Janis Joplin; he produced The Doors and Bonnie Raitt.

And he and I hit it off like wildfire, and the same goes for his amazingly talented son Dan, who later would help me produce some of my Americana music in the '90s. So to go back to 1989 . . . I'd auditioned for *Star Search*, which is like today's *American Idol*. I sang a very, very old song from the '20s, and I didn't look like the other typical people. And nothing really came of it. And I was working in an office – right, this crazy thing – in an office, and my phone rings. And this woman introduces herself and she says, I'm the assistant of Glenn Frey of The Eagles, and he and another guy who's a New York Broadway producer have a nightclub/restaurant in Aspen, Colorado, and they saw your audition tape for *Star Search*. And I'm thinking 'what?!'. They want you – they're going to make their restaurant called 'Andiamo' in Aspen – (where everybody goes in the wintertime, all the rich and famous people go there to ski and party) – and they just saw *The Fabulous Baker Boys* with Michelle Pfeiffer and the Bridges brothers, and they want to do a piano bar, and they are captivated by you. They've got three people that they're auditioning here in Los Angeles. Can you come tomorrow and sing/audition? And so I did. And I got the gig immediately. And my whole life changed. I was flown out to Aspen – you know, private jet – it was like a Cinderella story. Time goes by; so I sing out there and one of the important things that happens is that I meet Irving Azoff, who was the manager of The Eagles at the time. He'd heard me sing in Aspen, and Glenn Frey really believed in me, so Glenn helped tremendously after the Aspen shows ended.

Then I kind of lily-padded that exposure in Aspen along with Glenn Frey's help to jumping into the music scene in Los Angeles, and this would have been around 1990. So I had a meeting with Irving Azoff, where I told him that I felt I was somewhere on the spectrum between Melissa Etheridge and Sade. However, I didn't really have this – ALL – you know the 'complete package' or my own identity as an artist, singer/songwriter at the time – I could sing, but I didn't know who I was. I didn't know. So I couldn't tell them. That uncertainty put me in a very interesting and vulnerable position too, and looking back on this – well, I've never really talked about it like this before – so you guys are my first – and it's all flowing out of me because usually I can't remember. So I'm going to get quickly to the end. But I bring Paul Rothchild in as my producer, and he and I begin the work of trying to figure out what I want to sing that will be honest and ME. They throw some money at me; this is the music industry at the time: they throw some money at me. I go into the most amazing recording studios in Los Angeles – The Mamas and The Papas recorded at Ocean Way, A&M Records, Larribee studio, all this stuff started to happen. But again, I didn't have my identity. So I went into a song search, a process of searching for songs that I loved and felt a connection to. And, ironically, one of the songwriters we listen to was Sheryl Crow, and a bunch of other people were trying to find songs that resonated with me so I could come from a place of truth, blah blah blah . . . all of this at you know, age 26 (which many considered too old). And even then he wanted me to say I was 23 – it was fascinating. And during the course of all this, I started writing more of my own songs. But I went into development prior to this; I was sort of songwriting on the outskirts, but I was still trying to get this record deal.

We made my demo; time went by, and they decided to sign Sheryl Crow, because she was a songwriter and an amazing musician. And so, you know, her career went in the way you know; it was me or Sheryl Crow, and they picked/chose Sheryl Crow. . . . But anyway, so that ship sailed. And then

I started writing my own stuff. *And Paul wasn't into it*. And I remember I had done a show in a place called Genghis Cohen in LA, and I was doing my own stuff, and this guy Jac Holzman at Elektra Records – he was really digging it. But he thought I needed more time. So that was when they were putting artists – and this is all again – my whole . . . everything is up to men. At that time, it was all men making these choices and decisions. And I intercepted a letter that was addressed to me – my name but Paul's address: we both lived on Lookout Mountain up in Laurel Canyon, which had its own amazing music scene. Paul was at the top, I was at the bottom. And I intercepted this letter addressed to me at his address, and I decided to open it. And it was from the record company, basically offering me a development deal. So they were going to supplement my income as an artist, give me the stipend, and let me continue developing myself. *And Paul* . . . and I sealed it back up, and I took it to Paul's house. And a couple weeks went by and you [sic] never mentioned it to me. And I'll never forget this day, because I had him on the phone. He was in the recording studio with someone else at the time. So I said, "Paul, I gotta tell you, I found this letter". And he got so pissed. And he was like . . . he was on his back foot and he just attacked. And he just said, "I don't like the songs that you're writing. I don't like the direction you're going. It is not what we were looking for. It was not what", you know, blah, blah, blah . . . "And I don't think they should put this money into you. I don't think that they should invest in you". He no longer believed in my music – because he didn't have the control and, well, maybe my own songs just weren't his cup of tea.

And yeah, it was devastating. It was my last chance in a way, you know, not to like, pity party it. This is not . . . I haven't also talked about it. So, clearly there is a little emotion to it. So this – you know – this little dream went away, and also it felt like the betrayal of a friend. Because Paul and I went to the theatre together, and Paul and I went to the opera together; Paul and I, you know, would drink into the night and sing and cry, and his son is amazing, *his son is amazing* – Dan Rothchild – he's in Echo In the Canyon right now. We made music together and he seemed to really like my original Americana songs. We recorded several songs together. I love him madly.

Anyway, it was, it just was – my little thing, and it happened for a reason, because I'm also a super private person. And even though I sing and I do all this stuff, I'm also a little bit introverted. You know, maybe if you knew me well enough you'd be like – you'd be so [saying] – I mean, *you really are*. I don't need to wear the red dress. I don't need to do the – listen to me; look at me. . . . Even though I can front a band. I can tell a story; I can sing a song; I can be that entertainer, but in my own private life, I don't need to be the center of attention. I just don't need that limelight.

So I went to Nashville – I forgot all that part – I went to Nashville, on the radio, my whole country world that I made happen – I did get a small record deal. I got the deal, had songs and radio interviews, CDs in the record stores, all that stuff – I'll keep going and not get into the therapy! But I did find my own voice and I liked it. And I told my own stories in a way with the American fabric of Americana/folk, if you will, where I morphed into country/folk out of the rock. And I loved it, and I lived it for years and years and years. And I hooked up with other musicians and wrote many, many songs.

But anyway so the country thing – I was too country for LA and I was too folky/LA for Nashville, and this would have been in around '97; and then later on I moved away from it, because basically I'm 34, I got back from Nashville; I'm 34 – ran that track. Also, I didn't like all the attention afterwards, like men coming up to you and having their agenda in a bar in Nashville, such as "Oh my God, I'm in love with you, and I want you to meet my mom, and blah blah blah. . .", and my guitar player having to pretend to be my boyfriend. You know, it was a very interesting thing. But you're talking about a 30-year career in what I'm trying to shove in in 20 minutes – how I got to where I am. . .! But I came to LA, got into a job, because I didn't want to be a broke musician. So I got very practical. And then my work – I fell in love with my work – and did a sh** ton of movies and a sh** ton of work and walked away from the music, getting into sound editorial for film.

And we can get into that later – I don't want to just monologue for you – but I moved away from music. And then maybe halfway through my career, you know, working as an editor, I knew something was missing from my soul – this creative side too, that the editing wasn't satisfying – and that I wasn't playing music anymore. And some people I would meet at parties that I used to be in bands with, they'd say, "Aynee, how can you not be doing this?" And I would say, "well, you know, I met a girl, I got married and I got a house; I have a job. I'm, you know, blah, blah, blah, you know, that whole materialistic thing". And then I realized – you're right. So then I started to get in that groove. And then I ended up getting into a blues band. And that's when the blues band thing started because I could be any age – didn't have to be young and hot. I just had to sing and really have that together. And the blues was kind of jazz and country coming together. And it was this Bonnie Raitt thing and Susan Tedeschi and Aynee [Osborn Joujon-Roche] – this melding, and I was thinking, this is awesome, because it had the power but the finesse and the ease, and so I was digging that and I've done that for eight years, which brought me into where I am now.

And at that point I was 48 years old, and playing in a club from say 11 pm to 2 am, and – who am I? – driving home *drenched* from singing for three hours and feeling incredibly satisfied – and that you know had its trajectory and as bands go, as relationships go, we kind of had our climax and, it exploded and imploded when my brother got sick. And then after I lost my brother, I could not touch a guitar. I could not – you know, I wrote one song that nobody probably will ever hear called *Broken*. And then I didn't sing anymore, for a while, like over a year, year and a half, and then I slowly have gotten back into it, slowly, a little bit at a time, because it's all about our lifestyles, and our managing our stress, or whatever – that reflection of us – and also I was just, you know, healing. And part of that healing was me allowing myself to give myself a break, and to do whatever it is I wanted to do that I needed to do for me, because so much of my existence was output. You guys as Moms can understand that – as a mother, as a woman, whatever, giving, giving – and I almost, and it almost, killed me. To be perfectly honest, it almost killed me. And so climbing back from that, from the worst possible thing that you could imagine, you know, and coming back from that and how that's going to play into my music is still a story that's unfolding. You know, right now I'm singing – I've gone back to singing – not my stuff. It's too close. To be able to sing these classic songwriters and just have fun. . . . Just have fun again, right? I am also starting to play my own material with my writing partner Drew, and it feels good. I have a great home life with my wife, who is a light and my rock.

That's why I surf so much too, because it makes me feel closer to my brother. It resonates with me. And it's just pure fun. No thought. Just fun and joy and oneness with the planet, with the ocean. It's just the greatest gift in the world. I am so happy. I'm almost happier there than I am playing music.

PART 2 – WOMAN IN AUDIO

Interviewer

So what's the relationship between – or is there one? – between the sound editing work that you've done, and your original interest in music and singing? What's that segue?

Aynee Osborn Joujon-Roche

It was interesting because when I was making a record, when I first got into the music industry, it was analogue. So you're recording on this big two-inch tape, and it was visceral, and you were physically cutting the tape to edit out a breath that you didn't like, or whatever. And I was blown away by that process. When I started to work in sound, and I was helping to produce my own style of music, where we were taking control, it had become a digital world of sound. So now we're working in Pro Tools, but you're still seeing this visual representation of a track, and you're recording sound waves you can see going across this track. And so the music in Pro Tools, it's this linear thing, of this, for example, pink track, or orange track, or whatever; so it's really, it's almost like paint being dragged across the canvas horizontally with the soundwave going like this [gestures]. So it's really kind of cool visually. So when I first saw my voice being recorded that way, and the drums and how they looked, and the guitar, and all of this, I thought wow – that's pretty cool. That was my first introduction to sound being digitized and represented visually. And it was cool because it was multimedia, right? Plus, you're doing this creative thing that's energizing you and it's stimulating. So when I started to work for Warren Dewey, a sound engineer, sound designer – where he's designing sound, working in a digital format, with a huge mixing board still, it was the same kind of visual thing. But maybe in sound for film, as a sound effects editor, you might have that visual track going by, and instead of drums, it's the sound of crickets – because it's a scene that takes place at night in the forest. You might have another track of fire, crackling, and another track of the wind moving the leaves of the trees around. So you have that visual representation of the sound that way. It was an easy transition. And there's a ton of musicians who work in post-production sound, right? Because it totally translates across.

Interviewer

Can you say something about the changes you've seen in terms of the tools you use to produce audio since you first began?

Aynee Osborn Joujon-Roche

I would say that technology has made my job as an editor and musician much easier. For example, my cell phone is also a recorder – I can quickly record a song idea and then email that audio file to my writing partner, and he can then use that audio file as a foundation to start building a song. In the same way, recording equipment and digital editing tools make everything more streamlined. Fast internet speeds are also a great tool, making it possible to send large amounts of data over the internet – sharing songs, or sound for a TV show or film; it can all be 'uploaded' via fiber optics, making it possible to send huge amounts of this data, intact, to the studio servers in a very short amount of time. For example, I can send the dialogue for an entire episode of a TV show I work on in less than an hour. A few years ago, without fast internet it would take over 12 hours to upload this material. Before that, I would take a portable drive and have to physically drive to the studio and give it to the assistants personally. In this digital age, things can happen in seconds vs. hours, and that's a huge help in productivity.

Interviewer

Right. That makes sense. So back to the narrative about your career – you said you went to work for a sound designer?

Aynee Osborn Joujon-Roche

Yes, I was working for this sound designer. And he taught me a million different things. And there – this was a nice thing because it never ended in this betrayal [like] it did with Paul – as I said I worked for Warren Dewey, who had started also as a sound engineer for people like Aretha Franklin, doing concert mixes and amazing things. And he was a record mixer and came up with songs as well. So, we would sometimes do radio shows; we'd record the dialogue of a radio show, where you had, say, William Shatner, and these amazing actors, and we'd have a microphone on each guy and record them all. And we were still recording on digital audio tape, called Dat machines at the time. It wasn't even that digitized. Then we'd bring that back. We'd make this radio play, and even, say, how they did it in the old times with the guy holding the shoes, to make it sound like the footsteps were approaching or whatever – there's a door knock and glass breaking; we'd have the same thing. But we'd add it later; it exists in the library, a database that you could grab from and draw on.

And then I basically segued into working for a huge sound company in Hollywood called Sound Deluxe. And they did – you know – all of Oliver Stone's movies; they did everybody's films, mostly men's. So I basically worked my way up by working in a sound effects library, an encyclopedia of sound. So I would categorize, and I would organize, and I would edit and clean all these sounds from sound libraries from years gone by, and also [I was] creating new sounds. So I would go out with my microphone, and I would record a racing car going by, or we'd go out of the box and we'd go to, say, JPL – Jet Propulsion Laboratory in Pasadena. And we'd talk to some of the scientists there and go into their laboratories, to see if we could find new sounds to record. And there were animals to record, too; there were a million things like that. I worked my way up – I finally got into a union, where you have to have a certain amount of hours as an editor, a certain amount of hours working with audio on certain movies, because it's a whole political situation. I had to have someone write a letter for me. And I worked my ass off there for about five years.

And then they got bought by a bigger company. And it became less of a Mom-and-Pop type feel. And just as I got into the union, one of the men who owned the company, they basically let me go. They gave me severance. And they said – we don't need you anymore. Thank you so much, after I had produced a sh** ton of work, but they were downsizing and I didn't have the language, the self-love, to express myself and say, wait a minute, you guys are making a huge, huge – do you know what I could have brought to your company, the value? I wish, right, as a 55-year-old woman now looking back, I wish I'd had those – that language – but also with all of those heartbreaks or disappointments in this career, in this world that is dominated by men – music world and post-production – as are so many other fields, we find, it also led to another opportunity for me that was where I hit my stride. So it was 2002 when I joined the union and I was let go from Sound Deluxe. And a woman who is a dear friend of mine, we've known each other for many years – worked there. She was leaving at the same time to basically start her own company. And she said, because I was in tears, you know, and they knew that I had formed a department; I had made this thing – I had made, I had monetized their library; I had digitized their library, and continued to bring in money, and all this stuff that was going to go on in perpetuity. . . . Oh, by the way, as well, [it's] still going on today, over 20 years later. But she left and she said, come and work for me. Come and work with me. And here it was, this woman going out on her own with one movie called *Blue Crush* about women surfers, and that was it. And we were off to the races, and we haven't looked back since. There have been ups and downs, ebbs and flows – film segueing into television, which is fantastic. More women. And now there's a ton of women on our crews. And we keep working, and I've been working steadily for a really, really long time.

PART 3 – ON GENDER

Interviewer

Okay, the second question was around how you were supported along the way, and you've really covered that. The next question is about obstacles, and I feel that you've talked a little bit about that, but is there something more that you wanted to add? The question is – what obstacles did you face as producer/engineer/musician that may have shaped the future for you?

Aynee Joujon-Roche

I mean, I've had such cliché experiences with male chauvinism and male control. But I've also had, you know, the 'let me handle this for you sweetie' type thing. And, and men just sh** on you, thinking if I sh** on you, maybe it's going to make you better type thing; make you stronger. And, of course, sexual undertones/overtones always end up being a part of it. And, I've – you know, like many women I used it too, if I could, used it in my benefit. I did too. And you know, I think it's part of the learning process, and the experience that is reflected by our society. And that evolution of realizing – okay, I don't need to do that. And I'm not that cute anymore anyway, so maybe I . . . you know what I mean? So there were some obstacles like that, and the self-doubt and of jumping into an area where you think – I didn't go to school for this; I don't have a degree in this, where some people, you know, did and do; I don't have an engineering degree. So much of it again, comes back to when you finally realize who you are and what you bring to the table, and how sometimes we beat ourselves up about it. . . . And then that realization – the irony that happens later in life that you do [know who you are and what you are doing]. To find your voice and your self-worth and are able to stand up for yourself. If young women could find their voices and self-worth sooner – ah, the years of pain and BS that might be avoided.

But one other thing I wanted to tell you was a positive experience, also with men – is sometimes if you're lucky, and you're making music, it's one of the few things – also in Martial Arts – but making music sometimes – it's one of those beautiful things. Maybe you've experienced this, where you're involved in a creative process, and you're with the opposite sex, where you just become humans, musicians – making music, where I'm not looking at 'Oh, wow, those are nice breasts', or 'Wow, he's so tall and handsome, look at his jaw line'. No, we're just humans, in the same tribe, making music, and that's such a beautiful thing, when those outer things can fall away. So I've had some lovely experiences with other musicians making music, making sound for a film, making a soundtrack for a film, where for a minute, we take those hats off, of men and women.

Interviewer

What do you think contributes to that kind of experience, that kind of arena where people are just being themselves, where sexual politics or the attraction isn't a dominating factor?

Aynee Osborn Joujon-Roche

Well, so what do you think contributes to that positive? I think it's that the whole is greater than the parts. So, if that whole, whatever that is – that cake that you're baking – becomes the thing, this communal energy of making this cake, making this record, making the soundtrack, whatever, when you all – can get out of your heads, and just be in that creative process. Yeah, it's that moment of creation, right? That hopefully happens in any endeavor where you're making something or experiencing something – perhaps too, maybe you're not making it, you're not making the music, you're not making the cake, but you're experiencing whatever it is – the fire, the lightning storm, the things that are part of you, but they're also way bigger than you. It's about the purity of connection.

Interviewer

We can't categorize all men as a group or all women as a group. But if you were to give advice – let's start with male figures in the industry – if you were going to give advice to them about how to help the women in the industry feel included, and to be able to create that kind of natural ebb and flow that you just talked about, is there something you could say?

Aynee Osborn Joujon-Roche

Yeah, that's a tough one. It would be nice where, where things aren't qualified by gender? You know, to say – the cliché of she's such a great drummer for a woman. And she's this amazing drummer, she's an amazing woman/female, you know, female director. You know what? That part – the gender talk, the gender identification might be an interesting thing for them to try to think about more. And I think that it's happening too, because I was brought up in, came out, and was still working in the culture where men could still sometimes grab you in the butt or corner you – make you . . . for example, 'you're the secretary to this meeting. No, no, I'm a producer. I'm not. I'm not a secretary. Why is it, because I'm female?' You know, that kind of stuff, or the talking down.

Interviewer

Okay, so then, what about for the women, so younger women or women new to the business? Is there something that you would say to them, that would offer some insight into how to help create that kind of 'human' reality in the business that you just described?

Aynee Osborn Joujon-Roche

Boy, if it was a young woman, the more, the more work she could do *quickly* to somehow trust herself, to trust her gut, and to stand up for herself the way she might for her Mom. If somebody – you know – was downtalking your mother in the grocery store and embarrassing her, shaming her, or whatever, you'd stick up for your Mom or you'd stick up for your little brother, or your little sister or your best friend. Why can't we do that for ourselves in an educational, doesn't have to be a threatening, way; to be aware of your own self, and also your responsibility even about your tones – going back to the musicality of things, our tones – it's not what you say, it's how you say it. So, sonically, the way you might speak to somebody where they still feel like they're not being attacked; the man is not feeling like he's being attacked, but it's like – this would be so much more beneficial to all of us, if, you know . . . but so many people don't. That's the irony. That's, that's also the miracle of life, right, of aging, that you're not going to know everything at 20, nor should you have to. But, yes – I could have saved myself maybe some extra steps had someone said, you've got to believe in yourself and trust yourself, and you *know* what's going on, but you're ignoring the voice. Which of course the inner voice knows.

Interviewer

I hope that through this interview, some young women/women new to the music business will be hearing what you've just said. I hope they will trust themselves. And as you've said, in a professional and heartfelt way, the 'could you think about the way you've just said that' approach is so important – creating that trust.

Aynee Osborn Joujon-Roche

And this is so key, foundational, fundamental – 100%. But I also asked so many questions. And so all the men that I was around, that were so reclusive in their own little editorial worlds – but the minute I approached them with some curiosity – could you tell me why you like this or why you did it that way? Then they would say – oh, I would love to, and then they each give you little pearls of wisdom. And that also helped me in where I was going on my journey to educate myself to become that well-rounded person, and that also led to people going – she's pretty pleasant to be around; let's give her a call and get her on the show because she's easy to work with. And that's 90% of it. They assume you have the skills. So get the skills and know yourself and trust yourself. And then it's really about knowing when to stop talking and listen.

Because in that particular field, you could be with people in a dark room mixing a movie for 12 or 15 hours. And if you're a pain in the ass, on any level, you won't be getting that next job. The rub is, you get confidence with experience, but you need the experience to get the confidence. . . . As they say, fake it 'til you make it. I was thrown in the deep end, and I learned to swim and swim well – quickly.

Interviewer

So the next question is about the disparity in terms of percentages of men and women in music/audio production. And I understand that there are still far more men in the industry? Why do you think it is the case that there are fewer women working, and especially fewer women in the higher, more prominent positions?

Aynee Osborn Joujon-Roche

God, I wish I knew. Yeah, I really do.

It's not a sexy job. Not to say that women are after sexy jobs. But it takes a certain . . . I mean, now I'm talking about post-production, okay, post-production sound, which is mostly men. I mean, it's a technical job. There are a lot of female editors, also film editors, because, you know, it takes an incredible amount of patience. It's a super detail-oriented job, so you have to really be able to do that. And I almost look at it like virtual knitting. You're knitting this long, long scarf. And if you miss that one little thing, then the whole, you know, unravels. So it takes a great deal of patience. In so many ways women are so suited for it. And there are certain areas where in dialogue . . . although I started out in sound effects, but you know, dialogue is mostly women; sound effects and sound design, because it's gunshots and werewolves and who knows what, it could be mostly men. . . . So yeah, I honestly couldn't tell you why. It's just because there were more men in the workforce when this started to happen, and it just stayed that way. And it's slowly, slowly getting better.

Interviewer

And so you see it shifting?

Aynee Osborn Joujon-Roche

More women are coming in. Although I'm so sequestered I don't know a lot of the new young female editors. So I'm not sure about that. And in the music business, like any other business, too, there's a ton of men that are in power. But the music industry, which I don't have a lot of knowledge about, at this point, the music industry is changing dramatically too, because it's not like – oh, let's give you a record contract and make a record, as now people can make high-quality records in their bedrooms, and you can get it digitized and out there in the marketplace. Now the marketing – you need deep pockets for that sometimes, unless something goes viral. But yeah, the digital world – you know, my music is going to exist on Spotify and Amazon and iTunes, and all of that long after I'm gone. And that's really cool. You know, maybe there's 5,000 people that know of my music or perhaps 500 – whatever! The point is, it's out there in the world.

Interviewer

Okay, so final question, and we've talked a bit about this already, but what do you think the male music production population can learn from women working in the field? It's a pretty big question.

Aynee Osborn Joujon-Roche

So what comes to mind. Oh my God, the first word that comes to mind is *collaboration*. Collaboration. And to have that sort of inclusion as opposed to . . . sometimes men come from this egotistical place. That's just so – it's such a quick knee-jerk way to do it. Welcome to my recording studio. These are my guitars. That's my room. These are my mics. And what are we going to do with your song today? And it could really be, what are we going to do today? What are we going to do? Yeah, and here we have all these toys and all these knobs and it's going to be . . . but to make it that because women, you know, historically are much more communal. . . . And if that atmosphere was there, that sort of gentleness there – if men can embrace a little more of their feminine side . . . of tribal . . . of say, welcome to this space – a tiny bit of incense wafting through the room, carpets everywhere. We're going to make music together today. And I've encountered those men, too. And I love it. And you get up and you're in this vibe that is the most natural high when you're all vibrating on that creative level of like, yeah, yeah, right. Did you get – you know, oh, let's go again. Right. You know – we have to do it again. Right? Yeah, let's go back to the fifth measure, because . . . and they're reading your mind because they know/they heard/they felt/they experienced it. So you're all sharing that same thing. And even if it gets a little bit off-track, you knew exactly what to do to go back into that and in the corrective, collaborative moment, that's magical too. And there's a beautiful high there. And it's fun to be able to experience that with everybody you know – and the old guys with the old ego and it's my way or no way – they're going to be dying out on all levels, right? – the my way or no way. You guys are going to be going away. Just no room for tournaments anymore.

Interviewer

And I suppose just finally, because I always have a final question, but it's just to say is there anything more that you want to add about this topic? The book is entitled *Gender in Music Production* – is there anything more that comes to mind that you would want to add? Men and women will be reading this, men and women with different sexual identities, different roles/positions, people from different cultural backgrounds.

Aynee Osborn Joujon-Roche

Well, it's the dichotomy of embracing who you are, which also includes embracing your gender. And feeling empowered by that. As an example, to write a song from a female perspective of, say, watching your single Mom raising you and your brother – so that's going to have a female through line, and a feminine through line or a vulnerability. And so you know, okay, so because the idea would be to be gender neutral, and to go into a creative flow as human beings. That's while still honoring the idea of whatever might be happening in the project, while also the male, if the male is doing something, to be gender neutral, but also to embrace maybe his masculinity, of being able to comfort you at some moment or to be able to maybe put you back on track because women can go off on tangents. So then the gender could take place, come into play in a beautiful way. But then to remember to go back to that sort of neutrality perhaps, and always come from a place of trusting yourself, trusting your inner voice, because if you're not, then you really have work to do. You know, and if the making of the art helps you do that, fantastic. Start from there first because then we can heal it outside of ourselves if we work on healing it inside of ourselves.

Interviewer

Well, Aynee Joujon-Roche – the things you've said apply to my world too, and I just want to thank you for this amazing interview.

Aynee Osborn Joujon-Roche

Thank you!

CONCLUSION

This interview with Aynee Osborn Joujon-Roche adds voice to the period between the second-wave feminism of the 1980s and '90s and to the current moment, where culturally and institutionally, we are all seeing radical changes in terms of gender expression and gender equality. From my own vantage point as a 53-year-old woman, it is fantastic to see the way in which university students here in the UK express a range of gender identities, and raise points in class about queer identity, non-binary identity, and asexuality, for example. The popular MeToo movement inspired by Tarana Burke in 2006 has inspired professional counterparts; in my own institution, Student Services has developed an 'All About Respect' campaign, which particularly focuses on the need to report sexual assault. A young woman in my science fiction class just last week explained that she didn't know why some women she knew didn't call themselves feminists – she said she asked someone, 'do you think men and women should be treated equally?', and when the person replied 'of course', she said, 'well, that's feminism!' While editors such as Robyn Warhol and Diane Price Herndl (1997) have suggested through their title that in reality, the melody of voices which proudly foregrounds luminaries such as Virginia Woolf, Betty Friedan, Hélenè Cixous, bell hooks, Audre Lorde, Judith Butler, and Judith/Jack Halberstam is perhaps better represented as 'Feminisms', since feminists come from a range of different backgrounds and vantage points, my student nevertheless argues well. It is perhaps also true that the introduction of the more inclusive category 'gender theory', an umbrella term which includes theoretical essays on the topics of masculinity and non-heteronormative identities, has identified common ground between men and women, due to gender theory's simple acknowledgment that every human being (and every species on the planet) suffers in cultures where hypermasculinity dominates. We (and I am speaking as a woman, but also as a human being) needed the diverse language, the multifaceted discourse of feminism(s) to get to this greatly improved place in history, where it seems to me that finally our collective vision – where we come to perceive one another as people, first, and individual men and women, second.

Aynee's fluid description of a studio environment where men and women are going 'into a creative flow as human beings', while on a subtler level, accepting that there might be gender differences is for me, profound and deeply nuanced. Her comment 'if the male is doing something, to be gender neutral, but also to embrace maybe his masculinity, of being able to comfort you at some moment or to be able to maybe put you back on track because women can go off on tangents' is beautifully expressed. If a reader finds the phrase 'because women can go off on tangents' offensively stereotyping, I'm unsympathetic; of course this is not true of all women, but it is true of many. I'm happy to claim my own tendency towards self-interruptions, digressions, and trajectories. French feminist Hélène Cixous was perhaps the queen of the multi-leveled argument, and while few writers can compete with her erudition, she creates a meaning-making model which lends credibility to all habitual digressors!

Judith/Jack Halberstam famously explained that no one 'owns' masculinity or femininity in *Female Masculinity* (1998), and this observation is an important one for the context. In Aynee's description of a collaborative environment where 'the gender could take place, come into play in a beautiful way', it seems to me that the terms 'man', 'woman', 'masculine', 'feminine' have been thrown into free play, where all bets are off, and where the term 'collaboration' comes fully into its own. I can't help but return to my science fictional reference – in Piercy's *Woman on the Edge of Time*, the gender-fluid character Luciente tells protagonist Connie: 'strangeness breeds richness' (103). For me, this is where Aynee's journey, which contains many twists and turns along diverse and intersecting tracks, finally arrives. I cannot agree with her sentiments more thoroughly, and so to all who read these pages, here's to a future dense in collaborations rich and strange!

WORKS CITED

Beynon, J. (2002) *Masculinities and Culture*. New York: Open University Press.
Halberstam, J. (1998) *Female Masculinity*. Durham: Duke University Press.
Piercy, M. (1979) *Woman on the Edge of Time*. London: The Woman's Press.
Rich, A. (2003) 'Compulsory Heterosexuality and Lesbian Existence'. In: *Journal of Women's History*, Volume 15, Number 3, pp 11–48. Baltimore: The Johns Hopkins University Press.
Warhol, R. and Diane Price Herndl, eds. (1997) *Feminisms: An Anthology of Literary Theory and Criticism*. 2nd ed. New Brunswick, N.J.: Rutgers University Press.

Part Three
Personal Perspectives

11

Women in Audio

Trends in New York Through the Perspective of a Civil War Survivor

Svjetlana Bukvich

There has been very little written about forces that shape a woman music producer. More women are accessing audio technology today than ever before, and this has been attributed, in part, as a result of music technology becoming cheaper and ubiquitous in nature, rather than made possible by a cultural shift created consciously over time. In the US, there is currently a spotlight on women *en long et en large* – a loud buzz over 'who gets to say what happened in the bedroom' – which spills out into other narratives about women's lives: however, the 4% of women in the music industry who produce still haven't left the blind spot of the public eye. The rest of the world seems, for better or for worse, less impacted by this particular tension – the mild hysteria of the "me too" movement on one hand and the relative insignificance of its impact on the music industry on the other. In the US the invisibility/visibility pendulum for women is generally swinging harder – with yet-unseen results – and outside of it, there seems to be less drama and a sense that change comes at a slower pace. The "global" scene in music is becoming sonically hybridized, and yet, the behaviors in and around the recording business stay staunchly entrenched in the cultural folklore of those involved. Men and women are quietly suspicious of each other in contexts where music is produced, because they question each other's motivations for being there. To complicate things further, there are proven differences – perhaps less so in the US, but still pervasive – in pay scale for the same job performed. All this is accompanied with a lingering sense of confusion when it comes to understanding the gender roles they occupy in the ever-stimulating magical arena called the recording studio. I welcome this opportunity to bring attention to these matters by offering some hopefully useful solutions, and to shed light on the challenges men and women face when working together in sound, and beyond. Perhaps selfishly, I also welcome the opportunity to outline my own experience in the field, and create a profile that is identifiable within the history of women who produce. I especially don't wish to "add" a voice to anything. In the call-and-response dynamic, I want to be the call.

So, who are we, women producers? I'll narrow this question to where I now live – New York City. Like many other women producers here, I come from experiences that encompass other cultures. Those experiences are suffused with the unique pulse of sounds hitting the concrete, of ideas almost visibly expanding and contracting in a euphoric rush, of architectural structures that challenge the laws of . . . well, everything. These revelations are not unique to women, naturally, but the way women organize this data influences how a signal flow is routed in the studio, or the architecture of frequency equalization on an audio track. Women and men are not the same creatures, and that is okay. There are obviously differences in biology, but also assessable cultural norms and expectations that shape the way they look at things. They experience life differently. This should, by all measures and means, be a 'no-brainer', but somehow I needed to learn and re-learn that. In my youth, I continuously schlepped more synth poundage than my petite body structure could bear. This was to prove that I could handle my own gear in the same way the all-boy bands did. I don't do that anymore. Or, when I heard the mix of a song at 16 years of age while doing back-up vocals in Sarajevo (then Yugoslavia), and summoned the courage to offer a direction in the mix that had to do with what the song was about – the relationship between a boy and girl – and got slammed with: "It's time you start to shave your legs". I recoiled, said nothing, and finished the takes. I knew that nothing was 'wrong' with my legs and figured the comment was there to send me a message to stay in my place. I remember desperately trying to create a space between the oncoming feeling of hurt and the job at hand. Suffice it to say that today, with some 25 years of experience, and without the need of going into the validity of the response to my offering, I can retort that my advice back then was as sound as it would have been today for that very song, and that it would have benefited the production in unmeasurable ways. The difference is in the approach and the reply. Then, I felt fear. Today, I feel compassion toward that man and his words. There can be a sense of threat in the studio when a woman takes over, even fleetingly: deeper and more pronounced in music than in most other fields. Since men's work in the studio is often looked down on by the rest of society as not being a "serious" profession, they tend to hold their positions of power with extra zeal. This is probably true, or will be true for women too, and an interesting research topic in itself. What if roles get reversed in this way one day? Wouldn't we have the same problem? As kitsch as this may sound, have we not heard our own mothers warning against marrying 'musicians'? I have personally never met a woman, or man, who is *not* in music possessing an ability to distinguish the many levels of responsibility and income levels in the business of music.

Many women producers in New York, and other great cities – London, Berlin, Los Angeles – where they can realize themselves in music, have similarly tricky work memories from their home turfs and have needed to overcome them, thus taking the energy away from their work. In my experience, as suggested in the early part of the century by biologically centered gender theorists such as Simon Baron Cohen, and through societal conditioning, women I work with have brains that look at the universe in a relational way. Things don't stand as isolated islands: everything is interconnected; there are exponential possibilities in relational combinations in a single-family unit, neighborhood, workplace, city, country, world. Some of my fellow women producers in NYC have roots in tribal cultures on the verge of collapse, some come from ancient cultures older than Moses' tablets, some from sleepy bourgeoisie cultures, some have progressive working-class backgrounds, etc. These women are gloriously diverse, talented, and diligent and they are all here, ready to pounce, and change things. Paradoxically, they look at the world with these deep probing eyes and manifest meticulousness, inclusiveness (for the lack of a better word), and exceptional care in the studio and in the festivals/events they produce, and yet, they are looked at as curious, isolated islands. It's the women's festival, the women in audio, the women composers, the women who lunch.

Coming from Yugoslavia, a country whose model of living no longer exists due to the collapse of its experimental socialism and the brutal fight for money and territory helped by a new wave of tribal and nationalistic rhetoric, I have found myself increasingly re-connecting the broken pieces I took with me, consequently adding my "American experience" to my "new world order" fantasy. This fantasy includes, first and foremost, the re-ordering of the hierarchy in the "who sings under the dotted line" game in audio. I recently saw the Carole King musical *Beautiful* on Broadway. An anomaly for the times in which she lived – she signed her work as songwriter and producer – her early desire to author was met with "she writes girl songs", and was given a chance because of that. The ratio between the probability of her career longings and the amount of hit songs she amassed is soaringly high. King made a dent in the culture that didn't know "she" was possible in the role she secured. In my opinion, Kate Bush too falls into this particular category of trailblazers, and a few others. The situation is not all that bad, though. It is remarkably better than it was in the nineteenth century, for instance (yes, you can laugh here), and the very fact that I can write these words is a step forward toward having a music industry that co-creates with the other 50% of the planet. My desired scenario is this: the bastard proliferation of technology and its cheap price will inevitably pull in more women, and the pendulum will start to move toward their interest areas, therefore giving them more say. This could happen very dramatically in the fields of 3D sound and other immersive technologies, since women tend to live in that kind of world in their minds anyway. If they get to this field early enough, and claim these technologies for what they think is important for humankind, they'll reinforce the rule the big corporations will have to then accept. In traffic court, it's called 'the right of way'. NYC is a prime candidate for this to unfold. My feeling is that within the current political climate – and yes, all art is political – there is a sense of daring, hurriedness, and decadence, as when, I suppose, Rome was nearing its end. In a moment like this – in between radically different technological stages where nanotechnologies, Big Data, and talk of parallel universes are seeping daily into our cups of coffee – grand ideas can take hold, so stay alert! I, myself, don't intend to be left behind.

I'll narrow the question of who women producers are even further here and offer some images/situations that shaped me personally. When it comes to record production, I am interested in what with Simon Zagorski-Thomas calls 'active musicology', i.e. musicology that generates the kind of understanding that "helps us 'do' music more effectively". The diarized sections are mixed with the rest of the discussion here in an effort to do just that: to stretch the commonly accepted structure of an academic paper, and hopefully, offer a deeper insight into the subject matter – forces that shape a woman music producer – viewable from different angles and time-spaces. To me, this knowledge is important to be available because traditional epistemology, in the words of Alison Jaggar, is shaped by the belief that emotion should be excluded from the process of attaining knowledge. What I am inviting the reader to do here is experience alternative ways to interact with time and emotion in an academic-writing setting, especially when it comes to learning about women producers. What we *do* in our practice is handle time and emotion. But, I'd argue, we do it differently from men. What Australians call the 'fictocritical essay' is probably the closest inspiration for what interests me here: putting bits and pieces of my memories in relation to accepted authoritarian pools of knowledge about the matter. Another epistemologist who wrestled with the lack of critical thought that reflects women's position in the field, and whose ideas influenced my contemplations about women in music production, is Jane Tompkins. She offered that "an epistemology does not consist of 'assumptions that knowers make' in a particular field: it is a theory about how knowledge is acquired which makes sense, chiefly, in relation to other such theories". In other words, I'd like to offer a window through which I'd like to gaze myself, and invite others to see the view. My sense is that at this point in time, women who choose the path of music production all had to have had more than ordinary life circumstances, which in turn prepared them to face the challenges in the profession as it is today. Some created them consciously, knowing well that they'll need to learn from the results of their unfolding. I belong to this group, and offer this writing as a possibility for learning in the young field of theory about women producers.

SARAJEVO, YUGOSLAVIA, 1971

My sister is 13 and I am 4 years old. She is going to both general education elementary school and elementary music school, which was then common. Being her little sister, she teaches me *Für Elise* by L.V. Beethoven, two hands. As my feet couldn't reach the piano pedals, she tells me I could play loudly in places I want to. She takes me to her piano teacher to show me off. I play the entire piece and, before I finished, there are other teachers entering the room through the very, very tall, Austro-Habsburg wooden double doors, and then more teachers. I am making sound with my hands, just as I please, and they seem to be captivated by it. I enjoy myself immensely. I didn't know at the time that I would be, years later, playing the Minimoog synthesizer with the same kind of exhilaration, a kind of free abandon in the inclusion of images I felt were missing since I couldn't touch the pedals with my feet.

SARAJEVO, YUGOSLAVIA, 1974

Mali Slager is a children's festival of song, and I'm singing "The Strawberry Gardener" with the Sarajevo Symphony. It is the only song in minor key – this I know because I sneaked into the hall during rehearsals to hear all the other contestants. I am to sing in minor key with a choir! That is a pleasing thought. At home, the writer of the song comes with his black *achten taschen* filled with handwritten notes on staff paper. To this day I don't remember names well, but I know his: Edo Pandur. I think he is very important and interesting because he can imagine what his piano playing will wind up sounding like as an entire orchestra. I wish to tell him that I too sometimes imagine sounds and things while I play the piano. The pieces my teacher gives me have little stories hidden in them. His face is serious though, and I don't dare disturb the little time he seems to always have with me. I know I can sing the tone leading to the chorus really well (later on I discover it's *called* a leading tone in harmony), and I like how hearing me do it brings a fleeting smile to his solemn face. In the song, the lyrics talk about a butterfly who, riding as a knight on an early morning ray of light, lands on a poppy petal. My mother wants me to motion with my hand to the sky at that point in the song, since I was going to be on TV. During the live televised event, the live orchestra, the choir, the big microphone, and the dark chocolate-colored dress I didn't like all converge. I start to sing the second verse instead of the first, while motioning with my hand to the sky. I do a decreasing of volume in my voice and slow down the time in my head. I anticipate the entrance of the horns and land the remainder of the first verse correctly, switch the lyrics of the second so that they were not heard twice and that it all makes good sense for my Strawberry girl. That night, I learned that I can manipulate what is heard, and that the perception of sound differs from person to person. The audience apparently knew nothing of my mess-up.

ADDIS ABABA, ETHIOPIA, 1978

I am sitting on a tall wall that surrounds the elementary school – a slight little girl with golden hair. My dad was supposed to pick me up, but he isn't here. As the minutes go by, dread and then fear start to fill my chest. The sense of time seems stretched, unattainable. The schoolyard echoes with the rhythm of girls playing hopscotch: "A-B-C-D-College, A-B-C-D-College!" This was Addis's best school, with British and local kids in attendance, as well as some stand-alones like myself. I strain to focus on the quintuplets as they start to form sonic arcs with some counting in Amharic (ānidi, huleti, šositi, ārati, āmisiti) coming from the left field. Then there is the sound of the 1970 Mercedes my dad drove – off-white with navy leather seats – passing by with a Doppler kind of effect as it circles around and then vanishes through the school gate. In my mind all this is a language of salvation. Looking back, I didn't know if I heard music in all that clamor to suppress fear, or if my fear opened up a new kind of space for sound. All I knew was that I was going to see my dog Jimma soon, and mom too, and that all was good in the world. What I didn't know is that I became a producer on that day. All the sounds that I heard that afternoon – the mix of voices, languages, accents, and machine – was a language that would later become part of my own.

EDINBURGH, SCOTLAND, 1984

The George Heriot School Symphony is rustling in the pre-final rehearsal state of neurosis. I am playing Beethoven's Second concerto by heart that night in the hall larger than any of my biggest classical performance experience up until then. I am wearing an earring that says: "I'm a loud, super sparkly, prismatic, and fake aquamarine!" After a roaring premiere, the second violinist comments that the earring was too much, especially since I didn't come from affluence (in so many words), and was there on scholarship. I smile and say that, perhaps, she regrets not being the soloist herself with the freedom to set standards about proper attire. That same night, my gay Boy-George-with-raven-black-hair-looking school friend who came to the concert takes me to a bar stacked up with two floors – one for gay men and one for gay women. I had never been to a place like that. He buys me and his friends a drink (he exchanged music with me ever since I showed up in Edinburgh's oldest castle-school) and gives a speech. Some girls from school came, too. Everything is happy in a Harry Potter kind of way. That night I learned about inclusivity, the power of connection through music, the power given by music, and all that is larger than a lonely stint at a private school.

TROY, NEW YORK, 1993

Robert Ashley is fumbling through reels of tape he wanted digitized. He is preparing *Empire*, an excerpt from *Atalanta*, one of his several multi-disciplinary spoken-word TV-operas, which is to be produced at Rensselaer Polytechnic Institute (RPI), where he is visiting artist. I am his beaming student sound director, assigned to him in a program that favors an integrated approach to electronic arts, the first of its kind in American academia. No one really knows what we are doing: a Grass Valley video rig, early Kurzweil keyboards, and Studio Vision sequencing software with OMS are among the beauties we are installing in the studios as we relentlessly pursue the idea of converging technologies to fit our artistic visions. Things are discovered as we go: our teachers are open, curious, and gifted artists, and our efforts are reinforced by lofty postmodernist theories that seem to odiously lag behind. I have no knowledge of music production, so to speak. Previous to this, I had played with Atari computers in Sarajevo for the recording of my first symphony with the Yamaha DX 7 synthesizer, and had bartered my keyboard chops with boys in bands to let me record my songs, as long as they turned the knobs. So, I have to expand and adapt fast. In Ashley's piece we have video, pre-recorded sound, live performance, post-production, and words like Campbell soup to handle. Ashley is already highly regarded in his career, and yet, he appears to me as one of the most generous, patient, and open human beings I had ever met. And yes, he had that beguiling voice as a weapon. In a classic 'teach by example' fashion, Ashley had us experience the power of collaboration, insightful guidance, and grace. This discovery continues to serve me well as I find myself swimming upstream in many facets of collaborative sound work.

NEW YORK, NY, 1994

Looking Glass studios on Broadway proves to be a thriving artistic hive. The first Blue Man Group album was recorded there, as well as works by David Bowie and Goran Bregovic, among others, and of course, by Philip Glass, who is one of the co-founders and owners. What intrigues me so much as an intern there isn't Glass's compositional prowess as much as the fact that he owns a studio which expresses art that would otherwise not be heard, and that he uses electronic keyboards in new classical music requiring a conductor. His opera *La Belle et la Bête* based on J. Cocteau's film is being recorded here, and I am one of the assistants. His larger-than-life producer Kurt Munkacsi, responsible for the MIDI room and things of mysterious nature pertaining to the operation of the premises, teaches me one important lesson. The soprano is coming, the MIDI tracks need to be printed, all the audio interns are entangled with cables routing the MIDI from one MOTU Mark of the Unicorn rack-mount interface to another. (As a side note, I liked those interfaces because they had a picture of a unicorn that supposedly left a mark. I later wrote "Sabih's Dream", a piece about a flight on an Arabian horse, akin to those I used to have riding on a mountaintop back in the homeland.) Notwithstanding being inspired by my reverie, the MIDI isn't working for us, and the three interns are trouble-shooting for an hour to no avail. Enter Munkacsi, so tall that he can barely get through the door. Slowly, he walks to about the middle of the distance between the door and the gear, listens to our problem, and responds in an equally languid manner with about ten words. Then he swivels on one foot and disappears. The appearance was part comical and part surreal. His advice was gold, and we solved the puzzle, but what stuck with me more is the way in which it was delivered. It made the 'less is more' proverb more poignant to a budding know-it-all, and taught me about the importance of knowing how to balance coolness and efficiency in a stressful situation and to plain have a good time.

KINGSTON, NY, 2011

Incredibly, I just drove to Kingston to meet Tony Levin, the electric bass hero on the many records I adore. Next to me is Paul Geluso, the engineer from New York University's (NYU) Music Technology department, where I teach part-time. I feel grateful to Geluso for being my right hand in this – tracking Levin in his own house is something neither of us have imagined doing anytime soon. I don't know what to expect and am feeling a bit apprehensive. I had sent Levin the tracks with the bass part notated.

Yes, I came with a recommendation, but he tells me on the spot that he accepted to do this because I can write for bass and because he wanted to play on 'something different'. Generously, he pulls out more basses than I'd ever seen in one place: a couple of electric basses, a few uprights, and The Chapman Stick. Levin is completely prepared and full of ideas. The three pieces on my new album *Evolution* require 'big bass', and we follow Levin's lead to record with no frequency equalization, and no sound compression, using just one little pre-amp that he loves. His timing feels like pouring hot concrete over a bumpy surface, evening the bumps out. He seems to delight in odd time signatures and everything is going well. We utilize sliding and thumbing, but there is one section that just doesn't gel. What I wrote doesn't work so, quietly in panic, I decide to give up my structures and ask Levin to improvise with a bow on the Chapman. This technique produces a sound I have never heard before – a sound that, supported by the altered tuning in strings, so perfectly embodies the ideas of that particular section of the piece that I want to cry. That was one of the moments in my producing life where I tasted the power of making a decision that is right for the context. The decision was part-composition, part-production. Virgil Moorefield's book *The Producer as Composer* illustrates this profoundly: The role of the record producer was evolving from that of organizer to auteur; band members in some case became actors in what Frank Zappa called a "movie for your ears" (Moorefield 2005: 15).

As we parted, Levin shook my hand and said: "It was great. I hope we get to play again".

NEW YORK, NY, 2012

I'm producing my album *Evolution* at Stadium Red studios in Harlem. Many years in the making, this album contains *Before and After the Tekke*, considered by many to be my groundbreaking piece for violin, voice, piano, analog synthesizers, electric bass, and electronic sound in original tuning. Mixing with me are the classical producer David Frost (18 Grammys) and avant-garde jazz producer Tom Lazarus (7 Grammys). I'm excited, and hearing everything they do, and more. I ponder the hourly rate they must be getting with other, 'more commercial' projects, and the fact that I can shamelessly call the shots on every millisecond of sound on the tracks. Time comes to mix *Tekke*, and Lazarus kindly offers to do some premixing in order to save time. I gleefully accept and the next morning, solidly caffeinated, the three of us listen to the premix. My heart sinks. It is completely *not* the direction I want to go with the mix. Luckily, I had earned some feathers in the preceding sessions for the album, so I can speak. There's a pounding thought in my head: I am finally starting to feel validated in the studio (I embrace happily the question "have I arrived?"), but all this is also a favor to me bestowed by the two of them – helping an artist manifest her vision fully, for a price that she can afford. I don't wish to seem ungrateful. It is hard and we find a way. Lazarus and Frost are genuine gemstones, with music's best interest on their minds. We laugh eventually. This episode taught me that what a music producer really does is manage people, their legacies, personal histories, lives and wants, their talents and peculiarities. For that, one needs to walk the path of integrity, love, and respect, and, on that same path, to put the music first.

NEW YORK, NY, 2018

I am post–Thomas Dolby concert at the Cutting Room in NYC, a venue I hold dear in my heart. I managed, by a stroke of luck and some 150 email messages, to produce my album release concert here. Dolby plays through his oeuvre with nuance and humor, and with an informed technological wand in hand. I know his music and the innovation it brought about. After the concert I wait in line, holding my CD in hand. I approach Dolby, tell him I am a fan and why, and then throw in the names of friends and collaborators we have in common. His book *The Speed of Sound* is promoted on this tour. We exchange goods. I give him my self-produced CD, and he gives me his book – signed. A happy conversation ensues. It is a good night. I feel like I have 'arrived'.

NEW YORK, NY, 2019

The Dolby Laboratories midtown is a place I am initiated into by Mick Gochanour, a brilliant producer and director. I am in what seems like a tower, surrounded by the misty May clouds floating by the windows of the 28th floor studio. It's very zen-looking. The gear just got moved to another location, and what's left is the very essentials. I am getting a demonstration in Atmos, Dolby's cathartic new foray into 3D sound by none else than Darin Hallinan, the chief shaman. We fire up the engine. I feel like I'm hearing my music for the first time. My life is realizing itself.

So, this is my movie. And while some of these small, personal accounts may seem like a listing of milestones in my professional life, it is precisely these milestones that made me feel like I'm peeling away the onion of recognition; the insecurities about not having the opportunity to do the work drop away and instead of 'a woman producer' I become just 'a producer'. I am acutely aware of the general lack of female heroines, and like to tell stories that interest me in that regard. As someone born at the crossroads of East and West, I have always been interested in connections between fields that are not necessarily apparent "at first listening", and in finding/adjusting mediums through which these connections can take form – recording arts, concert stage, film, performance art, architecture, and dance. My heroine is omnivoracious and about connections. In the world I come from, men and women had equal pay and equal access to education. In Sarajevo, a mosque, a cathedral, an Orthodox church, and a synagogue all used to greet me while I strolled. Yugoslavia fell apart after a civil war, and the sense of unity in that part of the world shifted. My sense of myself as a person shifted, too. I evolved into a kind of a fractal split – an inhabitant of many worlds. However, I am not bitter about it. I have learned to adapt and to understand my place in history and purpose in life better.

Pre- and post-war history in Yugoslavia has undeniably influenced me as a producer. As with every societal upheaval in the past, the changing of hierarchical structures offered a unique look at what forces create us and change us. Having my nose "rubbed" in war – psychological and physical – felt like flying into an abyss, into another dimension from which to gather knowledge. That knowledge, however subjective, is part of my consciousness when I work. I left under tenuous circumstances, and I feel richer for those lessons. The experimental nature of society in Yugoslavia, coupled with the solid education I received there in classical music, the rich general education, and a coming-of-age within a vital rock scene all gave me a sense of curiosity about the world and the tools with which to explore it. My thirst for looking at things from different angles was intensified with my pursuit of electronic music and media in the US. I came of my own volition, on American stipends, and not because of war. I stayed because of it. In the 1990s, all I wanted to do was learn the technology so that I could fully produce my sound, and expand it into visions of genre-busting, hybrid performance. My influences back then were Meredith Monk, Laurie Anderson, Andrei Tarkovsky, Mesa Selimovic, Mikhail Bulgakov, Kate Bush, Led Zeppelin . . . and many more. This all happened at the time when multimedia in America was at its onset. Those were exciting times, and it felt like the war didn't actually slow me down. It taught me about the impermanence of things, yes, but I somehow continued to ride on the wave of innovation, fueled by the opportunity the US offered me. I clearly saw the historical chance for people like me to do new things and, potentially, leave things better than they found them. My influences now are stretched somewhere between early Igor Stravinsky and Jacob TV.

Women and men coming from different cultures have different ideas about the monetary value of the goods performed, patriarchal hierarchy, leadership, history, and politics. Post-Gregorian chant, music became a product and a powerful tool in the hands of, first, the church, and later, the executive producer, a male in most cases. Women producers don't make the same amount of money in music as their male counterparts. We all know that women who are successful in this field feel like they work twice as hard. Why? Because, in addition to the economy of things (related to their titular role in raising offspring), they need to overcome the still-prevalent understanding of them in a primordial sense – as muses whose purpose is to arouse magic. Yet, the (male) producer will get the credit and control the product and the money. So, there is plenty to overcome here – a few millennia?

So what are my solutions?

In my experience, when a woman walks into an all-male recording studio (or a man walks into an all-female studio, or a bunch of women walk into an all-male studio), there is a palpable vulnerability in the air. There simply aren't that many examples of the aforementioned scenarios, and people generally don't know what to do. The best result is achieved when there is openness in spirit and respect of each other's vibration. Most of the time, this has nothing to do with the actual knowledge of audio by anyone in the room, or talent. It's about acknowledging the roles people are in (she is here to do a mix, not to date, or he is here to sing, not to hang) and not confusing them with the personal set of cultural gender clichés.

In my work, I like to mix genres. I deconstruct, rebuild, repurpose, imagine, and re-imagine. What brings me joy is to see musicians from different backgrounds and with expertise in different genres share in the intimate space of a composition. To me, what they are sharing ultimately is an understanding of and a bending of time. That is one of the most precious gifts I gain from the mixing of genres. That understanding, however, comes from instrumentalists' – and my – experience with different musical languages (American minimalism, art and progressive rock, German classicism, Italian polyphony, Russian romanticism, jazz, ethnic idioms). Once I have a feeling for a musician or an ensemble, I apply their most appropriate language to the context of the piece according to my judgement and requirements. There's a definitive structure involved, rarely with improvisatory freedoms, and with notated parts and recorded electronic sound. As opposed to giving my music to anyone who wants to play it, I like to know who I'm composing for, if possible. This aesthetic probably resembles a rock ensemble practice more than that of an ensemble in the classical field, even though I work with classically trained musicians most of the time. This is possible now for me because in my work I assume the role of the producer as well as composer. This means that I can fully take control of all aspects of performance, timbre, timing, microphone placements, and the ultimate intention in the sound of a piece. I am drawn to technologies that hold a promise for the attainment of 'flow', or of uninterrupted and immersive experience in art. As children, we have a glimpse at that freedom, and then we take on heavy specialization. I aim to reinstate some of that early flow ("tok") by creating a language which expresses the experience of life now and reads much like strings of micro-euphorias, vectors, prisms, and vines intertwining, branching out, accelerating. When I look around, things seem to come out of one another, much like in a fractal sequence, or a cell splitting. I understand music's elements as something organically interconnected. (I talk about this in the book *In Her Own Words: Conversations with Composers in the United States* by Jennifer Kelly.) Music genres don't hold my interest because I see them as stultifying, as something that breaks things apart, and as mere steps toward human emancipation, not their final destination.

I don't fantasize that all this should be known to a guy who's folding cables with me so that he can gauge what to do and what not to do. I expect, though, that with global access to data, men and women will be more open to that which they know little about, and will embrace the learning. Men and women in the music industry *will benefit* from creating opportunities for working together. In music lies freedom, and together, we are better positioned to free the world from the shackles of prejudice, pitch/tuning domination, gender bias, genre compartmentalization, tribalism, doctrine, and, ultimately, poverty. It is a win-win situation. I believe that it is where we could head as a species. The way new space can be created is by imploding models that don't work, and music is probably one of the fastest highways that can teach us that.

SELECTED RESOURCES

Baron-Cohen, Simon. *The Essential Difference: Male and Female Brains and the Truth About Autism*. New York: Basic Books, 2003.

Baron-Cohen, Simon. *The Science of Evil: On Empathy and the Origins of Cruelty*. New York: Basic Books, 2011.

Jaggar, M. Alison. *Love & Knowledge: Emotion in Feminist Epistemology. Gender/Body/Knowledge: Feminist Reconstructions of Being and Knowing*. New Brunswick: Rutgers University Press, 1989.

Kelly, Jennifer. *In Her Own Words: Conversations with Composers in the United States*. Urbana, Chicago and Springfield: University of Illinois Press, 2013.

Moorfield, Virgil. *The Producer as Composer*. Cambridge, MA: MIT Press, 2005.

Tomkins, Jane. *Me and My Shadow: New Literary History, Vol. 19, No. 1*. Baltimore: The Johns Hopkins University Press, 1987.

Zagorski-Thomas, Simon. *The Musicology of Record Production*. Cambridge: Cambridge University Press, 2014.

12

Three-Pronged Attack

The Pincer Movement of Gender Allies, Tempered Radicals, and Pioneers

Julianne Regan

This chapter aims to highlight the role of gender allies and tempered radicals, as they attempt to change a culture wherein they are considered dominant. It acknowledges some of the self-limiting behaviors exhibited by women in the field of audio engineering and production and also serves as a reflection upon the pursuit and attainment of gender equality in the recording studio, with consideration of the significant role of these gender allies. Factors potentially contributing to the dearth of women in audio practice will be addressed from an autoethnographic perspective and will draw upon the documented experience of music educators, electronic music pioneers such as Delia Derbyshire, and self-producing artists such as Björk and Grimes. It also considers anonymous qualitative data gathered via a questionnaire completed by members of social media groups populated by female professionals in the fields of audio engineering and sound production.

Since becoming a music practitioner in 1981, I have always worked and created in a predominantly male environment. The first band of which I became a member was Gene Loves Jezebel, a post-punk outfit from Wales. I was the only female therein, but this five-piece enjoyed an overall androgynous aesthetic. Fronted by twins, whose prettiness bristled with a muted masculinity, the band became known for its angular dark pop music and left-field, anti-machismo lyrics such as 'I'm just shaving my neck for maximum effect' (Aston and Aston, 1982). The experience afforded me a false sense of security about the music industry, in that I found myself in an egalitarian meritocracy, in which my gender was inconsequential. My role in the band was that of bassist. I played a Fender Musicmaster short-scale 30-inch bass, but only because I was a skinny 19-year-old with a slight frame. Almost as a counterbalance to the perceived concession of using a smaller than average bass, I used a Carlsbro amp combo. I think it was a Stingray, but can't be sure, as it is long since stolen. Whatever the model, it was heavy in sound and stature, and I took great satisfaction in that. My time with this band – aside from the toxic sibling rivalry – was a positive experience, and I was always treated as an individual, and never as merely a 'girl'. We enjoyed many tea-fueled deep and wide conversations allowing me to grow as a person intellectually, and we had a solid work ethic that allowed me to flourish as a musician.

My first encounter with sexism in the world of music occurred as I was sound-checking for a gig. This was my first gig, not only with Gene Loves Jezebel, but my first gig *ever*. I was nervous but determined. A member of the all-male in-house tech team approached me and leered, "My, that's a big instrument for a little girl like you", before proceeding to ask me which of the men in the band was my boyfriend. I gave no verbal response, but was angered by having been singled out as the lone female, and also embarrassed by the implied sexual connotation of the remark about the 'big instrument'. But what irked me most was the assumption that I didn't know much about bass playing and was quite probably there as a token appendage of an assumed beau. I knew as much as I needed to know about bass playing. I knew how to choose, change, and tune my strings; I knew how to get the best sound from my amp; I knew how to play bass for the genre of music I was involved in, *really well*, because in my teen bedsit, I had taught myself by listening closely to Jah Wobble while playing along to PiL's *Metal Box*. However, I failed to challenge the behavior of the stage technician because at that point in life, I was a shy and sensitive individual with a very rudimentary idea of what feminism actually was, and indeed, if I were a feminist or not.

Being the bass player was a role in which I felt comfortable. I was under no pressure to look or dress in any certain way, and I could maintain a low and sexless profile towards the back of the stage, in black combat trousers and jacket bought from the legendary Laurence Corner Army surplus store. This happy anonymity was forfeited when I became a singer, which would eventually happen three years after leaving Gene Loves Jezebel.

My personal experience of music production began with a Fostex 250 porta-studio, which although limited in function, was state of the art in the early 1980s. I used to clean the heads with Johnson's Cotton Buds and nail varnish remover, but then, didn't we all? Regarding music technology courses, as far as I knew, they either did not exist or I had no financial access to them, and so consequently taught myself how to record and mix, by using the manual and my ears.

A number of generally enjoyable music projects started, floundered, and ended, yet the Fostex was a constant throughout. I also used the Roland SoundMaster Memory Rhythm SR-88 drum machine. Both were analogue, and using them felt intuitive and straightforward. Then came the digital revolution.

I bought a Yamaha RX-11 MIDI drum machine, which could be played live (real time) or programmed (step). Again, it had a user-friendly interface, just a set of buttons to press for each sound required, be it snare, hi-hat, and so on. This machine would go on to be used in the band All About Eve in 1984, in preference to the complications a human drummer might have brought.

It was a fellow band member who discovered how to download data from the machine onto a cassette tape so that we could play live to a drum backing track. It seems that what I called the 'blip tape' triggered and controlled the RX-11. I didn't understand. This prompted the beginning of my deferring to others concerning recording technology. One of 'the boys' could deal with it, and not because I was lazy, but because I felt out of my depth.

Once the band was signed to a record label, we increasingly found ourselves spending time in recording studios, with two-inch tape, Studer machines, SSL desks, and multi-colored patch-bay spaghetti; again I felt as if I were in back in a technological comfort zone. Mixing was often a 'all hands on deck' endeavor, with time cues scrawled on bits of paper and each band member assigned a number of faders to ride. As time progressed, automated Flying Faders took over the task, and computers found their way into the studio control room. However, it wasn't until our third album, recorded in 1991, that a MIDI keyboard controller and computer was used, with Cubase software running.

We worked in prestigious residential studios, including The Manor, Ridge Farm, Chipping Norton, The Mill, and Hook End, which offered cutting-edge, top-end equipment. With the exception of a manager at Ridge Farm, the only other women present would be cooks, cleaners, wives, and girlfriends. I had become desensitized to the lack of women in my immediate environment. In fact, the presence of certain wives and girlfriends could sometimes be quite irritating as it often changed the dynamic of what was essentially a place of work. Cleaners were seemingly under the radar, with only the starched white sheets and the next morning's disappearance of copious numbers of empty bottles to indicate their presence. The cook at Ridge Farm became a friend, perhaps because, like me, she was able to become 'one of the lads' when appropriate; or rather, she was very comfortable in male company, without there being any sexual tension or undercurrent at play.

At no point in my career as a recording artist did I ever encounter a female engineer, programmer, or producer, on more than just one occasion. The band needed a remix of a single and was given a choice of three or four names. One of the names was female, and for that reason, I insisted that we employed her – positive discrimination. But it is quite telling that I can't remember her name.

The session was strained. Technically she was as knowledgeable and creative as any man I had worked with, at that level, and of that there is no question; but the atmosphere was decidedly strained. We were not a particularly bawdy group of individuals, but we did share a sense of humor that could be a little crude and improper. I can only imagine that she felt uneasy, an outsider, as I was more closely allied to my male band members than to her. However, it cannot be determined how much of this unease could be attributed to her being female and how much to her personality or her sense of professionalism. Personally, although having attended an all-girls school from the age of 11 to 18, and having had no male siblings, I had grown to be very comfortable in all-male company. Dr Jim Dickinson, a music educator and senior lecturer at University level, says:

> some women either have or adopt what are considered to be male characteristics – language, social behavior etc. to fit the group dynamic – or they have these character traits already – I have observed both. If

this does not occur, the dynamic of a group of men can change when women are present – sometimes this can be a positive influence. In my experience, the intent (rightly or wrongly) is to allow the female(s) to integrate into the group more easily.

(Dickinson, 2019)

Gender fluidity is currently a significant focus point. Blurred boundaries are being acknowledged and gender traits challenged perhaps more than ever before. A questionnaire I had designed with the aim of gathering qualitative and quantitative data for this essay attracted kindly alerts to how some of my questions were not inclusive. I asked how traditionally female considerations such as menstrual cramps, PMT, PCOS, and so on might impact upon performance at work for female producers, engineers, or programmers, and it was pointed out to me that this made no consideration of transitioning individuals whose performance might be negatively affected by, for example, hormone treatments. My attempt to remain robustly inclusive, while inviting males, females, male or female identifying, non-binary, or transitioning individuals to partake in the research, failed. I semi-abandoned the exercise.

However, much of the data collected was valuable, primarily in demonstrating that many of the males or male-identifying respondents are frustrated that gender is still such an issue, with all genders expressing that they felt frustrated at what they believed to be a protracted rate of progression. These individuals are gender allies, instrumental in encouraging and hastening the demise of gender inequality. In addition to assuming the position of gender allies, a number of them will potentially be tempered radicals. Meyerson and Scully state that tempered radicals are committed to the industry wherein they work, even when they support an ideology that may not resonate with the culture of that industry, and whose radicalism "stimulates them to challenge the status quo" (Frost in Meyerson and Scully, 1995: 585). Their temperedness arises from how they are "angered by what they see as injustices or ineffectiveness" (Ibid.).

Dickinson has many years of music industry experience as a performer and producer, and would seem to exemplify Meyerson and Scully's definition of a tempered radical. While acknowledging that women in the music industry still encounter some obstacles that men tend not to, he states that the current challenge is to:

make females feel confident that they can achieve the same or better results than men. The lack of numbers is simply to do with a wider social perception that technical roles are for males – in the same way there are still very few male nurses or air stewards as these are viewed as requiring 'softer communication skills'.

(Dickinson, 2019)

There is some resistance amongst my female peers, and indeed females I teach, that this perception of a difference in approach in the studio might overly 'feminize' them. However, Dickinson has guided hundreds of music students through Production Technology modules at both the FE and HE levels, and concludes that the female students he has taught differ in their approach, compared with males, where males tend to focus more closely on technical elements of production. He states:

> This does not necessarily mean that the end product [for female students] is any worse – in fact it is often better. They tend to take a more holistic approach – considering the musical/emotional side of music production as well as the technical – and this can often benefit the record – whereas sometimes male students obsess about tiny technical details while ignoring fundamental issues regarding the music and performance etc. There are always exceptions.
>
> (Ibid.)

Despite the best efforts of gender allies and tempered radicals, there is clearly some way to go in eradicating the tendencies in some individuals to adopt traditional roles, whether that adoption be as a result of outside pressure or self-limiting behaviors.

Respondent X, a classical, jazz, and contemporary musician who self-produces, pursued a BTEC Music Production course at college, where she found her lecturers to be less than enthusiastic, encouraging her to focus on working on her vocals and to aim to become a "really great female bassist". She states that her lecturers' argument was that she should follow a trajectory in which she already had experience, and that she "wouldn't feel very comfortable working in an environment without any other women" (Respondent X, 2019). Unfortunately for Respondent X, the environment in which she studied seemed lacking in a culture of gender alliance, and she did not flourish therein.

Respondent Y, an EDM artist, producer, and programmer, with a bachelor's degree relevant to her profession, says that Production classes at university felt like a 'boys' game', and that she and other females in her cohort became disinterested in lessons as they tended to lag behind the males.

> Boys answered the questions and took control of the sound desk when we were working in the studio. Girls accepted that their job was to be recorded and not record. Retrospectively, it's quite funny how the girls just floated into the live room and the boys would make a beeline for the sound desk.
>
> (Respondent Y, 2019)

When asked what might have impeded the success of women in the studio, one respondent offered the succinct response: "Bullshit in my own head". Perhaps 'nonsense' can replace the word 'bullshit' in this statement, as a combination of Imposter Syndrome, ruminative thinking, and a natural but detrimental response to negative experiences. It is of course important to acknowledge that Imposter Syndrome is far from an exclusively female malaise. Clance and Imes used the similar term, 'imposter phenomenon', to describe "an internal experience of intellectual phoniness" (1978: 241). They state that although their clinical experience tells them that this occurs more often and more acutely with women than with men, but with appropriate psychotherapeutic interventions and a willingness to change, women can free themselves "of the burden of believing [they are] a phony and can more fully participate in the joys, zest, and power of [their] accomplishments" (1978: 246).

Evidently, there are numerous and complex reasons why women may not find a career in a still male-dominated industry attractive, with sexual harassment being one. Artist and producer Grimes (aka Claire Boucher) says she has found herself in situations where she has been told by male producers: "We won't finish the song unless you come back to my hotel room" (Grimes in Hiatt, 2016).

A number of my questionnaire respondents agreed that there had been inappropriate sexual advances made towards them, in the working environment of the studio, with one stating: "In my 10-year career as an engineer, I have had SIX men hire me to try and sleep with me".

From responses to my questionnaire, it is also clear that the role of prime carer of children and also of elderly parents still tends to fall to females over males, instigating career breaks that make it difficult to re-establish a reputation and/or client base. However, it is necessary to consider the role of self-limiting behaviors.

Lilith Lane is one of two female engineers currently working at Newmarket Studios in Melbourne, and on the subject of the perpetuation of gender stereotypes concerning certain tasks, she states that she had recently witnessed a situation where a light bulb had gone out:

> some women expect that it will be a guy that fixes that . . . women, and the woman herself are as equally capable. . . . In reverse, it could be that a male feels it is his role to fix that problem and prevents the woman from doing so.
>
> (Lane, 2018)

Just as I deferred to a male band member concerning data being downloaded from our Yamaha RX-11 drum machine, perhaps I might have been capable, but he felt it was his responsibility, and so I was prevented from learning what was actually a simple task. A scenario such as this can initiate a mild sense of learned helplessness in an individual, where discrepancies between belief and behavior warp into a self-fulfilling prophecy of failure.

Lane goes on to say that a cumulative effect of these instances can result in a "confidence gap for women and technology as well as society's confidence in women in technical roles" (Ibid.).

Among gender allies, men who wish to guard what has for so long been their domain are still to be found, and the adage 'When you are accustomed to privilege, equality feels like oppression' applies. Once a former co-writer of mine discovered that I had learned how to use Logic 5.5 through observing him working with it, he repined that I wasn't going to 'need' him anymore. On some level, my learning seemed to have emasculated him, or at least deprived him of a responsibility that had either consciously or subconsciously become a power.

ROLE MODELS FOR WOMEN IN PRODUCTION

Dickinson notes that there is an increasing number of role models for women in production, something borne out by conversations with some of my own students. However, rather than these role models working in commercial studios with clients, they tend to be artists who are self-producing, for example, Björk, Grimes, and production pioneer Kate Bush.

Grimes has railed against the music industry's view of her as a 'female musician' with a 'girly voice', saying: "I'm a producer and I spend all day looking at fucking graphs and EQs and doing really technical work" (Grimes in Friedlander, 2015).

Similarly, Björk's talents as a producer were undermined in 2001, even though she had been working with music technology for decades. Of her 2001 album *Vespertine*, Björk (2015) says that she spent three years composing 80% of the beats, and in an expression of the meticulous nature of her work with microbeats, likened it to creating a "huge embroidery piece" (Björk in Hooper, 2015). However, although electronic duo Matmos only became involved in the final weeks of the recording, according to Björk, they were "credited *everywhere* as having done the whole album", although their contribution was to overlay songs with percussion. She went on to state that Drew Daniel of Matmos was keen to clarify that this was not the case: "in every single interview he did, he corrected it. And they don't even listen to him". This was not to be an isolated incident, as 2015's *Vulnicura* was originally misreported as having been produced by Arca, when in fact, it was co-produced by Arca and Björk (Ibid.).

Björk seems torn concerning a need to be accurately credited for her production skills, stating: "You're a coward if you don't stand up. Not for you, but for women", and then conversely states that she has considered providing a 'map' of her albums, in order to clarify "who did what", yet is concerned that it may come across as overly "defensive", concluding "it's pathetic" (Ibid.).

Kate Bush, however, seems at peace with her place in the producer's chair with little need or desire to defend her position. It could be said that as a national treasure of the UK, she is beyond reproach – certainly since 1982's self-produced *The Dreaming*, her studio skills are rarely questioned – and as a creator, she chooses her allies carefully and says that she has been fortunate that: "people that I've worked with have been really supportive and embracing of trying to make the project as good as we could" (Bush in Myers, 2016).

Although my own success in the music industry reached nowhere near the glittering heights as those mentioned previously, as an artist, since 1985, I have worked in studios at the side of producers and engineers, including Paul Samwell-Smith and Warne Livesey, and learned from them, and have also had a home set-up from the days of Porta-Studios up to the present day's Logic and iMac.

My first co-production credit came in 1991, for my band All About Eve's fourth studio album, *Ultraviolet*, and the second came in 1996, for a solo album I recorded under the name Mice. In 2009 I received an audio engineering credit for work on tracks I wrote with The Eden House, a kind of gothic dreampop supergroup, and my most recent co-production credit was for a 2011 album of cover versions I recorded with Wayne Hussey, primarily known for his work with The Sisters of Mercy and The Mission. All of those credits have been as co-producer and/or engineer; I have never been asked to actually produce other artists, whereas my male counterparts have. In short, I've never been 'hired'. To call attention to this is not a case of sour grapes, but simply an observation, a reflection. I can only surmise that I have not been invited to produce anyone because:

1. Perhaps I'm not as good at it as I think I am.
2. I'm female.
3. I'm backwards in coming forwards, I hide my light under a bushel, and other clichés that add up to my being too hesitant to truly succeed in the music industry.

The most probable reason is a hybrid of 'b' and 'c', with perhaps a sprinkling of 'a'. However, it may just be that I'm not sufficiently ambitious or driven and prefer to be immersed in my own work rather than act as a hired hand for other artists. If that were the case, then I'd certainly be in good company because, as far as I am aware, Kate Bush has not yet produced for any other artist, other than one track for Alan Stivell, even though she has surely been asked numerous times.

Bush's *Hounds of Love* album had been released in September 1985, and two years later, my own band had been signed; we were looking for a producer for our debut album. Our A&R executive was in full agreement when I suggested that we approach Kate Bush. I'm not aware of how that approach was made, perhaps it was management to management. The response came back that although she liked the demo she had been sent, Bush was too busy. It transpired that between 1987 and 1989, *The Sensual World* was recorded. Regarding her polite refusal, I am under no illusion that this might not have just been because she was busy, but perhaps because producing other artists was not something she wished to pursue.

While Bush, since early in her recording career, seems to have been in the position of not having to defend her role as producer, the robustness of the skill sets of Grimes, Björk, and others have not always been entirely acknowledged. It is clearly not a new phenomenon, as demonstrated by the blighted career trajectory of composer, audio engineer, and producer Delia Derbyshire.

Derbyshire attended Barr's Hill Grammar School for Girls, leaving in 1956. I attended the same school, leaving almost a quarter of a century later. I went on to study Fashion Journalism at the London College of Fashion but dropped out to join a 'rock-and-roll' band. Nowhere near did

I live up to the standard set by Derbyshire, who had gone on to read Mathematics, and then Music, at Girton College, Cambridge.

Interviewed for *Surface* magazine in 1999, by Spacemen 3's Sonic Boom, Derbyshire explained that following her graduation, the advice she was given from a careers officer was that, as she was "interested in sound", she would be wise to pursue a career in "deaf aids or depth sounding" (Derbyshire in Boom, 1999). Undeterred, she went on to apply for a position at Decca Records, without success, having been told "they didn't employ women in the recording studio" (Ibid.).

She eventually found employment with the BBC, where according to Derbyshire, she was seen as something of a 'rebel' or at least an 'independent thinker': "I think I was a post-feminist before feminism was invented!" (Derbyshire in Cavanagh, n.d.). Described in *The Quietus* as a "spacescape-faring Queen of the Unheimlich" (McNamee, 2009), Derbyshire eventually become a pioneer in electronic music and, most famously, as part of the BBC Radiophonic Orchestra created the original theme to Dr Who from a score by Ron Grainer. However, BBC policy at the time means that "Grainer – unwillingly – is still officially credited as the sole writer" (Pidd, 2017). According to Kember, aka Sonic Boom, Derbyshire met with opposition in pursuit of her creativity, and on the subject of her realization of Grainer's Dr Who theme, she worked alone and nocturnally: "People would try to fuck her over by monopolizing the equipment, so she would work late to get around all that" (*M-Magazine*, 2013).

Derbyshire died in 2001. On receiving a posthumous PhD from Coventry University on her behalf, Derbyshire's long-term partner Clive Blackburn said that technology was "finally catching up with what she managed to achieve manually in the 1960s using the most rudimentary of equipment" (Pidd, 2017).

Shortly before her death, Derbyshire looked set to enjoy something of a renaissance, with '90s electronic music practitioners. Spacemen 3's Peter Kember, aka Sonic Boom (2018), says that he encouraged Derbyshire to become musically active again, but with limited success; however, she taught him much about sound and its structure, and told him that her work at the BBC had attracted the attention of

> Syd Barrett, Paul McCartney and Brian Jones. I think that gives a hint as to what an impact she was making culturally – which resonated right through electronic music via White Noise, the Silver Apples, Kraftwerk and Aphex Twin.
> (Kember in Coney, 2018)

The pioneering spirit of ahead-of-their-time women such as Derbyshire, and of co-founder of the BBC Radiophonic Workshop, Daphne Oram, undoubtedly etched cracks into the glass ceilings of the latter half of the twentieth century. The influence of Oram is such that the PRS Foundation has created The Oram Awards, in an effort to "celebrate role models for the next generation", stating that she had been instrumental in "establishing women at the forefront of innovation, in newly emerging audio technologies, in the UK and around the world" (PRS, 2019).

Thanks to pioneers such as Björk and Grimes calling out sexism and demanding credit for their work, and Kate Bush who became powerful enough at a young enough age, i.e. to perhaps become immune to it, inoculated with her own visionary confidence, the place of women in the studio certainly includes the producer's chair. Building on the creative audacity and boldness of women such as Daphne Oram and Delia Derbyshire, they make a nonsense of the concept, indeed the myth, of male superiority in the recording studio.

Rayona Sharpnack, founder of the Institute for Women's Leadership and the Institute for Gender Partnership, writes: "we will look back and see that while the first wave of gender equality was driven by women, the second was spearheaded by men" (Sharpnack, 2015).

Sharpnack's statement might present itself as a bitter pill if perceived as male apologism, but in the context of the far-reaching and prevalent issue of toxic masculinity, and the conundrum and challenge it poses to good men, to decent men, it is prudent, and indeed fair, to give these allies the opportunity to help facilitate women's continued permeation of the music industry.

In her 'Billboard's Woman of the Year Award' acceptance speech, Lady Gaga identified a common enemy in the traditional model of the music industry, when she said: "It's like a fuckin' boys club that we just can't get into" (Lady Gaga in Sullivan, 2015). The 'we' in that statement is not exclusive to women.

Male allies who reject the patriarchal 'no-girls allowed' den that recording studios have often been, and who are frustrated and appalled by the primitive and primordial attitude still exhibited by some of their male peers, can combine with female pioneers in a powerful pincer movement that promises, at an accelerated pace, to afford women their deserved equality.

I consider my colleagues Dr Jim Dickinson and Stuart Bruce both as gender allies and tempered radicals, as they are deeply invested in nurturing students of all genders in the art and science of audio engineering and production. Bruce, a celebrated producer as well as a university lecturer, is perhaps best known for having engineered the Live Aid charity single *Do They Know It's Christmas*, yet has also recorded Kate Bush, whom he considers a "phenomenal producer" (Bruce, 2019). Born in 1962, when gender equality was in an embryonic state, he ruefully admits to having occasionally 'mansplained' in the studio:

> It's difficult because it was ingrained in me. I went to an all-boys, English public school and grew up with the attitude that we were "superior in all ways". Luckily, my mother was a feisty working woman who would never readily defer to a man.
>
> (Ibid.)

Bruce has worked alongside many female engineers and producers across his 35 years in the recording industry and speaks of them with respect and admiration. Regarding the students now in his charge, he ensures that they all feel that they have a "level playing field" and are afforded respect as individuals looking to enter the recording industry. Determined to help redress the gender balance in the profession, he states:

> There is still an imbalance, but each academic year sees an incremental increase in the number of female applicants and subsequently the number of students we take on. My hope is that we eventually achieve at least a 50:50 situation.
>
> (Ibid.)

Much hope is pinned upon this younger generation of women, unjaded and uncynical digital natives, be they self-taught or products of a largely inclusive FE and HE education system where production, engineering, and programming are part of a syllabus, more than ready and able to assume a confident, creative, and authoritative demeanor at the helm of a control room. Personally, I am at a stage in my career where the desire to produce other artists has long slid off my list of ambitions. However, I have a sense of gratitude toward the allies I met along the way, and to those I still encounter: men who see a female presence in the studio as not just natural, but positive, and who generously share their skills rather than insecurely guarding them.

I acknowledge that for the most part of my recording career I was a principal songwriter and lead vocalist, and that afforded me queen bee privileges. It wasn't easy to intimidate me, though many tried. From the small girl on a stage with apparently too big an instrument, I have arrived at a place where I produce and engineer my own music, predominantly for my own pleasure, or co-produce and engineer music amongst a selective coterie of longstanding and trusted allies. I can't say that I'm fortunate, because I've had to work hard at threshing some irritatingly misogynistic chaff out from the wonderful wheat, but it has been incredibly worth it. Treasure your allies.

REFERENCES

Aston, J. and Aston, M. (1982) *Shaving My Neck*. London: Situation Two.

Boom, S. (1999) *Delia Derbyshire; Electronic Music Pioneer*. Available at: www.delia-derbyshire.org/interview_surface.php (Accessed June 12, 2019)

Bruce, S. (2019) Conversation with Julianne Regan, October 10, 2019

Cavanagh, J. (n.d.) *Delia Derbyshire; On Our Wavelength*. Available at: www.delia-derbyshire.org/interview_boa.php (Accessed July 10, 2019)

Clance, P. R. and Imes, S. A. (1978) 'The imposter phenomenon in high achieving women: Dynamics and therapeutic intervention'. *Psychotherapy: Theory, Research & Practice*, 15(3), pp. 241–247. Doi: 10.1037/h0086006. (Accessed November 10, 2019)

Coney, B. (2018) *The Strange World of . . . Pete Kember*. Available at: https://thequietus.com/articles/24930-pete-kember-sonic-boom-spacemen-3-spectrum-interview (Accessed July 22, 2019)

Dickinson, J. (2019) Conversation with Julianne Regan, July 10, 2019

Friedlander, E. (2015) *Grimes in Reality*. Available at: www.thefader.com/2015/07/28/grimes-cover-story-interview (Accessed June 22, 2019)

Frost, P. (1995) 'Tempered radicalism and the politics of ambivalence and change'. *Organization Science*, 6(5), p. 585. Available at: www.jstor.org/stable/2634965 (Accessed July 1, 2019)

Hiatt, B. (2016) *Grimes on 'Art Angels' Follow Up, Why She Loves Tool*. Available at: www.rollingstone.com/music/music-news/grimes-on-art-angels-follow-up-why-she-loves-tool-162756/ (Accessed June 10, 2019).

Hooper, J. (2015) *The Invisible Woman; A Conversation with Björk*. Available at: https://pitchfork.com/features/interview/9582-the-invisible-woman-a-conversation-with-Björk/ (Accessed May 25, 2019)

Lane, L. (2018) *The Confidence Gap: Talking Audio Production with Recording Engineer Lilith Lane*. Interview with Lilith Lane. Interview by Monique Myintoo for *Collarts*, 13 May 2018. Available at: https://blog.collarts.edu.au/confidence-gap-talking-audio-production-recording-engineer-lilith-lane (Accessed July 1, 2019)

McNamee. (2009) *It Started with a Mix – The Best of the BBC Radiophonic Workshop on One Side of a C90*. Available at: https://thequietus.com/articles/01010-the-best-of-the-bbc-radiophonic-workshop-archive-on-one-side-of-a-c90 (Accessed June 3, 2019)

M Magazine. (2013) *Interview - Sonic Boom*. Available at: www.m-magazine.co.uk/features/interviews/interview-sonic-boom/ (Accessed July 27, 2019)

Myers, O. (2016) *Kate Bush Speaks*. Available at: www.thefader.com/2016/11/23/kate-bush-interview-before-the-dawn (Accessed May 20, 2019)

Pidd, H. (2017) *Doctor Who Theme's Co-creator Honoured with Posthumous PhD*. Available at: www.theguardian.com/music/2017/nov/20/delia-derbyshire-doctor-who-theme-co-creator-posthumous-phd (Accessed July 29, 2019)

PRS. (2019) *The Oram Awards*. Available at: https://prsfoundation.com/funding-support/funding-music-creators/next-steps/the-oram-awards/ (Accessed June 10, 2019)

Respondent X. (2019) Anonymous questionnaire response to Julianne Regan, June 27, 2019

Respondent Y. (2019) Anonymous questionnaire response to Julianne Regan, June 27, 2019

Sharpnack, R. (2015) *When It Comes to Gender Partnership, Aristotle Had It Right*. Available at: https://womensleadership.com/comes-gender-partnership-aristotle-right/ (Accessed July 28, 2019)

Sullivan, L. (2015) *Lady Gaga Accepts Woman of the Year Award at Billboard Women in Music: Read Her Full Speech*. Available at: www.billboard.com/articles/events/women-in-music/6813845/lady-gaga-billboard-women-music-full-speech (Accessed July 30, 2019)

13

Gender in Music Production

Perspective Through a Female and Feminine Lens

Louise M. Thompson

INTRODUCTION

The majority of critical and scholarly discussions on gender and music production illustrate the gender inequalities and masculine dominance in music as an omnipresent feature. Paula Wolfe states that female producers are marked primarily for their absence in the field, and have been under-researched, along with analyses of their practice (Wolfe 2012). Feminist research in the field has found that the masculine-led music industry is largely ignorant in regards to women's productions, which has led to research on women's contributions to the field critiquing the inequalities in the workplace, rather than offering an analysis of the works and practices of female producers. Sally Macarthur explains that "Women's productions are not discussed in mainstream music production discourse or tediously analyzed in comparison to men's music, nor given the same attention as exemplified in performances and broadcasts" (Macarthur 2002: 2). This chapter is a step towards correcting this discursive trend, through reflection on the contributions of female producers in music production. This reflection includes explorations on gender performativity and gendered modalities of music production. This chapter takes a broad definition of the music producer, encompassing both the studio producer and the electronic music based artist/producer. As electronic music has become part of the wider popular music industry, this elucidation situates this chapter within the current music industry context, drawing on semi-structured ethnographic interviews with nine female producers from North America, Mexico, Australia, and The Netherlands. This chapter provides a trans-local perspective on the female producer and an examination of music production through a female lens. Through this we are able to gain a deeper understanding of the feminine culture of music production and the feminine aesthetic of music production, gaining a fuller understanding of women's participation in the field.

In line with the feminist ethic this research is undertaking, a suitably accommodating methodology has also been chosen. Ethnography is a qualitative research method originating from sociology with the study of values, beliefs, behaviors, or language within a distinct group in society.

Many feminist musicologists use this method of research, including Dr Jodie Taylor, Rebecca Farrunga, Magdalena Olszanowski, and Elle M Hisama (Taylor 2012; Farrugia 2012; Olszanowski 2011; Hisama 2014). This approach is useful as it validates the personal perspective from which I am starting this research. As my practice is music production, my immersion in the field aids in gaining an understanding of this culture from this female perspective. This method allows for free-flowing peer communication and qualifies narrative accounts. The ethnographic interviews, while semi-structured, seek to discover the gender performativity and gendered modalities of music production. The interviewees describe gender performativity and gendered modalities of music production in regards to their works and practice. This study concludes that while the experience of one's gender may be preconscious in a woman's approach to music production, the feminine manifests as a distinct aesthetic in all cases herein.

In order to clarify my use of the terms *feminist* and *feminine*, I adopt Grosz's definitions of these as referred to by Sally Macarthur: "According to Grosz the four criteria for a feminine text are the sex of the author; the content of the text; the sex of the reader; and the style of the text" (Macarthur 2002: 149–150). Grosz explains that "Feminine texts are those written from the perspective of a feminine experience or composed in a style culturally represented as feminine [while] feminist texts self-consciously challenge the methods, objects, goals, or principles of mainstream patriarchal canons" (Macarthur 2002: 149–150). Hélène Cixous further defines the term feminine in the binary system of cultural representation:

> Traditionally, the question of sexual difference is treated with coupling it with the opposition: activity/passivity. Philosophical discourse both orders and reproduces all thought. One notices that it is marked by an absolute constant, precisely this opposition activity/passivity. Moreover, woman is always associated with passivity in philosophy.
> (Cixous 1997: 149)

Cixous uses examples of the binary opposites to define the cultural representations of masculine/feminine such as: sun/moon, culture/nature, father/mother, head/heart, intelligible/palpable. Cixous explains that

> These sexual differences are not distributed on the basis of socially determined sexes [man or woman]. To avoid the confusion: man/masculine woman/feminine, for there are some men who do not repress their femininity, some women, who more or less, inscribe to their masculinity.
> (Cixous 1997: 152)

It is then essential to recognize that not all women are feminine or feel comfortable expressing feminine behavior. Charlotte Greig affirms:

> We can't talk about women as if we were a subgroup of humanity that speaks with one voice. We don't, and there's the same level of

diversity and conflicting points of view amongst women songwriters as you'd get anywhere else in music.

(Whitely. Ed. Greig 1997: 168)

This is the same in music production with a level of diversity and conflicting perspectives amongst female producers. It is important to acknowledge that not all of the female producers I interviewed feel their practices are feminine or see themselves as feminists. Some female producers do not acknowledge feminism as a cultural framework embedded in their music production, though I am viewing their discursive and material contributions through their embodied gender and a female lens. Many have the perspective that when you focus on gender it separates the female producer, trivializing their practice. I am making female producers the object of my study, invoking Magdalena Olszanowski's assertion as follows:

As a separate tradition is not isolationist; rather, it is a strategy in recovering them, in making them an object of discourse. Separation is a means of offering women visibility that they would not otherwise possess and enabling discussions that could not otherwise proceed.

(Olszanowski 2011: 10)

Sally Macarthur suggests that gendered divisions need to be drawn between the work of male and female producers because of the difference between men and women on the whole due to the different ways in which they have been socialized (Macarthur 2002: 2). Rebecca Farrugia indicates that social constructs of gender significantly impact the ways in which men or women may or may not seek out and find their voice in all creative fields (Farrugia 2012: 68); therefore, they additionally impact how women express and interact creatively with music production. Further research is needed to understand the female process and practice of music production to gain insights into combating the gender inequalities in the field; therefore, a discussion about this division is necessary.

GENDER PERFORMATIVITY IN MUSIC PRODUCTION

Susan McLary explains that music does not passively reflect society but serves as a public forum in which models of gender are "asserted, adopted, contested and negotiated" (McLary 1991: 8). Norma Coates refers to Teresa De Lauretis's analogy "that gender is the product of social technologies, including forms of popular culture, institutionalized discourses, critical practices and practices of daily life. Gender is at the same time constructed and always under construction" (Whitely. Ed. Coates 1997: 52). Coates further compares De Lauretis's analogy that these social practices construct, indicate, and secure a cultural form as associated with a particular gender. Judith Butler's belief is that the social practices also allude to and reinforce acts, gestures, enactments, and other signifiers, which express that gender or perform a particular gender. Coates explains that Butler's interpretation of the performance of gender is an expression, a non-tangible fabrication, manufactured and sustained through corporeal signs and discourse: "That the gendered body is per-formative suggests that it has no ontological status apart from the various acts which constitute its reality" (Whitely. Ed. Coates 1997: 52). Coates compares the tangible embodied gender of female and male with the per-formative non-tangible fabrication of femininity and masculinity, denoting that gender can be performed. The following analysis uncovers this gender performativity in female producers' works and practice.

Fieke Van Den Hurk[1] performs femininity in her practice by capturing the emotion in the production. She explains that her approach is by going with "a gut feeling" rather than constructing a mix from "a technical" mindset. She likes to capture the feeling in the production and then reflect on what she has composed or captured to know how to proceed with the production. Van Den Hurk feels that she contributes femininity to her artists' projects with this approach of "capturing the emotional content," though she always focuses on the "artists' creative direction and intention" (F Van Den Hurk 2019, interview communication, 27 March). Sophie Botta[2] focuses on capturing the emotion in a recording, as songs aim to express a feeling. She says, "This is capturing the heart of the song". Botta feels a certain responsibility to her clients, and she mentions "the importance of understanding what their expectations" are and what they want to convey "in a song, musically, emotionally and conceptually" (S Botta 2019, interview communication, 17 March). Missy Thangs[3] also follows her "gut instinct" in her practice instead of following the conventional methods. The "instruments" that she "chooses to put in the foreground of a mix" display her unconventional approach, illustrating her feminine performativity. She can be "quite playful with the board/console while she mixes", often "dancing while mixing". Thangs also has a feminine performativity in the way she talks about her approach to her practice. Thangs explains panning a guitar in terms of "sweeping a guitar to the left" (Missy Thangs 2019, interview communication, 31 March). Amelia Warren's[4] production style "doesn't fit into a paradigm of someone else's making." Using her gut instinct, Warren describes capturing the emotion of a recording by "knowing when someone has put their heart and soul into a piece. The whole composition tells the story, not just the vocals" (A Warren 2019, interview communication, 12 April).

Patty Preece[5] from The Ironing Maidens is "creatively influenced by the state of the world" and comes from a political stance to her practice. There is feminist performativity in this approach, with the intention to "bring women to the forefront in the home and in the studio". This celebration of women addresses the visibility of women in the workplace. The Ironing Maidens' motivation for getting into the studio is the "inequality of women". Their recent song 'Pick you shit up' was "inspired by the recent census results where women are still doing the majority of the housework and domestic duties within the home" (P Preece 2019, interview communication, 2 April). Feminist musicologist Sally Macarthur explains: "feminist texts self-consciously challenge the methods, objects, goals, or principles of mainstream patriarchal canons" (Macarthur 2002: 149–150), due to the majority of discourse composed by men. The Ironing Maidens have made "electronic instruments and midi controllers out of irons" (P Preece 2019, interview communication, 2 April). The use of a household domestic device reconstructed into a music technology device is an example of feminist commentary and performativity in Preece's practice.

I observe Catharine Wood[6] as having a feminine performativity in her practice as she produces her personal productions with "a confessional singer songwriter" style. At times she writes music "just for her and may never be released or may not be released for some time after composition" (C Wood 2019, interview communication, 26 March). This style of composition and genre of writing is a more personal style and has been described as feminine by Greig: "Singer-songwriters, perhaps implied an arty, feminine kind of solipsism not present in the term folk singer" (Whitely. Ed. Greig 1997: 174). Wood illustrates feminine performativity in this approach to practice.

When working with artists, Karina Rivero[7] concentrates on creating a welcoming, homely environment for the artists to perform in, showing a feminine performativity with this approach to practice. She brings a focus to "logistics and catering" for recording sessions unlike her male colleagues, and says "the males usually go in and hit record" (K Rivero 2019, interview communication, 7 February). Van Den Hurk believes her gender influences her behavior with her artists. The difference that she can see between her and the male producers and engineers in the studio that she works alongside is that the male producers behave with a hierarchy over their artists and clients, behaving with a "bossy" personality in the studio. The males are more controlling of the sessions: "they don't feel into what the artist is doing", and they follow a conventional studio process. The males take less care of keeping the energy in a session: "they just go-go-go and at times you have to tell them to take a break" (F Van Den Hurk 2019, interview communication, 27 March). Botta finds if things aren't working in the studio, she also "takes tea breaks" (S Botta 2019, interview communication, 17 March). Preece says women create a "nicer vibe" when they are in charge in the studio. She mentions working with a student who had to record vocals for a song competition: "The guy that was receiving the tracks also came into the studio. He was given the brief that the female artist is sensitive so sit back in the studio and chip in when he needs to". There was a young female student with Preece running the computer while she worked the console. She commented that, "The guy went on to take over the computer from the female student" along with communications with the artist, and that "he took over the whole session and made the artist uncomfortable by being really pushy with her". Preece explains that, "He got a great sound but at the cost of making the artist uncomfortable, along with the student and teacher" (P Preece 2019, interview communication, 2 April). Feminine performativity is illustrated in the approach to practice with making the artist comfortable, taking regular breaks, and not behaving hierarchically with the artist. Arguably, this allows for a much more comfortable session, leading to a better performance and a better recording experience.

Feminine performativity in record production is evident in many female producers' works. Many female producers focus on the melody and melodic content of the production. Riveros's mixes often illuminate the higher frequencies, and she has had her productions described as "sometimes lacking in the lower end". She believes this is due to women's influences in the childhood and teenage years in regards to sound. Most females "play melodic instruments" and therefore, focus more on these areas (K Rivero 2019, interview communication, 7 February). Warren also enacts this feminine performativity; the focus of her productions weighs on her melodic compositions (A Warren 2019, interview communication, 12 April). As an artist in her personal productions Wood says she is "very particular with melody and has a unique sense of harmony with composing chord progressions". She mentions she is "firstly melody oriented then lyrically oriented". Lyrics take her a long time to write, and it is "the melodies in her music that connect them together as her works" (C Wood 2019, interview communication, 26 March). Missy Thangs' personal productions are feminine sounding in regards to the melodic content. She explains that these "melodies create a vibe that is really quite beautiful". Her songs "don't have a lot of dissonance and are usually written in the minor keys" (Missy Thangs 2019, interview communication, 31 March). Many believe minor keys are associated with the feminine, corresponding with theories on aesthetics in music. I would like to acknowledge that these theories are described in mainstream musical discourse, which like all mainstream discourse has produced and reproduced the feminine in a subordinate position to the masculine. Philip Stoltzfus explains that "musical expression can provide orientation for the entirety of the inner life", including the characteristics of gender, heard in the metaphors of the masculine and feminine in the major and minor keys (Stoltzfus 2006: 81). These metaphors come from different theories on sonata form, with the first theme denoting the masculine, written in a major key. The second theme portraying a feminine character is written in a minor key. Macarthur describes Karl Marx's concept on sonatas as one that depicts the first theme as masculine, "constructed decisively and completely with energy and vigor" in contrast to the second theme: "tender feminine themes, dependent and determined by the preceding masculine theme" (Macarthur 2002: 90). Most take for granted these aspects of musical practice as simple elements that structure his or her musical and social world, though McLary states:

> They are perhaps the most powerful aspects of musical discourses, for they operate below the level of deliberate signification and are thus usually reproduced and transmitted without conscious intervention, for it is through these deeply engrained habits that gender and sexuality are most effectively and most problematically organised in music.
> (McLary 1991: 16–17)

Preece says that The Ironing Maidens' music is "more groove-based with no aggressive synthesizer lines. The melodies are really beautiful with a prominent vocal" (P Preece 2019, interview communication, 2 April).

These female producers interviewed describe a feminine performativity in their works by focusing on the melodic content, while other female producers focus on producing a smooth and sensual mix. A Hundred Drums[8] performs her femininity when she is DJing her songs. She says she has a "smooth and sensual" approach to her mixes, "unlike a guy who may do a hard drop" (A Hundred Drums 2019, interview communication, 8 March). Ania Grzesik[9] also has a similar approach to her works, with "no sharp edges" in her mixes and a "fluid" nature (A Grzesik 2019, interview communication, 15 March). Van Den Hurk likes it when things "stay organic and not so clearly defined". She likes it when sounds have a fluid sonic quality (F Van Den Hurk 2019, interview communication, 27 March).

Feminine performativity in music production is embedded in the communication of the musical story and the delivery many female producers offer through their works. Van Den Hurk says her music "has a feminine sound and thinks it would be hard to escape, as she is a female writing it, expressing a female story" (F Van Den Hurk 2019, interview communication, 27 March). Rivero believes female producers express emotions and convey ideas or messages differently in a mix. She thinks that everything that defines her emotions as a woman is expressed through her arrangements, mixes, and sometimes even the mastering process. She explains that "If it's your voice that's being expressed and you're a woman, it will come through" in your work (K Rivero 2019, interview communication, 7 February). A Hundred Drums performs femininity in her record production in the way she encodes her energy into her productions; she says she is "transmitting" or communicating the story with the way she weaves musical elements and the choice of language used in the lyrics of her works. She thinks that many males may tell their story differently, possibly with the use of "dominating" or aggressive sounds (A Hundred Drums 2019, interview communication, 8 March). Botta performs femininity in her works by composing on "personal themes and experiences" (S Botta 2019, interview communication, 17 March). Missy Thangs also has this approach to her works, composing from her "experiences and feelings about what's going on around her". Missy says that due "to being a female she is naturally going to bring" the feminine experience or "a feminine quality to her work" (Missy Thangs 2019, interview communication, 31 March). Wood also composes her personal productions on her own experiences. She comments on "craft versus being an artist". When you're thinking about yourself as an artist and you're creating, "whether it's painting, drawing, or music, the thread that goes between your works is you as the artist. Therefore that's what people recognize in your work" (C Wood 2019, interview communication, 26 March). Warren finds that she composes music based on her "emotional experiences rather that what's happening in the outside world". She believes there are "unintentional nuances in her music that would express her gender. She is a female experiencing life as a female and composes music based on her experiences" and believes that "other females would decode these experiences when listening to her music and relate to them". When she hears female productions, she can hear "the femininity in the recordings and productions". Her older compositions are "like diary entries" with her femininity expressed in the lyrics composed from the feminine perception. She says she likes to "sum up her emotional experiences; if she can't do this with words then she knows she has to do it in a song" (A Warren 2019, interview communication, 12 April). Botta composes from "her personal experience as a way to come to terms with events", expressing how she feels as a way of processing. She believes "music works from the inside out", and the connection to the composer or producer is unavoidable. Botta believes as female producers we express our personal experiences or emotions through the musical product, whether it's a personal production or for a client. The composer's/producer's voice can be heard in the production; for example, "a chord progression or certain riff is chosen as it relates to the producer's taste" and expression (S Botta 2019, interview communication, 17 March).

Preece also performs gender in the communication of the music, producing female stories with her duo The Ironing Maidens. The Ironing Maidens' songs express the female domestic duties in the house to address the gender inequalities of females in the home and in the field of music. They "use repetitive lyrics" in the song 'The Dangers of Ironing' to "express the repetitive monotony of housework". The "narrative of the song communicates that the character/protagonist of the story dealing with mental problems, due to this monotony". They express these feminist stories in music with the ultimate goal of addressing the gender inequalities of females in the domestic sphere. They are trying to "address issues of sharing the load in the home with housework, and male privilege socially and culturally". The Ironing Maidens "perform on stage with ironing boards, and electronic instruments and synthesizers made out of irons" (P Preece 2019, interview communication, 2 April), showing another aspect of feminine performativity in their works with the use of a domestic object coded as traditionally female and used in the performance of their music. Female producers who perform their music on stage have to take their personal presentation into consideration, unlike male producers who can perform in a T-shirt and jeans. This is largely because women are judged not only on their music but also on how they look. Preece and her partner are "conscious of not playing into traditional assigned gender roles". As Preece is "more masculine and her partner is more feminine", Preece wears pink and her partner wears blue to complicate the way in which they portray gender (P Preece 2019, interview communication, 2 April). Warren's onstage presence illuminates her gender when she performs, as she likes to "wear a dress and makeup, and [she] curls her hair" (A Warren 2019, interview communication, 12 April). A Hundred Drums likes to dress up and "set a vibe" when she plays live. The way she "moves when she is DJing her music onstage" also expresses her femininity due to the fact that the female body moves differently from a male's (A Hundred Drums 2019, interview communication, 8 March).

Feminine and feminist performativity in music production is evident in many female producers' practice with: following their gut instinct instead of conventional methods of constructing a mix; focusing on capturing the emotion in a recording; having a socio-political creative influence;

utilizing household domestic devices in the production of music; drawing on a confessional singer-songwriter style of composition; and by making the artist comfortable and not behaving hierarchically with the artist. Arguably, feminine and feminist performativity can be enacted in many female producers' works through their melodic content, their production of a smooth and sensual mix, their personal/political lyrics, and their physical performances on stage. The following analysis focuses on the gendered modalities of music production with the aim of offering a further understanding of the female producer and her practice.

GENDERED MODALITIES OF MUSIC PRODUCTION

Gendered modalities of music production define the modes of music production that exist and are experienced and expressed by many female producers. These modalities of music production are from a tangible, embodied perspective, in correspondence with a person's ontological gender: for example, female, male, or intersex, unlike gender performativity, which is from a non-tangible perspective irrelative to a person's embodied gender: for example, a female can have a masculine performance and vice versa. Consequently, a tension exists with using the terms feminine and feminist to enable a discussion of embodied experiences. Sally Macarthur constructs this embodied perspective as an "autonomous human subject that is variously labelled male or female in order to ground what is being spoken about" (Macarthur 2002: 18). Macarthur comments on the perspective of gender as embodied due to its tangible and social construction:

> For all the slipperiness entailed in the notion of a feminist style, it is also imperative to conceive of feminist aesthetics as being grounded and embodied . . . As notions of transcendence, composer, feminists, women, feminine principle, and so on are all constructions of a real, material, social world.
>
> (Macarthur 2002: 20)

Gendered modalities are a tangible social construction, so to analyze these from an embodied perspective is imperative. Eva Riegers' notion is that gender is one of the most important constructs of human behavior. Therefore, gender will influence the way in which men and women produce music (Macarthur 2002: 13). Macarthur states: "Rieger, Citron, and McLary want to claim a space for their women subjects, to suggest that even while working with inherited paradigms and stylistic norms, women compose music differently than men do" (Macarthur 2002: 19). The following analysis discusses the gendered modalities of music production in relation to the practice and works of female producers. Not all female producers feel their practices are feminine though I am exploring the way in which feminine behavior is something celebrated and reclaimed by many women interviewed.

The majority of music production practice is done in recording studios, a sterile space. Though many female producers' approach to practice starts with making this space an environment to inspire creativity. Riveros's approach to the recording sessions is by concentrating on creating a "welcoming homely space" for her artists (K Rivero 2019, interview communication, 7 February). Botta has a studio attached to her flat with its own entrance, though at times she hires another studio for clients to work in as well. She also keeps a homely environment in the studio with plants and ornaments, though she says "it's a mess with guitars everywhere" (S Botta 2019, interview communication, 17 March). Van Den Hurk works from her own professional, acoustically built studio within her apartment where she records many bands and artists. Her studio has a "cozy homely atmosphere. It is light filled with a big window and a plant across the window still with a beautiful view of the garden outside the window". She says, "it's quite different to most male producers' studios or man caves" that she's been in. She "put a lot of thought into how she wanted her studio to look" when she was building it and really cared about the "aesthetic" of the studio (F Van Den Hurk 2019, interview communication, 27 March). Wood, like Van Den Hurk, also has this approach to practice and likes to work in a "light filled environment, not a dark room or cave". Her creative environment is her own studio, which has been "professionally built with acoustic treatments" (C Wood 2019, interview communication, 26 March). Missy Thangs doesn't like working in a sterile environment as it "puts a damper on her creativity". She cares deeply about creating a vibe in the studio and uses "incense and lighting to set the space" for her artists (Missy Thangs 2019, interview communication, 31 March). Preece produces music for her duo The Ironing Maidens from her home studio, which is "a light filled space surrounded by nature with coconut palms and passion fruits growing outside. It is a creative space for jamming and production with a pop theme created with large colorful Lego bricks" (P Preece 2019, interview communication, 2 April). A Hundred Drums feels that "colors and visualization help invoke her creativity". Her studio is within her home. The environment is "filled with visual art, instruments and is royal purple to stimulate her mind and creativity". She has an "altar in her studio with candles, animal bones, and crystals" to further stimulate her environment and creativity. Her studio is more than just a music studio; it is her "creative space", the one room where she is "100% raw and in full empowerment of her self-expression". When she enters this space, "the valve to her creativity vessel in her brain completely opens" (A Hundred Drums 2019, interview communication, 8 March). For many, the female producer's practice starts with making the studio an environment to inspire creativity, setting a comfortable space with use of color, ornaments, and nature. The prominent characteristic to the female producers' studio practice is having the studio in the environment of their domestic homes, a feminine space, which has significantly contributed to their development of production skills. Mary Celeste Kearney believes this is due to the fact that "the power and control associated with the producer (Bayton 1998: 6) appears undiminished, even in the digital age (Wikstrom 2009), the ability to avoid such 'grappling' via self-production, therefore, is welcomed by the women here" (Whitley. Ed. Kearney 1997: 218).

Another female mode of music production in practice is related to cleanliness and organization to create a pleasing and calming atmosphere in the studio. Rivero notices that many female producers that she has worked with "are gentler with the equipment and have a cleaner studio environment" (K Rivero 2019, interview communication, 7 February). Grzesiks' practice is to prepare for her creativity by "ensuring that all the technical aspects" are ready so she can enter her creative space. This "allows for her process to flow without getting caught up in the technical elements". The female modality of her music production is evident in her practice with more mindfulness, attention to detail, and organization (A Grzesik 2019, interview communication, 15 March). Van Den Hurk's studio is very clean and organized. She has "a system for everything" and her studio set up in a particular way for efficiency, suiting her workflow. Van Den Hurk has had someone comment that, "you can clearly see it is a woman's studio". She notices that "the studio gets messier sooner when there's a male engineer" in there (F Van Den Hurk 2019, interview communication, 27 March). I'd like to acknowledge that not all men are messy and not all women are tidy, though for many women, creating an aesthetically comfortable, pleasing, and calming atmosphere is conducive to productive music-making.

Botta's approach to practice is to help her clients be comfortable and prepared for the recording sessions so that they capture the desired emotion in the recording. She does this by allowing her artists to rehearse to "get used to using microphones and the studio environment before hitting record" (S Botta 2019, interview communication, 17 March). Van Den Hurk uses her intuition in the studio when working with an artist or client to "pick up on their emotions and stresses" and make them comfortable. When the artist is comfortable, it makes a better recording and session. She does this not only in her actions, but also with how she chooses the "process to flow". She is conscious of how she communicates with clients. She says, "some clients perform better when you are warm and with others you need to be strict". She sets the studio space to suit her client by "dimming the lights or lighting candles to make the artist more comfortable" (F Van Den Hurk 2019, interview communication, 27 March). Preece is more intuitive in the studio. She thinks about every session in the studio as an experience: "It doesn't matter if you don't get what you set out to achieve the first time, as each time you can reflect on what happened and go back and change what's needed. This helps artists to grow and learn their own processes" (P Preece 2019, interview communication, 2 April). Warren also feels as a female producer her approach to practice can be "more accommodating in the studio than her male counterparts" (A Warren 2019, interview communication, 12 April). Missy Thangs believes as a female producer she is "very adaptable and can mold to suit" her artist's style, and connect with the artists. She's "very malleable and soft, nurturing" and personal to get to the heart and capture the emotional intention of the music, though she can "be a hard ass if need be". She works with her intuition and "finds out the artist's intention" for the music and works with this. She likes to "hear what they are doing and find opportunities" in their music to inspire her further with the production (Missy Thangs

2019, interview communication, 31 March). Another female modality of music production in practice is the way that Botta thinks and feels about creating music for someone. Botta feels that most of her clients want to create music that is "really personal to them, songs about breakups" and current life events. She thinks "as a producer they have entrusted their baby to you. It's special, it's precious to them and the producer is the one that has to breathe life into it" (S Botta 2019, interview communication, 17 March). Angela McRobbie attests that "[t]he role of the music producer within the popular music industry has been recognized as a profession strongly associated with notions of power and control" (McRobbie 1978: 66–96). These women show that in a female modality of music production, the power and control lie with capturing a great performance. Some of them do this with what might be called a feminine approach to their practice, by making the clients comfortable with a gentle manner, and by using their intuition to read the client's needs, intentions, and processes.

Gendered modalities of music production can be seen in many female producers' practice with their workflow. Rivero believes women have different workflows in the studio "as females think differently to males" (K Rivero 2019, interview communication, 7 February). A Hundred Drums believes females convey things or communicate musically, "due to thinking differently" than men (A Hundred Drums 2019, interview communication, 8 March). Missy Thangs believes that her "interface" of practice is very "effeminate due to women thinking differently to men" (Missy Thangs 2019, interview communication, 31 March). Warren has a unique workflow with the way she "arranges music, as she doesn't write linear". She doesn't start at the intro and work forward (A Warren 2019, interview communication, 12 April).

The Do-It-Yourself (DIY) approach to practice is another female modality of music production. Warren uses DIY in her practice with use of "a USB microphone in a closet with blankets to deaden the room". She also uses methods of self-teaching with certain gear, though she prefers to have someone walk her through something rather than to read a manual. She enjoys gaining knowledge from video tutorials and gets "more benefit" from these than a manual (A Warren 2019, interview communication, 12 April). Grzesik also uses DIY in her practice with methods of "self-teaching" (A Grzesik 2019, interview communication, 15 March). Preece uses a DIY approach in her production by "fixing our audio gear and making our own cables, instruments, synthesizers and controllers". The Ironing Maidens also "make their own website and do their own marketing". Preece also uses methods of self-teaching for learning new gear and techniques (P Preece 2019, interview communication, 2 April). Kearney defines the history of DIY as an "anti-corporatist ideology, which grounded various leftist movements, committed to creating non-alienated forms of labor and social relations" (Whitely. Ed. Kearney 1997: 215). Further explaining, feminist DIY has been a major access point for women to create music by providing "safe women-only spaces for the learning of skills as well as rehearsal and performance, challenging ingrained technophobia and giving women the confidence to believe that, like the boys, they can be music-makers rather than simply music fans" (Whitely. Ed. Kearney 1997: 216).

Feminine modalities of music production are evident in the gentle approach that female producers have to their works. Preece's feminine modalities of music production are notable in The Ironing Maidens' music as it is "less aggressive and more about the story in the lyric and political message" that they are sharing. Preece has a gentle approach to processing, including refraining from over-compressing the drums. She explains, "All the production in our music supports our message of the social inequalities of women in the home and in music production" (P Preece 2019, interview communication, 2 April). Van Den Hurk doesn't like to overly compress her sounds and likes them to be more "natural and organic," compared to some men's productions which are polished and heavily compressed and "shape the sound heavily, harsh and squeezed into shape" (F Van Den Hurk 2019, interview communication, 27 March). Warren also has these feminine modalities of production in her works, with gentle processing on the drums, and she regularly uses a wet reverb. She also has a gentle approach to her arranging style and is "not aggressive with muting things" in the mix (A Warren 2019, interview communication, 12 April). Rivero explained that another feminine modality of music production is having less aggression or a gentle touch with the faders. "Women have a different touch on the fader, it's not aggressive" (K Rivero 2019, interview communication, 7 February). Grzesik says that typically, males may have an aggressive approach to production and their studio processing, whereas "women may have a lighter touch" when it comes to this. She also adds: "Then how much of this has to do with the fact that men can get away with more". She gives the example of times when she has gone out on a limb and taken chances with being more aggressive with her processing and "guys have been like, yeah, I don't think this works here". She says, "Who knows if a guy tried this same thing he could have gotten away with it". Although this is of course speculative thinking, this appears to make her think that women's approaches have less to do with our gendered natures and potentially more to do with our timidity as female producers. We have a subordinate position in the field and studio environment, which impacts our creativity due to "wanting to fit in, be hired again, and please people" (A Grzesik 2019, interview communication, 15 March). A Hundred Drums says she has "some songs that are feminine sounding and others that aren't feminine at all". She doesn't believe that the process of music production is a gendered process, or that she does anything different in the studio compared to males. She has collaborated with male dub-step producers and recognizes similarities that can be explained as "genre-based processing". Though by contrast, she believes her female modalities of production are subtle in the finer detail of her productions, what she calls "the essence of it" (A Hundred Drums 2019, interview communication, 8 March).

Gendered modalities of music production were revealed in many female producers' works through their gentle approach to processing, and were evident in many female producers' practices. Their feminized approaches were demonstrated through the creation of homely spaces to create in, cleanliness, organization, a unique workflow in the studio, and the use of DIY culture. The approach to practice these women have shown with their clients in regards to using their intuition to read what the client wants and needs, and their softer personality styles as producers, are arguably gendered modalities of music production.

CONCLUSION

This study concludes that while the experience of one's gender may not be conscious in an approach to music production, the feminine manifests as a distinct aesthetic throughout this analysis. Many participants found it hard to think about music as gendered. Wood remarked that "music is math and emotion" and doesn't believe in the cultural anthropological perspectives to music theory. She thinks "of herself as human" and found it hard to judge the femininity in her practice (C Wood 2019, interview communication, 26 March). This demonstrates the fact that women's approaches to music production are diverse and that not all female producers will or will seek to express femininity through their practice. Farrugia notes the disadvantages that female electronic music producers face when their music is categorized as women's music, which results in harsher criticism and potentially distances male fans and lessens the amount of airplay as electronic music is mostly consumed by and generated through male DJs (Farrugia 2012: 68). Macarthur explains that some believe the label 'women's music' illustrates the perspective that it is separate from men's music, illuminating the perspective that men's music is simply music (Macarthur 2002: 2). Many women want to gain recognition for their music production practice, but they want it to happen without reference to their gender, with the belief that relating gender to practice diminishes the impact of their work. Like myself, Macarthur was also aware of demonstrating the difference of the feminine aesthetic to celebrate validity in the practice of females. She says: "It becomes obvious that I want to demonstrate its difference in order to celebrate its worth. On the other hand, it is apparent that I am painfully aware of the deficiencies involved in such an argument" (Macarthur 2002: 3). I believe these deficiencies will dissolve along with the gender inequalities in music production with further research and representation of the female producer and her practice. Celebrating the feminine aesthetics of music production and focusing on the representation of females in the environment of music production will eliminate the sense of devaluation that female producers feel with the intersection of their gender and practice. By acknowledging that women, just like men, bring a range of skills into the process of music production, those of us who feel passionate about this subject will help balance the female-to-male producer ratio and generate more awareness of women's productions.

Many believe the representation of music production in the media leads to the belief that production is a male domain. In the majority of discourse around the subject and on social media, the representation of a producer is male, with images of "a male behind the console or with a boom mic" (K Rivero 2019, interview communication, 7 February). There is a lack of articles written about female producers, and even the high-achieving female producer lacks publicity in the media. Botta noticed that articles on the male producer of the year are prevalent in the media and feels "there is a natural gravitation to write about male producers" (S Botta 2019, interview communication, 17 March). Lieb comments on how important media representations of the female are to be interpreted as a reality for the audience:

> Historically, women have been underrepresented in the music industry and in society (e.g., women only got the right to vote in 1920 – less than 100 years ago). This makes the few representations of women that we see all the more important from a social influence standpoint.
> (Lieb 2013: 163)

Preece believes "as an educator that research into the female producer and her practice needs to be researched and published in the academic world," leading to a larger representation of females in the field. That will "benefit the culture of music production with assets to everyone in the industry, males included" (P Preece 2019, interview communication, 2 April). Grzesik feels "the more that we see women involved in this industry, the more women will feel comfortable joining the industry" (A Grzesik 2019, interview communication, 15 March). Providing the dominant ideology and identity of a music producer as a male not only marginalizes over half of society, but it can also lead to issues in regards to the creation of identities for female producers in the industry and those wanting to access the industry. Marian Wright Edelman explained: "You can't be what you can't see" (Marian Wright Edelman; Spelman College 1959). We need more female role models of music producers to be represented in the media and in music production discourse. This will address the gendered assumptions and "inspire more women to the field" (K Rivero 2019, interview communication, 7 February), giving aspiring women producers someone to identify with.

ACKNOWLEDGEMENTS

Thanks are due to Dr Jodie Taylor for her mentoring and support with guiding me toward relevant information sources, giving me feedback on drafts of this chapter, and for inspiring conversations on gender and music. I would also like to acknowledge the help of my colleague Wayne McPhee from the Masters in Creative Media program at SAE Institute Australia for his feedback on the drafts of this chapter, along with the interview participants for their time and openness in interviews. A final thank you to my daughter Charli Gaiter, for her patience and being my foundation in life.

NOTES

1. Fieke Van Den Hurk is a studio producer/engineer from Graveland, a small village half an hour from Amsterdam, in The Netherlands. She has a Masters of Music degree and has worked for Wisseloord Studios since 2012 as an engineer.

She also has her own studio called Dearworld Studio. Van Den Hurk has worked with many known artists including Ozark Henry, Candy Dulfer, Eivør, Anna Rune, Kingfisher Sky, and the Rembrandt Frerichs Trio (F Van Den Hurk 2019, interview communication, 27 March).

2. Sophie Botta is an artist/producer and studio producer/engineer from Sydney, Australia. Having graduated from a Bachelor of Music with a major in Audio Engineering, Botta has worked across the live and studio scene on a variety of projects in a diversity of genres including jazz, pop, and world music (S Botta 2019, interview communication, 17 March).

3. Missy Thangs is a studio producer from North Carolina, USA. With a degree in Recording Arts, she started out working on her own productions, and then went on to writing and performing with No One Mind and Birds of Avalon. Thangs currently works at Fidelitorium studio as a producer and engineer and has worked with many renowned artists including Ian McLagan, Ex Hex (Merge Records), Avett Brothers, Birds of Avalon, Las Rosas (Burger Records), Pie Face Girls, Jenny Besesztz, The Tills, and Skemäta (Missy Thangs 2019, interview communication, 31 March).

4. Amelia Warren is an artist/producer/DJ from Austin, Texas, USA. Amelia produces her own productions, along with working with clients to produce their works. She works as a freelance engineer and producer at Dub Academy Studio and South by Southwest Studio. She has a Bachelor degree in Songwriting and a Masters of production, technology, and innovation, which was completed at the University of California, Berkeley (A Warren 2019, interview communication, 12 April).

5. Patty Preece is a studio artist/producer/DJ from Cairns, Australia, producing in a duo called The Ironing Maidens. An award-winning electronic duo, The Ironing Maidens are putting domestic labor, technology, and the history of women in electronic music center stage with their live electronic performance art piece. Preece is also a lecturer at Southern Cross University in Cairns, Australia in the music tech field. She entered the field of music production through the health field and played drums in a band. She has completed a Diploma of music production and a Bachelor Degree of Audio Production (P Preece 2019, interview communication, 2 April).

6. Catharine Wood is a studio producer/composer/engineer from Los Angeles, USA. Beginning her career in audio in 2005, she has worked in audio post-production for commercials, engineering on the first Apple iPhone commercial, and as a mix/mastering engineer, engineering over 500 commercially released songs. Her company, Planetwood Studios, specializes in producing singer-songwriters and providing engineering, production, and composition services to the TV and film industries. Wood is also a Grammy Voting Member and Producers and Engineers Wing member (C Wood 2019, interview communication, 26 March).

7. Karina Rivero is a producer and a recording, post-production, and mastering engineer, from Mexico City, Mexico. She has worked in a recording studio/post facility for the last four years and has a Bachelor and Honors of Music Production (K Rivero 2019, interview communication, 7 February).

8. A Hundred Drums is an artist/producer/DJ from Grass Valley, Nevada, USA. She has been producing electronic music for the past five years. A Hundred Drums is a self-taught producer, though many mentors have helped guide and

support her along with further developing her self-taught skills. She has performed at many renowned events including Coachella, Lucidity, Enchanted Forest, and Bamboo Bass Festivals (A Hundred Drums 2019, interview communication, 8 March).

9. Ania Grzesik is a studio producer practicing in the New York City Metropolitan area, USA. With over 17 years of professional experience, her current practice is focused on freelance podcast production and sound design, though she has previously worked in recording studios and live sound in theatres and music events (A Grzesik 2019, interview communication, 15 March).

FURTHER READING

Macarthur, S. (2002) *Feminist Aesthetics in Music*. Westport, CT: Greenwood Press. ISBN 0-313-31320-2. (Musicology resource on feminist aesthetics in music.)

McLary, S. (1991) *Feminine Endings*. Minneapolis, MN: University of Minnesota Press. ISBN 978-0-8166-4189-5. (Musicology resource on feminist aesthetics in music composition.)

Mills, S. (1995) *Feminist Stylistics*. London: Routledge. ISBN 0-203-40873-X. (Musicology resource on feminist aesthetics in music.)

Rodgers, T. (2010) *Pink Noises: Women on Electronic Music and Sound*. Durham, NC: Duke University Press. ISBN: 978-0-8223-4673-9. (Musicology resource on gender and electronic music production.)

Whiteley, S. (1997) *Sexing the Groove: Popular Music and Gender*. New York: Routledge. (Popular musicology resource on music and gender.)

REFERENCES

Bayton, M. (1998). *Frock Rock: Women Performing Popular Music*. Oxford: Oxford University Press. ISBN 9780198166153.

Cixous, H (1997) 'Sorties: Out and Out: Attacks/Ways Out/Forays'. In Schrift, A. D. ed. *The Logic of the Gift: Toward an Ethic of Generosity* [Chapter]. New York: Routledge. ISBN 0-415-91098-6.

Farrugia, R. (2012) *Beyond the Dance Floor: Female DJs, Technology and Electronic Dance Music Culture* [Article]. Bristol: Intellect. ISBN 978-1-84150-566-4.

Hisama, E. M. (2014) 'DJ Kuttin Kandi: Performing Feminism' [Article]. *American Music Review*, Vol XLIII, No 2. Conservatory of Music, Brooklyn College of the City University of New York, NY, USA.

Lieb, K. (2013) *Gender, Branding and the Modern Music Industry*. New York: Routledge.

Macarthur, S. (2002) *Feminist Aesthetics in Music*. Westport, CT: Greenwood Press. ISBN 0-313-31320-2.

McLary, S. (1991) *Feminine Endings*. London: University of Minnesota Press. ISBN 978-0-8166-4189-5.

McRobbie, A. (1978) 'Settling Accounts with Subcultures, A Feminist Critique'. In Frith, S. and Goodwin, A. eds. *On Record, Rock, Pop and The Written Word*. New York: Pantheon Books.

Olszanowski, M. (2011) 'What to Ask Women Composers: Feminist Fieldwork in Electronic Dance Music' [Article]. *DanceCult*. Retrieved from http://dj.dancecult.net

Stoltzfus, P. (2006) *Theology as Performance: Music, Aesthetics, and God in Western Thought*. New York: TT Clark.

Taylor, Dr J. (2012) *Playing it Queer*. Bern: Peter Lang AG. ISBN 10: 3034305532

Whiteley, S. (1997) *Sexing the Groove: Popular Music and Gender*. New York: Routledge.

Wikstrom, P. (2009) *The Music Industry, Music in the Cloud*. Cambridge: Polity Press. ISBN: 978-0-7456-4390-8.

Wolfe, P. (2012) 'A Studio of One's Own: Music Production, Technology and Gender' [Article]. Retrieved from www.arpjournal.com/asarpwp/a-studio-of-one%E2%80%99s-own-music-production-technology-and-gender/

Wright Edelman, M. (1959) 'Mariane Wright Edelman Quote' [Website]. Retrieved from https://quotes.thefamouspeople.com/marian-wright-edelman-2254.php

Part Four
Industrial Evolution

14

Addressing Gender Equality in Music Production

Current Challenges, Opportunities for Change, and Recommendations

Jude Brereton, Helena Daffern, Kat Young, and Michael Lovedee-Turner

INTRODUCTION

Although music production encompasses a broad selection of disciplines, job roles, practices, and methodologies, it is arguably one of the least gender-diverse occupations. The ratio of male to female music producers is estimated to be 47 to 1 (Smith et al. 2019).

The purpose of this chapter, however, is not to provide a detailed analysis of gender representation in music production, nor a comprehensive discussion of where biases have been formed or perpetuated. Whilst we will touch on the factors which contribute to the current state of the industry's bias, we will predominantly explore the potential for changing the landscape of gender representation in music production by capitalizing on current positive work in the sector. We will also present a number of recommendations to make a rapid, significant, and sustainable change for the better.

There still prevails a stereotypical view of the Music Producer as a lone practitioner, an audio engineer controlling huge amounts of complicated-looking technology, with buttons, sliders, LED displays, and many cables. Indeed, this stereotype is historically heavily gendered and bound together with notions of male dominance in technological expertise, control, and creativity. Wolfe (2019) provides a wider exploration of gendering in the music industry and the implications of the patriarchal framework this creates.

A web-based image search on the term *Music Producer* returns hundreds of photos in which the technology (mixing desk, multiple loudspeakers, PC/laptop screens, keyboard) are front and center under the control of a sole (male) audio engineer. Very few images depict people working in a team, and even fewer show obviously female faces. Those that do include female producers/engineers are usually found on specialist websites aimed at encouraging women into audio engineering.

Despite such pervasive, narrow stereotypical images, music production involves much more than studio recording and requires a wide range of technical, artistic, and social skills. Ian Shepherd (2009) suggests that music producers embody a variety of different skill sets and identifies seven types of music producer, including engineer, artist, musician, and mentor.

With this in mind, we use the term *music production* in this chapter to span a wide variety of roles within a number of disciplines such as audio engineering, creative music technology, sound recording, electronic music composition, and game audio as well as audio for emerging immersive technologies (e.g. virtual and augmented reality). As such music production covers a broad range of activities all of which combine creativity and technology, arts and engineering; "from studio based roles, to digital processing, composition, mixing, live sound engineering, recording processes, audio for games" (Murphy *et al.* 2006).

CURRENT DATA AND CHALLENGES

The proportion of women working in the music industry rose from 45.3% in 2016 to 49.1% in 2018 (UK Music 2018). Although this headline data seems encouraging, when looking at the type of roles women are working in, the picture is quite different. Approximately 21.7% of artists in the music industry are women, but only 12.3% of songwriters, 2.1% of music producers, and 3% of engineers/mixers in popular music are women (Smith *et al.* 2019) – so women are excluded from crucial roles in the industry.

The low representation of women in the creative roles in the music industry is perhaps surprising given that, at all stages of education in the UK, equal (and sometimes higher) numbers of women than men study music.

The percentage of female students within GCSE Music (Level 2) cohorts in England has fluctuated between 48% and 57% since 2002 (Figure 14.1,

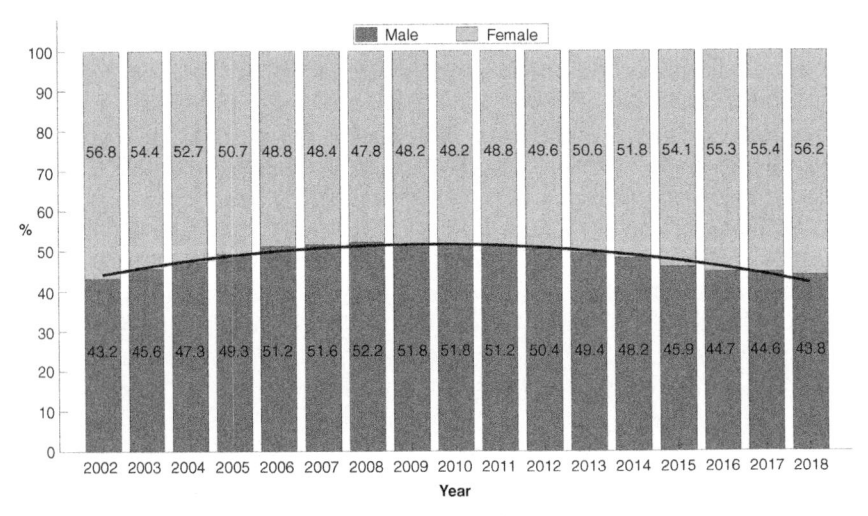

Figure 14.1 Percentages of Male and Female Students Taking Music GCSE From 2002–2018. The polynomial trend line indicates male percentage

Source: Data from Gill 2010, 2016a; Joint Council for Qualifications 2019a.

Addressing Gender Equality in Music Production

data from: (Gill 2010, 2016a; Joint Council for Qualifications 2019a). The percentage of female students with Music and Music Technology A Level (Level 3) cohorts has fluctuated between 39% and 54% since 2002 (Gill 2012, 2016b; Joint Council for Qualifications 2019b). Note, the government statistics combine the numbers for Music and Music Technology A-Level courses.

After some years of near parity between 2004 and 2013, it is worrying that the gender gap between pupils' take-up of Music seems to be widening once again. Indeed, the traditional gender divide in subject choices seems to have deepened since 2013, with the proportion of girls taking GCSE Music (Level 2) increasing whilst the proportion of girls taking Design and Technology has fallen from 67% to just 33%.

In England at least, some attribute this widening gender gap to the introduction in many schools of the EBacc (Knott 2018), which has seen an overall 10% drop in arts GCSE entries, against a 1% overall increase in total GCSE entries across all subjects.

An overall decline in student numbers for Music Technology A level (Level 3) is similar. Between 2008 and 2018, the number of students taking the Music Technology A Level dropped from 3,422 to 1,397 (Pearson Qualifications 2019). A survey of 464 secondary schools in England between 2016 and 2019 underlines this reduction, with 31.7% fewer schools offering A-level Music Technology over the time period studied (Daubney and Mackrill 2018).

Splitting the overall total by gender paints a different picture to that of GCSE Music data. In contrast with Music courses, Music Technology courses see much higher numbers of male than female students (Figure 14.2), with a female percentage of between only 16% to 25%,

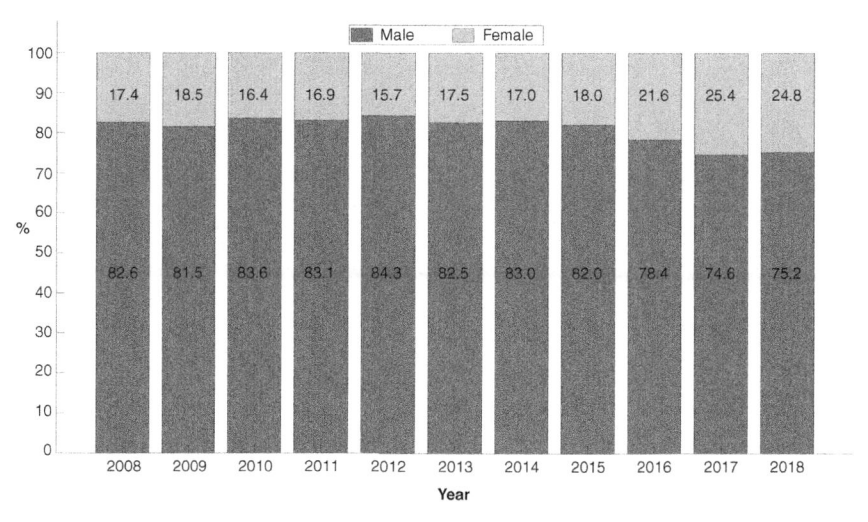

Figure 14.2 Percentage of Male and Female Students Within the Edexcel (Pearson) Music Technology A Level From 2008–2018

Source: Data from Pearson Qualifications 2019.

although this is somewhat offset by the reduction in numbers of male students more so than female students (Figure 14.3), which is not accounted for by general decline in A-Level cohort size. The more vocational BTEC First (Level 2) Music Technology courses are also dominated by male students (63%), a gender imbalance which deepens as students progress to Level 3 BTEC National qualifications, where 80% of the students are male (Ruthmann and Mantie 2017). Such a gender imbalance is of course not restricted to Music Technology, with most IT and technology subjects at Level 3 (A Level) seeing in the region of 80% male students.

The pattern of male dominance in Music Technology at Level 3 continues as students move into Music Technology degree programs. In UK Higher Education, *Music Technology* acts as an umbrella term for a variety of differing approaches, all of which involve music (or audio) and technology. Music Technology degree programs have proliferated in the UK after the introduction of the first Music Technology A level in 1998; Born and Devine (2015) estimate that numbers taking Music Technology degrees rose by nearly 1400% between 1994 and 2011. A search in July 2019 for undergraduate degrees in music technology on UCAS (the Universities and Colleges Admissions Service – a centralized service for university admissions in the UK) returns 324 degree programs hosted by 93 different institutions.

The large numbers of Music Technology degree programs reflect the wide variety of programs in established Higher Education institutions which cover some aspect of music and technology, e.g. audio engineering, music production, music technology, creative music technology, to name just a small selection. Often the differences among programs center on the balance between *making music with technology* and *making technology for music* (and audio). Boehm et al. (2018) have noted a reduction in

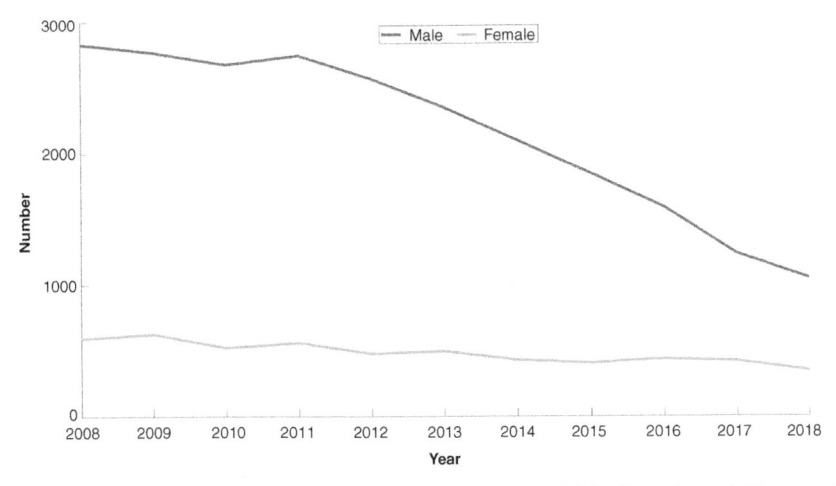

Figure 14.3 Numbers of Male and Female Students Within the Edexcel (Pearson) Music Technology A Level From 2008–2018

Source: Data from Pearson Qualifications 2019.

the number of 'Audio Engineering' oriented degrees over the last decade, alongside a dramatic increase in 'Music Production' type programs. The broader context of this divide and its implications for program design is considered through a study of one UK Music Technology masters program in Brereton et al. (2019).

Obtaining gender-disaggregated data on those studying Music Technology in Higher Education in the UK is difficult because of the variety of Music Technology degrees offered; some fall under Music and Arts HESA codes, and others might be coded under Engineering Technology HESA data (HESA 2019). However, a study by Born and Devine (2015) obtained data for Music and Music Technology degrees between 2007 and 2012 and found that, whereas traditional music degrees saw a gender profile in line with the overall student population, music technology degrees were overwhelmingly male: in 2015 only 12% of Music Technology students in Higher Education were female. This is somewhat unsurprising, given the larger numbers of male students at Level 2 and Level 3 in Music Technology.

Girls and women are excluded from the technological aspects of music-making from an early age – as Ruthmann and Mantie (2017) note, since the reforms to the UK national curriculum in 2014, music technology is largely absent in the Key Stage 2 curriculum (for ages 7–11), meaning that the opportunity is lost to introduce pupils of all genders to music technology at this crucial age where we understand that gender norms begin to be established. Music Technology as a school subject somehow falls into the male-dominated *technology* domain, rather than a more gender-balanced *music* domain, despite the fact that the most recent A-Level Music Technology curriculum emphasizes that music is the focus and "technology is the servant of music, not an end in itself" (Edexcel 2013).

A study by Mathew et al. (2016) estimated that only 7% of members of the Audio Engineering Society (AES) were not male – this data could only be estimated since the society does not systematically record the gender of members. More recently, Young et al. (2018) proposed a new method to collect accurate gender data on those presenting (paper authors, keynote speakers, and workshop leads) at AES conferences. Between 2012 and 2016, presenters at AES conferences were 88.98% male, 9.09% female (1.82% unidentifiable, 0.11% non-binary). The authors have recently updated the data[1] to include conferences from 2016–2019; a small change is evident, with participants within this second time period being 85.6% male, 12.1% female (1.75% unidentifiable, 0.55% non-binary). However, it is too soon to see whether this is the beginning of an improvement in the gender balance or just "noise" in the data over time (Figure 14.4).

In order to better gauge whether the positive movement in some of the gender balance data for education and the industry is reflected in the aspirations of youngsters, we ran a small survey of female UK secondary school pupils (aged 13–18). All were involved in making music or music technology related activities, and we asked which job roles in the audio industry they were aware of, and which they might be interested in pursuing.

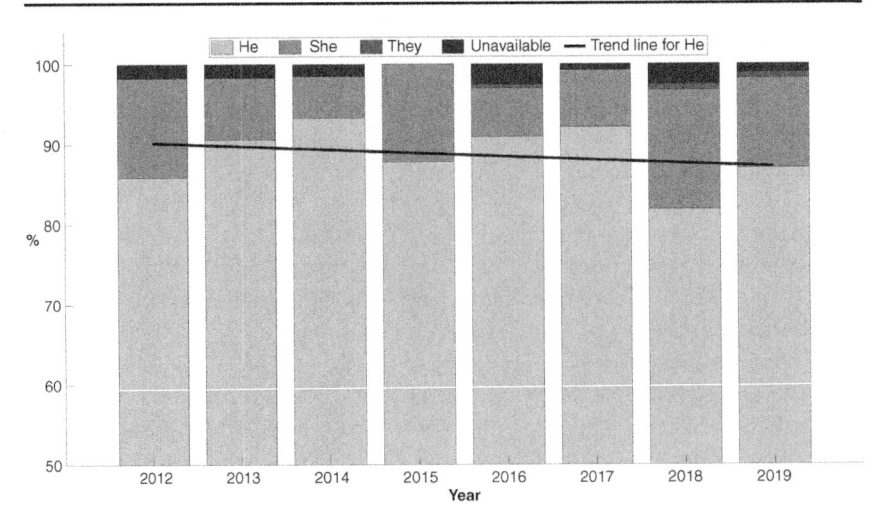

Figure 14.4 Gender Representation of Presenters at AES Conferences, 2012–2019

Source: Data from Young *et al.* 2018 and https://tibbakoi.github.io/aesgender/.

Of the 57 respondents, 39 were involved in instrumental music lessons, three in ensembles, 13 in school music lessons (both GCSE and A Level), but only one in studying Music Technology (A Level). Nine respondents reported some involvement in additional music technology activities, such as helping with sound or lights for school productions or mixing performance soundtracks.

Eighteen job roles were listed in the survey, covering a variety of aspects within audio technology, music production, and the audio industries (seen in Figure 14.5). The respondents had heard of the majority of roles (most dark grey in Figure 14.5), with "producer" (82%), "composer" (82%), and "software developer" (67%) the most recognized.

However, the numbers of respondents potentially interested in taking up each job role is substantially lower (middle grey in Figure 14.5). Only 49%, despite being involved in music or music technology, said they were maybe or definitely interested in a job in the music industry (indeed, only 4/57 said "yes definitely"). Results for these 28 respondents have been extracted and plotted in the lightest grey in Figure 14.5: "Adjusted Interest". In alignment with the roles most heard of, producer and composer have the largest percentages: 57% and 25%, respectively.

It seems that very few respondents were interested in any of the roles associated more with engineering (only 7% were interested in "mixing engineer" or "research engineer", and only 4% "acoustic engineer").

All nine respondents who had some form of music technology experience (whether through formalized education or additional activities) reported "yes" or "maybe" interested in a job in the music industry/music production. These respondents appear to have heard of more of the job roles than the general respondents; however, the number of respondents interested in each role is still low, with the exception of "producer" (see Figure 14.6).

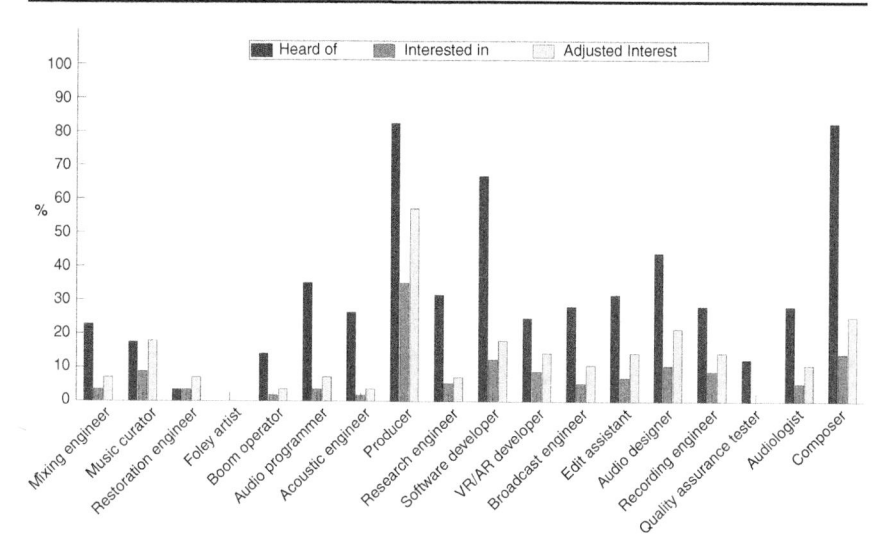

Addressing Gender Equality in Music Production 225

Figure 14.5 Percentage of Respondents Who Had Heard of the Listed Job Roles and Who Were Interested in Each Role (calculated as the fraction of total respondents). Adjusted interest is those respondents who said "yes" or "maybe" to interest in a job in the music industry (calculated out of 28).

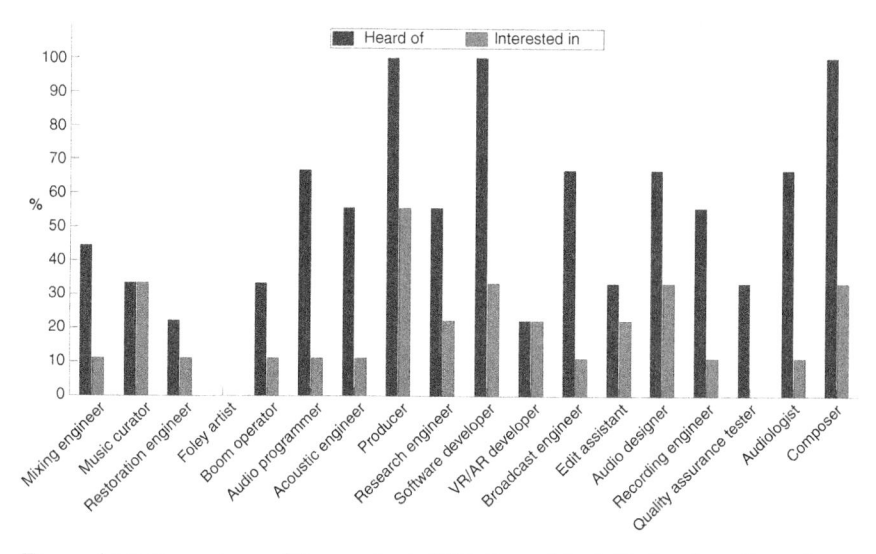

Figure 14.6 Percentage of Respondents With Some Music Technology Experience (nine respondents) Who Had Heard of the Listed Job Roles and Who Were Interested in Each Role.

"Composer" does not feature highly in this subset of respondents, where it is one of the more popular in the general set. This is to be expected, as the data can be interpreted as it is likely that the other respondents are studying music rather than music technology. It seems that, for the female students we surveyed, having some experience of music technology at school raises awareness of the range of job roles available under the wide umbrella, but does not necessarily translate into an active desire to pursue these roles. Negative perceptions of the industry may be driving this trend.

Attempts to gather and publicize data on gender in the audio and music industry are relatively recent (for a good example, see Women In Music[2]), and data collected invariably re-emphasizes the small proportions of women/non-binary participants in music technology and audio-related subjects at school, college, and in tertiary education – there are few surprises.

OPPORTUNITIES FOR CHANGE

As data attests, audio engineering, music production, and music technology have traditionally been male dominated, a position which unfortunately seems to be resistant to speedy change. In this section we look briefly at some of the current interventions which seek to address gender balance in music production, note some of the successful initiatives and bank of knowledge that has been drawn together on gender equality in STEM disciplines, and discuss the opportunities for change unique to music production which can be exploited.

Interdisciplinary Nature of Music Production

Music production and music technology are inherently interdisciplinary pursuits, with methods, approaches, and knowledge shared among the separate disciplines of music, art, engineering, creativity, performance, analysis, and technology. It might, therefore, seem odd that music production has become so male-dominated when the related fields are often praised and noted for their truly interdisciplinary nature.

This nature of music technology and music production is, potentially, a real draw for those looking to work across traditional arts/science disciplinary divides. A review in 2007 identified a large number of sub-disciplines involved in sound and music computing including music, music composition, music performance, science and technology, physics, maths, psychology, and engineering (Bernardini and de Poli 2007; Serra et al. 2007).

However, as Boehm (2007) and Boehm et al. (2018) argue, Music Technology moving from interdisciplinary to truly transdisciplinary and, as such, transforming into a single discipline in itself has not materialized. Reasons for this are many and related to external motivations and pressures on both school and tertiary-level education. Since 2012, STEM (science technology engineering and maths) subjects have been better funded by government. This has led some institutions to adapt Music Technology programs to increase their focus on the science/technology element of the discipline, potentially as a means to attract greater numbers of students, exploiting the perception that employment prospects are increased for those studying STEM rather than arts-based subjects.

What is more, the traditional idea of a music producer as closely aligned with that of an engineer is very often one that prevails, despite music production encompassing a huge variety of skills and not just relying on the mastery of technology.

There is a danger that the interdisciplinary nature of music production means that those in the industry and education are not fully involved in initiatives to address gender equality that are led by and housed under single disciplines of music and engineering, such as the PRS Keychange Manifesto (PRS Foundation 2018) and the Engineering UK campaign on diversity (This is Engineering 2018).

Campaigns are, however, becoming more popular, especially as they learn to exploit the potential of social media, and are playing an important role in setting the scene for significant change by raising awareness of gender issues.

Increased Awareness of Gender Issues

The need for work to address gender equality in many areas of life seems to be receiving renewed interest in the media, with blockbuster films highlighting the role of women in space science, the #MeToo movement, issues around the gender pay gap highlighted in the press, and high-profile public figures giving support.

The results of a non-diverse engineering sector on our modern world are also increasingly being discussed. Ely (2015) presents a series of examples where design does not meet the diverse needs of end users as a result of the non-diverse workforce: the design and testing of car seats, seat belts, power tools, and medical research have all historically failed to include women.

For the audio industry, the gender data gap is also an issue since much of the seminal established literature in areas such as audio perception has been based on male participants. There is now an increased awareness that participants in perceptual studies should be gender balanced and results reported by gender where possible. This is especially important in particular areas, for example, in research on spatial sound perception and where researchers are trying to tailor sound reproduction based on the size and shape of the listener's head.

An increase in the participation in research studies by female students and researchers will help to raise awareness of the interdisciplinary research being undertaken in audio and might help to address the gender imbalance in the field in the future.

Many departments in higher education institutions in the UK have engaged with the Athena SWAN charter, which promotes interventions to tackle gender inequality in academia. It originated within science, but has since expanded to all academic disciplines. As such, there is now perhaps a wider understanding of some of the social and work-based barriers that those in under-represented groups face in any arena, including imposter syndrome, stereotype threat, unconscious (and conscious) bias, the leaky pipeline, in-group socialization, nepotism, homosociative habits, the glass ceiling, lack of role models, prevalent gender schemas, and sexism (Valian 1999).

The music industry has similar problems to those found in academia; a recent survey of women in the US music industry (Prior et al. 2019) found that the majority had experienced gender bias, and that nearly half felt that they were given less recognition than their male peers and that their careers had been held back because of gender. Women of color in particular felt less well supported in the workplace.

Much of the academic literature in this area seeks to address the problems of women leaving a discipline (known as the "leaky pipeline" (Ceci et al. 2009)), often in disciplines with equal numbers of male and female undergraduate students (such as chemistry), or with greater numbers of female students at the undergraduate level (such as biology; see, for example, Burrelli 2008). But music production and music technology attract low numbers of female students to start with, and although efforts to retain women within the industry are valid, we need first to better understand why women are not involved in the discipline throughout school, university, and into work.

Fixing the Industry

Although much effort has been expended to encourage female students to consider STEM careers, and within that music technology more specifically, all of these efforts will be wasted if we don't make a concerted effort to address the culture of the industry itself.

We must address the justified perception of the music industry as heavily male-dominated and toxic. As we noted earlier, although women are now equally represented within the music industry workforce as a whole (Prior et al. 2019; Smith et al. 2019), they are still under-represented, overlooked, undervalued, overcriticized, excluded, and sexualized as performers, composers, and producers. They face entrenched sexism, a lack of role models, and multiple barriers to entry that means that any female musician or producer must fight significantly harder than her male musical peers for recognition (Prior et al. 2019; Smith et al. 2019).

A clear and very common example of this hostile environment can be seen in the online comments made on one video which is part of a series of over 500 video interviews with music producers in their studios. Here, Catherine Marks, winner of the Music Producers Guild 2016 "Breakthrough Producer of the Year Award" and 2018 "UK Producer of the Year Award", talks about her approach to recording and production. One of the comments posted reads: "Hot and an engineer toooo!!!!" (We looked at many of the interviews with producers and could not find a similar comment directed towards a male producer.) Female music producers and women working in the audio industry also face subtle (hidden) as well as overt (open) discrimination and gender bias; this is the case too in other areas where a discipline is male dominated (Heilman and Welle 2005; Jones et al. 2016).

There has been plenty of recent research on women in science and academia over the last 20 years, as well as organizations working in this space to champion gender equality in science and engineering, such as the

Women in Science and Engineering (WISE) campaign[3] and the Women's Engineering Society (WES)[4].

A growing awareness of the need to counteract the male domination of the music industry has led to the creation of websites and networks to highlight the contributions of women both historic and current, such as Her Noise,[5] the Delia Derbyshire Archive,[6] She Is The Music,[7] the Women's Audio Mission,[8] and Pink Noises.[9] These organizations not only organize face-to-face physical meetings but also act as an online source to collect material and share information. This is crucial in order to highlight not only the hidden history of women in music technology, but also in retaining current contributions by women to music production, electronic music, DJing, songwriting, sound art, and audio engineering for the future.

A number of women-only groups have also been established across the world for those working in the discipline, such as the Yorkshire Sound Women's Network,[10] Women in Music,[11] SONA WOMEN,[12] and Women in Sound Women on Sound.[13]

These groups provide a safe space for women and non-binary people to meet together, offer mutual support, and take part in activities in a space free from discrimination, gender bias, imposter syndrome, and stereotype threat.

There are also good examples of interventions, focused mentorship, and sponsorship for female engineers and producers such as Spotify (2018), which provides mentorship and work experience in three recording studios aimed at eventually improving the numbers of women working in the industry.

Such organizations and schemes often face criticism for being "exclusive" rather than inclusive, as they mostly exclude those from the majority group. This can be seen as discrimination against the majority group; however, this fundamentally misses the point of minority-only spaces. Designating a space for a specific minority group allows members of that group somewhere where they can feel less guarded, more supported, and often safer than they do in the space where the majority has control. A female-only space allows women to own the space rather than it being a men's space that the women have to then come into.

Women-only groups and interventions aimed at raising the profile of women in STEM and in the audio industry are undoubtedly helping to build confidence amongst those in the minority group and working hard to encourage women to participate. If they continue to grow and diversify traditionally male-dominated spaces into those where all are welcome and included, then surely we should hope that, eventually, the need for women-only safe spaces within the industry becomes redundant.

Inspiring the Next Generation

As the data outlined herein shows, women are not taking qualifications in technology-based subjects at the same rate as men. Seemingly the divide between arts and science subjects at school is as marked as it ever has been, and gendered expectations of school students means that a higher proportion of those studying arts subjects are women. In the UK, students specialize early in their academic career, with most choosing three or four A-Level (Level 3) subjects.

Of those students who combine AS level subjects across disciplinary divides, girls are much more likely to choose combinations which include Arts and Humanities (37.7% of female 2012–13 AS level students compared to 19.7% of male), and boys are more likely to choose Arts and Science subjects in combination (9.0% of male 2012–13 AS level students compared to 7.9% female) (Sutch 2014). Given that most Music Technology degree courses require skills in both music and science/technology subjects, and music qualifications are often gained outside of the school curriculum, it is clear that girls, with a lower take-up of maths and science, are choosing the "wrong" subjects at school to easily progress to Music Technology degree-level study.

In fact, the early specialism roughly at age 16 prevents many school students from retaining a broad-based education, which is beneficial for such interdisciplinary subjects (McLeish 2019). Indeed, many students don't study any science subjects at all between ages 16–18. One attempt to encourage students to study a broader range of subjects at Level 3 (age 16–18) was the introduction of Core Maths A level. It was hoped that this would attract those who needed to continue with some maths as a facilitating subject in order to progress to numerical subjects (such as physical sciences and engineering) at University level. Unfortunately, this radical move hasn't been effective in opening up otherwise closed-off paths to HE, since many of the most selective Universities in the UK routinely do not accept it as part of their entry requirements.

The gendered expectations of school subjects has long been a source of concern for those working in computer science and engineering areas. Indeed, there has been substantial research on the reasons that girls don't consider these subjects at university-level study. For example, girls don't think that computer science/IT is relevant to them, as a narrow interpretation of what computer science is and what computer scientists actually do has been presented at school. Many girls have negative experiences in a male-dominated classroom; being one of few girls studying the subject leads to social isolation and insecurity, and girls choosing to study computer science must make a positive, conscious decision against most gendered norms and peer pressure (Goode et al. 2006).

Interviewing teenage girls about their perception of engineering as a career, Andrews and Clark (2012a) highlight three areas where gendered expectations against women in engineering are experienced: in the transition from primary school to secondary education (where science becomes less practical, more theoretical), in perceptions of science at secondary level, and in perceptions of engineering as a career choice.

Such findings are important for music technology/music production, since the results focus around working with technology and technological expertise (rather than the creative use of technology). If female school students don't feel that they belong in audio engineering/music technology, then it will be difficult to establish a critical mass of female students participating in the subject. This leads to a worrying cycle of low participation, which in turn feeds into future attitudes towards STEM subjects.

It is interesting in itself that, although actual GCSE results show that female students slightly outperform male students in maths and science (68% of girls achieve grades A–C, compared to 65% of boys), when pupils are asked which subjects they are best at only 33% of girls chose STEM subjects as opposed to 60% of boys. Indeed, despite similar performance in Maths and English, boys are much more likely to state that their best subjects was Maths, whereas girls chose English (Department for Education 2019). Schools and teachers will need to work hard to dismantle some of this gendered perception around subject disciplines, although this is quite a task given that subject choices seem to align with more widespread societal "Gender Schemas" as first proposed by Valian (1999) and further investigated by Hewlett et al. (2008): "In white middle-class society, the gender schema for men includes being assertive, instrumental, task oriented, and capable of independent autonomous action" – which aligns nicely with technological expertise. The gender schema for women is different: "it includes being nurturant, expressive, communal, and concerned about others" – perhaps better suited to the performing and expressive arts (Hewlett et al. 2008).

Many organizations across science and engineering are investing considerable time and funding into interventions and outreach activities to inspire and encourage greater diversity in the next generation of those working in STEM.

The Athena SWAN charter has been a driving force for addressing gender balance in Higher Education in the UK. The Women's Engineering Society, The Institute of Physics, and the Royal Society of Chemistry have undertaken sustained efforts to address gender balance in their respective disciplines. Some of this work is through outreach activities in schools targeted specifically at female pupils, and raising the profile of the diversity of those already working in the field to act as potential role models for future generations of scientists and engineers. This means there is now a substantial body of work on interventions to address gender imbalance, and a growing understanding of what works and what doesn't work (see, for example, Valian 1999; Bohnet 2016; Savonick and Davidson 2016; The Center for WorkLife Law 2018).

Off the back of this, there is already much work being undertaken by a variety of groups in the UK and beyond to raise the profile of women in music production and to inspire and encourage the next generation. Many of these groups and initiatives are highlighted elsewhere in this book and online,[14] so we will not list them here.

In many ways, the multifaceted nature of music production, the blurring of lines between traditional technology-based and artistic roles, and the changing nature of job roles and multiple career options may be confusing for students entering further study and aiming for a role in the industry. The current data also suggests that the perceived strong alignment of music technology with engineering and technology disciplines is the main reason why the subject is heavily male-dominated.

With both the creative industries and the engineering sector recently renewing efforts around aspects of diversity, inclusion, and outreach work, there is huge potential for the audio industry to capitalize on its interdisciplinary nature and learn from other fields and those working to improve gender balance.

Outreach, Inclusion, and Belonging

While there is an abundance of outreach work to encourage underrepresented groups to participate in STEM (and in Music Technology in particular), this does not alter the industry's inclusion record to date, and we end up with a situation which one commentator called the "all space camp and no space suit problem".[15]

Coined on the occasion of the infamous cancellation of the first NASA all-female spacewalk due to a lack of appropriate equipment, this term represents the push to get girls and women interested in STEM, or into STEM careers, without acknowledging or addressing the reasons why girls and women are then pushed away. Much time, effort, and resources are spent on encouraging more women into the industry, only for them to then find that they are not welcome because of sexism and a toxic male-dominated culture within the industry itself.

Potential of Future Technologies

The increase in the availability and affordability of technology for making music over the last few decades has the potential to increase the diversity of those participating in music technology. It has been shown that the proportion of students from lower socio-economic groups studying Music Technology is much larger than traditional Music degrees, where the potential entry costs might be much higher (e.g. thousands of pounds for a classical music instrument and private tuition) (Born and Devine 2015).

Indeed, since the 1990s there has been a huge increase in the range and number of music technology-based degrees in Universities and Colleges in the UK, which Born and Devine (2015) hail as a move away from the more historically oriented traditional music degrees. But there is also the danger that Music Technology degrees are spaces which exacerbate the gender imbalance problem, since they often "participate in feedback loops whereby existing ideologies of gender and technology, and social class differences, are being reinforced or even amplified" (Born and Devine 2015).

But it seems that the democratization of music technology hasn't so far resulted in music technology being used by all and equally by men and women. In many ways, the market for consumer music technology has mirrored the marketing of personal and home computers which targeted dads and boys during advertising campaigns in the 1970s and 1980s. The stereotypical image of the computer gamer as male is still prevalent in popular culture. Such images feed the common attitude that it's okay for teenage boys to isolate themselves in their home or bedroom studios surrounded by technology, but not so much for teenage girls.

However, there is evidence to suggest that the gender balance of those using and purchasing music technology is beginning to change. For example, a recent survey by Fender found that 50% of beginner and aspirational guitar players are women (Duffy 2019), and there is evidence that in the USA more women are buying and selling vinyl (a traditionally stereotypically male domain) (Mixmag 2018).

We are also on the cusp of exciting new developments in virtual and augmented reality technology, and recent media coverage has suggested that women are playing a leading role (Onanuga 2019). However, it is vital that well-known issues around the non-inclusive design and development of technology (Faulkner 2000) are avoided. There is a huge opportunity to all of us working in music production and music technology-related fields, to harness the potential of the fourth industrial revolution. Women's participation in designing and shaping immersive and interactive technologies and experiences should be cherished and nurtured.

Music and sound play a key role in the development of XR (virtual reality, augmented reality, mixed reality) technologies. Nevertheless, there is a danger that the typically male-dominated technology sector will steal the ground. If we are careless, there is a real chance that women will be prevented from participating in and benefiting fully from the presaged fourth industrial revolution.

What Doesn't Work

Identifying potential reasons for poor non-male representation can only be a positive step forward. It provides the basis for the development of practical strategies and policies that target the heart of the problem, and therefore attempts to address the symptom rather than the cause. However, the complex and heavily encultured nature of gender and its representation in society can make well-intentioned interventions highlight the fundamental issues rather than provide a positive framework for change.

In 2012, the European Commission launched "Science: It's a Girl Thing" to improve the number of females choosing STEM subjects between the ages of 13–17. The campaign was data-driven in its approach, acknowledging the need to address education and the pipeline of decision-making for STEM subjects, with an aim to make science appear "cool" to females. However, the promotional video for the campaign received attention for all the wrong reasons, and was quickly removed from its website after criticism for its stereotyping of the sexes (Collins 2012) and a defensive if not critical public response from the advisory group for the project (Sánchez de Madariaga et al. 2012).

Another example, this time of a commercial company drastically missing the mark, was the initial release of the Midiplus MIRROR, for which the product web page[16] contained the phrases: "an audio interface specially for females", "through a female perspective and taste, we attempted to integrate mirror light", and "twirl the colorful eye-shadow shaped knobs, glance at the equalizer LED", amongst others. Many within the audio industry were outraged at the stereotyping and patronization running throughout the product description.

At the time, professional musician Catharina Boutari commented to Gearnews:

> It feels like I'm being treated as a puppy that wants to play and not as a musician being talked to at eye level. And that it, in my opinion, is what we women demand: talk to us at eye level, treat us as pros or future pros, design advertising where professional female musicians present gear in a professional way to other women, and you're gonna sell your shit.
>
> (Gearnews 2018)

As a result of the backlash, Midiplus released a statement on their website[17] apologizing for any offense caused and stating: "we have no intention of offending anyone. We have the deepest respect for people of all genders, ages, and orientation". What used to read "designed for females" now reads "designed for live-streaming". They claim the product was aimed at those wanting to livestream karaoke performances, the majority of whom are kids and young adults, and that this had been lost in the translation.

These examples highlight a fairly nuanced problem facing the industry. Proactively addressing the issue of under-representation of minority groups requires careful thought and the balancing of (often misunderstood and uninformed) representation of common belief systems and societal cultures and norms.

The Problem With ED&I

There has been some criticism of equality, diversity, and inclusion (ED&I) initiatives in recent years, and frustration that outreach activities, diversity interventions, and equality charter awards seem to have made little impact in many disciplines.

In most STEM workplaces (including industry and academia), the "leaky pipeline" – the drop-off of female participation moving from junior to more senior levels – still persists, with few institutions making any significant change to the numbers and ratio of women in more senior leadership positions (Hewlett et al. 2008).

There is often an over-reliance on diversity training packages, whose effectiveness has not been demonstrated. Dobbin and Kalev (2016) argue that diversity training designed to reduce bias in recruitment (hiring) practices doesn't work, as the effects tend not to last beyond a couple of days. More seriously, they can often activate a backlash from employees who don't like being told what to do, or what to think, by senior managers. This leads to small rebellions against rules brought in to engender equality of opportunity, and accusations of being terrorized by the "thought police" (Dobbin and Kalev 2016).

It is also arguable whether outreach activities designed to attract those from under-represented groups into a discipline – in our case designed to attract women and girls to consider study and careers in music production/audio engineering – are actually working. Despite years of work on attracting women into engineering, the Royal Academy of Engineering and the IET (under the umbrella of Engineering UK) have not seen any substantial increase in the proportion of female engineering students. It is still stubbornly low at 15% and hasn't improved since 2012 (National Centre for Universities and Business 2019).

Indeed, when considering STEM workforce and looking at the rate of publication in scientific journals by gender, Holman et al. (2018) predict that at the current rate of progress and without some impactful interventions, it would take 258 years for the gender ratio of senior physicists to achieve parity! This is just one example in a discipline where gender equality work has been established for a number of years (Institute of Physics 2019).

Within music production, if we simply calculate the linear trend line using data on paper authorship at AES conferences since 2012 (Figure 14.4), it would take until the year 2105 to reach 50% male authorship (note, this is calculated using only eight years of data).

STRATEGIES FOR DRIVING CHANGE

Considering the current gender bias in music production across all levels of engagement from education into professions, alongside the existing interventions and initiatives attempting to elicit a change in attitudes and action, this section outlines the most important steps identified by the authors required to drive change in the sector.

Deliberate effort is required by all involved in education and industry to ensure that music production can fully realize its potential as an exciting interdisciplinary dynamic sector, combining art, science, and technology, for participation by all to benefit all.

Understand the Problem: A Data-Driven Approach

Gather Data

A data-driven approach to both understanding the problem and designing solutions is fundamental. We have presented data by gender in this chapter to illustrate the current state of music technology education and the audio engineering industry. However, much of this data was difficult to obtain since it is not readily available, partly as a result of the interdisciplinary nature of the field and the variety of routes into music production. It is imperative that data is collected and made available throughout processes across the industry. Whether concerned with education or professions, this needs to include gender data from expression of interest stage (where possible) through to recruitment, retention, and progression.

Recommendation: Gather and publish data on women's participation in music production, audio engineering, and music technology education.

Understand the Causes of Under-Representation

Collecting quantitative data is only the first step in starting to address the issues behind under-representation. Numbers only tell us so much; qualitative data is also needed to reflect the lived experiences of women in the industry and the variety of career journeys they've taken. Only with both quantitative and qualitative data can we start to properly understand the barriers to inclusion.

It is easy to assume that women don't participate in particular subjects and industries through a personal choice or a lack of interest, but this may be in effect confusing the symptom for the cause. In fact, there are multiple larger factors related to biological and societal gendered expectations that mean not all choices are freely made (Ceci and Williams 2011), including familial influences and parental expectations, peer norms, the awareness of being outnumbered in a male-dominated environment, and the perception that engineering just isn't for girls (Andrews and Clark 2012b; Dasgupta and Stout 2014). This also extends to awareness of work environments, including whether the hiring process is biased, whether an industry or department is inclusive or not, and whether one would be able to maintain existing work and family commitments (Dasgupta and Stout 2014).

Work such as this should be extended to music production to examine whether these factors are also at play here, and to what extent.

Recommendation: Undertake and publish more research on the causes of women's under-representation in music production (this book is a good start!).

Find Out What Works

There are a number of professional bodies and organizations who have been championing gender equality in STEM disciplines in higher education (e.g. Athena SWAN, Royal Society of Chemistry, Institute of Physics, Royal Academy of Engineering) over the last 20 years. The Audio Engineering Society has also recently strengthened its efforts with the formation of the Diversity and Inclusion Committee.[18] Music production is an interdisciplinary field, and is therefore not covered by one professional body alone – but here is an opportunity to learn from work on gender equality in the STEM disciplines.

Unfortunately, there is little systematic research and reliable evidence on what actually works in changing culture and fixing gender imbalance. Many outreach activities raise the profile of the discipline, but it has not yet been shown that targeted outreach activities don't just appeal to those who might otherwise be interested even without the intervention.

Nevertheless, researchers are now beginning to collect and evaluate data on the effectiveness of gender equality interventions. Dobbin and Kalev

(2016) suggest that the most effective interventions include: engaging managers in solving the problem; promoting the desire to look and be fair-minded; using targeted recruitment; mentoring; and making training in diversity voluntary rather than mandatory. Dasgupta and Stout (2014) identify a number of evidence-based recommendations for tackling under-representation of women in STEM fields including: educational collaboration between schools and universities, informal STEM workshops, mentoring, networking, blind review of applications, and supporting work-life balance.

There are now a number of data repositories which can be consulted, such as the Harvard University hosted Gender Action Portal (Harvard Kennedy School Women and Public Policy Program 2019). There are also books (for example, Bohnet 2016) and some freely available online resources with suggestions for evidence-based interventions for changing culture and addressing gender inequality, such as The Center for WorkLife Law (2018), to which those in music production education and industry can refer.

Ultimately, improving the diversity of any organization should not just be seen as a tick-box exercise. It must be regarded as the most important key to best practice in effective management and human resource policy and implementation. But to do this effectively, organizations must heed suggestions as to which inventions actually work.

Recommendation: Undertake and publish research on effective interventions for addressing gender inequality in music production.

Change the Culture

Many gender equality initiatives and outreach schemes focus on encouraging women to get involved in STEM or to adapt their approach to working life and their own values in order to "fit" the prevalent male-dominated industry culture. Such approaches can only have short-term and somewhat fleeting success since ultimately "the very nature of an occupation is discerned and organized around social identities like gender and race" (Ashcraft and Ashcraft 2015). Women should not be required to change themselves in order to fit into the majority-lead culture – the so-called glass slipper problem. Rather, the culture of the music industry system needs to change itself so that all are able to participate and thrive.

Set Target/Inclusion Goals

It is clear that, whilst current interventions might be making positive changes, if current trajectories continue we are many generations away from a gender representation which reflects the wider society. One more proactive approach is to set targets for inclusion which stipulate specific numbers or percentages of non-male participation. This may include specified numbers (beyond a target of one) of non-males on strategy/advisory boards, expert panels, within professional teams. and on recruitment short lists. Whilst somewhat controversial, a precedent is beginning to be set by large corporations, for instance, with ITV recently announcing a veto on all-male writing teams (Guardian 2019). Whilst this may initially be set quite low based on the current under-representation of females in the industry, setting short-term to long-term targets which are regularly reviewed could ensure a consistent increase in the trajectory of improved gender representation.

Even within the music industry this is being explored: the Recording Academy (USA) introduced The Producer & Engineer Inclusion Initiative in 2019, which stipulates that at least two females should be included in any hiring considerations. More than 200 artists, labels, producers, and management companies have signed up so far. A webpage dedicated solely to facilitating the process of identifying working female producers and engineers explains why this sort of intervention is necessary:

> We aren't here to tell anyone who to hire, but we have seen repeatedly that the simple act of making sure diverse candidates are always seen and considered makes it more likely that women will get the opportunities they previously have been denied.
> (Recording Academy Task Force 2019)

Recommendation: Set challenging targets for diversity for organizations working in music production and related education.

Majority Group Driving Change

Expecting females to take the lead in designing and delivering effective initiatives for change has been shown to be counterproductive (Ovseiko et al. 2017). When women are given sole responsibility for inclusion agendas, their time is taken up on the gender agenda, which prohibits progression within their actual field. For quick and meaningful change to happen, males need to not only be on board as allies but also take proactive roles in pushing the agenda. This includes listening to under-represented groups and understanding their stories (evidenced through the data), but not expecting them to fix the problem. Campaigns such as HeForShe[19] (the United Nations' global movement for men and people of all genders to stand in solidarity with women) represent a strong step forward in this regard, and represents a positive attitude amongst males to take a proactive role in changing our culture and understanding the need of males to be seen to be allies. Dan Rowson commented on Twitter:[20] "If you're not actively challenging and dismantling inequalities, you're playing a part in their continuation".

It has been shown that women working in academia already get less credit for doing the same amount of work as their male peers and are overlooked (Macaluso et al. 2016; Feldon et al. 2017). Expecting women to then also take on the extra burden of fixing gender equality within the industry can be massively detrimental to their careers, and only exacerbates the gender imbalance problem. Having to act as a role model, mentor, and networking facilitator in addition to working hard to gain recognition is too much to ask from the small numbers of women in the industry. It is just one of the many "double-binds" that those in a minority face in the workplace (Ballenger 2010). Additionally, this perpetuates the culture that women take on more of the academic "housework" than their male colleagues (Guarino and Borden 2017).

In line with our previous recommendations concerning the introduction of targets, a powerful message here would be for administrative responsibilities and roles (such as chairing of equality and diversity committees) to be more equally shared between women and men, and for those men to be visible in advancing the agenda. This also extends to the running of targeted outreach activities: whilst outreach specifically targeting women is positive with clear benefits of such activities being led by women, this again creates an extra burden on the minority. Fixing the gender problem should not, and cannot, be left as the sole responsibility of the minority group.

Recommendation: The majority group in music production (men) take responsibility for driving change, as well as acting as allies to their female colleagues.

Work to Break Down Sexism in the Industry

The way that the music industry thinks about and treats women is the biggest barrier to their inclusion and progression in music production. Women in the industry report that their skills are discounted or dismissed by colleagues, their work is not taken seriously and not recognized, and they have to prove themselves in order to earn respect (Prior et al. 2019; Smith et al. 2019). Continuing to raise the profile of women in the industry, offering mentoring, and raising awareness of gender issues are good initiatives, but more must be done within the industry to address stereotypical views of women.

In addition to feeling professionally overlooked, women are typically stereotyped and sexualized: they report being "subject to innuendo, undesired attention, propositioned, valued for their appearance, and even an awareness or fear of being personally unsafe in work situations" (Prior et al. 2019; Smith et al. 2019: 9) or even subject to "constant sexual harassment" (Prior et al. 2019; Smith et al. 2019: 19).

With regards to the higher education sector, the 1752 Group recommend a list of strategic priorities to start to address sexual misconduct (The 1752 Group 2017). The first of these is to establish an enforceable code of conduct which makes professional boundaries clear. A mechanism for those affected to report sexual misconduct/harassment is key. It is also crucial for workplace leaders to raise awareness of the very damaging cumulative effect of gender microaggressions, the prevalence of subtle and not-so-subtle gender bias, and to provide guidance for those who witness gender discrimination on how they should respond (Basford et al. 2014; Jones et al. 2016). Those who experience sexism (in whatever form) must feel like action will be taken if they report it.

Recommendation: Adopt a zero-tolerance approach to sexual harassment and gender microaggressions and implement a robust and sensitively handled reporting mechanism.

Code of Best Practice for Music Production

We are presented with a timely opportunity to build on the positive forward steps being made across our culture in gender diversity, alongside the initiative and campaigns that directly target the music production industry that have been discussed so far.

Recommendation: Establish a code of best practice for the music production industry which addresses the following:

Continue Effective Interventions

Whilst we have highlighted that perhaps not all interventions perform as desired, there is evidence that some do have an effect, including visibility of role models, networking, and mentoring schemes.

The lack of role models in male-dominated sectors is often quoted as one of the barriers to participation in STEM (Valian 1999; Dasgupta and Stout 2014; Prior et al. 2019; Smith et al. 2019). Increasing the visibility of role models helps to break down stereotypes and inspire the next generation of women to choose to work in music production.

Targeted schemes which provide structured mentoring and networking for women and other under-represented groups have been shown to be effective in supporting career progression (Dasgupta and Stout 2014; Dobbin and Kalev 2016). Approximately 54% of women reported that access to networking opportunities had a positive impact on their careers (Smith et al. 2019), and 92% of mentored women felt that it had contributed to their careers (Prior et al. 2019). Some great examples of current mentoring schemes are those run by organizations such as She Is The Music (USA) and the Equal Studio Residency Program for women run by Spotify (Spotify 2018). Effective interventions must be supported and encouraged to continue.

Recommendation: Continue effective interventions to support gender equality at all levels.

Intentional Diversity

A code of best practice and policy on inclusion and diversity is a good first step for organizations wishing to improve the gender balance in music production. However, policies are too often written and stay on the shelf, little read, and never fully implemented in day-to-day operations.

A key recommendation from women in the US music industry surveyed in 2019 is to increase the focus on diversity in the workplace: "Focusing on increasing diversity through providing equal opportunity was the most common recommendation from women about how to improve inclusion in the music industry" (Prior et al. 2019)

We therefore urge all in music production to practice "intentional diversity". This means:

- Being open about the need for greater diversity in music production.
- Being honest about the current state of industry.
- Being realistic about the difficulties and challenges involved in changing cultures.

Establishing a safe space for discussion of issues around diversity and inclusion is a key first step to making sure that minority voices are heard, respected, and included in conversations about strategic change. Working in this way means that the "double bind" that women are often the ones left to fix gender equality (as discussed earlier) can be avoided.

Responsibility for delivering equality and diversity objectives should be included in role descriptions of all staff, and a commitment to the equality agenda must be an explicit expectation which is outlined in job adverts, person specifications, and interview processes. Accordingly, effective training programs should be included as part of continuing professional development to support this key element of everyone's role.

Diversity and inclusion needs to not only "be on the agenda", but also, importantly, it needs to be seen by all to be on the agenda, with those in positions of power, influence, and leadership driving the fight for change. A recent excellent example of intentional diversity is the BBC Research and Development division policy on diversity in public speaking,[21] which aligns with the BBC's overarching diversity and inclusion policy, but also gives more specific advice and recommendations for those working in R&D on how to foster inclusion and be a part of positive change.

Of course, working to change the culture of any industry or workplace is a huge task, and there is no silver bullet or one key intervention which will see rapid results. Establishing a more inclusive space for all is hard work, and all involved need to be honest about this challenge. With time and sustained effort, the explicit task of being more inclusive will become commonplace, and the focus will turn less on outreach and inclusion, and instead to making sure that all those involved in the industry find it to be a place where they feel they belong.

Recommendation: Be open, honest, and proactive in efforts to increase diversity and inclusion.

Move From Outreach and Inclusion to Belonging

A number of campaigns are now embedding active gender diversity across programs without explicitly targeting only females, often through diversifying the image of engineering. The children's *Hello World* magazine,[22] for example, which can be utilized by schools as part of outreach activities, is carefully edited and curated to make sure that diverse images are used to break down stereotypes of computer science. To this end, it always includes diverse images of representation of females in school with cross-curricular contents, including projects which establish computing as an interdisciplinary field which includes technology, art, making, computer science, and STEM/STEAM education. It is freely available online and has a global reach (Hello World 2019).

Extending this idea, Engineering UK's "This is Engineering" campaign[23] seeks to highlight that engineering is for all, and works to break down outdated and stereotypical ideas about what engineering is and who engineers are. Whilst one goal of the campaign is to attract females into engineering, it has a broader remit to change the stereotypical image of engineering. By portraying diversity in all aspects, it hopes to encourage representation across all traditionally under-represented groups.

A social-belonging intervention trialed at a US university engineering program (Walton et al. 2015) improved academic grades by about 10% for female students in this male-dominated discipline. This demonstrates the importance of efforts to improve a sense of social belonging in music production and related subjects beginning early in education. Social marginalization contributes to gender inequality not only in the workplace but also in educational settings, and it is essential that interventions start there.

Recommendation: Encourage diversity in all forms.

Commitment to Diversity Supported Through Accreditation

Many UK Music Technology degrees are accredited by a relevant professional body, either the Institution of Engineering and Technology (IET) or Joint Audio Media Education Support (JAMES)[24]. Neither of these bodies makes any explicit reference to equality, diversity, inclusion, or issues around widening access to the discipline.

A recent discussion document produced by David Ward (JAMES founder and Executive Director) and Phil Harding (JAMES Vice Chairman) highlights the opportunity for music technology to be more fully embedded in the school curriculum; one of the key considerations here is not only a diversity of musical genre, but also diversity of those involved in making and teaching music with technology: "This could dispel the outdated notion that only boys are interested in technology and sciences, as an example, as well as the lack of music education in deprived areas".[25]

Recommendation: Professional bodies who accredit music technology degrees should include diversity and inclusion considerations as part of the accreditation criteria.

Education

A sense of belonging is established early in educational settings, with female pupils feeling marginalized at a young age from engineering and technology-based subjects (Colley et al. 1994; Cheryan et al. 2009; Armstrong 2011; Born and Devine 2015; Department for Education 2019). As such, all those involved in audio/music technology education, at the primary, secondary, and tertiary levels, need to actively engage in the process of change.

Strong Foundation of Gender Diversity for the Next Generation

Addressing the gender imbalance in subjects studied at school will allow us to harness the potential of music technology as an interdisciplinary field (combining arts and science) to establish greater representation and diversity of those coming into the music production industry in the future. In 2015, Born and Devine argued that degree-level education in music technology has the potential to produce graduates who are "equipped for and can be inserted into a host of new technical and professional occupations in the burgeoning, intersecting fields of music, sound and audio, IT, design, and the other media and arts" (Born and Devine 2015).

However, it is clear that those entrusted with education in music and music technology in schools need to address the gendered expectations and stereotypes around music technology and STEM education. As highlighted in our earlier section, many of the issues around representation at all levels of the music industry are rooted in school and the factors which drive female students away from participation in music technology.

We can only achieve this change through some dedicated partnership working between those involved in school education, further and higher education, working together with industry partners. Good practice on gender equality has been established through the Gender Action Schools Award,[26] which works to put gender equality at the heart of schools' policy and everyday practice, in order to challenge gender stereotyping.

Building on such initiatives but specifically focusing on music and music technology would enable music production to benefit from the experience of others working in this space, for example, school-industry partnerships supported through the Women's Engineering Society and the Royal Academy of Engineering.

A new program has recently been announced, which will investigate effective interventions in engaging female students in computer science at GCSE and A Level, which is a joint initiative funded by the Department for Education in England bringing a number of key partners together.[27]

A similar joined-up dedicated campaign around music production/music technology would ensure that the interdisciplinary nature of the subject could be understood and embraced by all: this would enable both schools and the industry to break down inequalities, to inform students and teachers about the numerous exciting and stimulating roles in the music industry, and in turn helping to change the industry itself.

Recommendation: Formation of a national education-industry partnership to promote gender equality in music technology.

Establish a Gender-Inclusive Curriculum for Music and Music Technology

As yet there is little guidance for school teachers or those in further and higher education (post-16 education) on how to ensure that music and music technology teaching is inclusive. Recently published guidance on inclusive curricula for engineering and physical sciences (e.g. Hughes (2015) can offer ideas transferable to music technology. Much can be learned from those at HE who have recently established more inclusive engineering programs (Salminen-Karlsson 2002; Mills *et al.* 2008; Peters 2011), which involve critical evaluation of all aspects of the curriculum including subject content, teaching styles, marketing materials, wall displays, etc.

Change will not be effected merely by a change in curriculum content since "gender inclusivity has to be thought of as a continuous process" (Mills et al. 2008), but industry partners and higher education educators working together with schools, teacher trainers, and exam boards must be the starting point for establishing a more gender-inclusive music technology curriculum at all levels.

Recommendation: Guidance provided for educators at all levels on inclusive curriculum and teaching practice.

CONCLUSION

We have shown that gender inequality persists in education and the industry around music production, despite the great potential and opportunity for change towards a more gender-diverse discipline, which is ripe to be harvested.

Nevertheless, without strong commitment, keen awareness, and positive engagement this exciting opportunity, signaled by emerging signs of change, and arising from the potential for a closer interdisciplinary working afforded by the development of new technologies, could be squandered.

It is worth keeping a future vision in mind – where would we like to be in 5, 10, 15 years' time? Perhaps in a world where an all-female spacewalk isn't news, where an all-female music production team isn't news, where we don't need safe spaces for women to work in music production, and where an interview with a female music producer doesn't garner comment on their level of attractiveness.

To achieve this vision we need to be bolder and adopt active approaches to gender diversity so that the future of music production is one where it's equally as likely that today's children become the next generation of Music Producers whatever their gender.

NOTES

1. https://tibbakoi.github.io/aesgender/
2. www.womeninmusic.org/stats.html
3. www.wisecampaign.org.uk/
4. www.wes.org.uk/
5. http://hernoise.org/
6. http://deliaderbyshireday.com/dd-archive/
7. https://sheisthemusic.org/
8. www.womensaudiomission.org/

9. www.analogtara.net/wp/projects/pink-noises/
10. https://yorkshiresoundwomen.com/
11. www.womeninmusic.org/
12. https://sonawomen.co.uk/
13. http://wiswos.com/
14. For an extensive but not exhaustive list, see: https://drlizdobson.com/2018/02/18/feministsoundcollectives/
15. https://twitter.com/kejames/status/1111122027652546561
16. https://web.archive.org/web/20181104004611/www.midiplus.com/html/mirror.html
17. www.midiplus.com/html/mirror.html
18. www.aes.org/community/diversity/
19. www.heforshe.org
20. https://twitter.com/d_rowson/status/1129080507415838722?s=12
21. https://downloads.bbc.co.uk/rd/pubs/bbc-rd-diversity-public-speaking-policy-v1.0-jul-2019.pdf
22. https://helloworld.raspberrypi.org/
23. www.thisisengineering.org.uk/
24. www.jamesonline.org.uk/
25. www.jamesonline.org.uk/resources/JAMES-Music-Technology-in-Education-2018-discussion-Doc.pdf
26. www.genderaction.co.uk/
27. https://blog.teachcomputing.org/new-gender-balance-in-computing-project-announced

REFERENCES

Alex Ruthmann, S. and Mantie, R. (2017) *The Oxford Handbook of Technology and Music Education.* Oxford: Oxford University Press.

Andrews, J.E. and Clark, R.P. (2012a) 'No one said girls could do engineering': A fresh look at an old problem. *In: International Conference on Innovation, Practice and Research in Engineering Education (EE2012)*, Coventry, UK, 18–20 September 2012.

Andrews, J.E. and Clark, R.P. (2012b) Breaking down barriers: Teenage girls' perceptions of engineering as a study and career choice. *In: Presented at the SEFI 40th Annual Conference 2012, European Society for Engineering Education (SEFI).*

Armstrong, V. (2011) *Technology and the Gendering of Music Education.* Farnham, Surrey and England: Routledge.

Ashcraft, K.L. and Ashcraft, C. (2015) Breaking the 'Glass Slipper': What diversity interventions can learn from the historical evolution of occupational identity in ICT and commercial aviation. *In:* V. Schafer and B.G. Thierry, eds. *Connecting Women: Women, Gender and ICT in Europe in the Nineteenth and Twentieth Century.* Cham: Springer International Publishing, 137–155. doi: 10.1007/978-3-319-20837-4_9.

Ballenger, J. (2010) Women's access to higher education leadership: Cultural and structural barriers. *Forum on Public Policy Online*, 2010(5), 1–20. ERIC. Available from: https://forumonpublicpolicy.com/vol2010no5/archivevol2010no5/

ballenger.pdf; https://forumonpublicpolicy.com/vol2010no5/womencareers 2010.html

Basford, T.E., Offermann, L.R. and Behrend, T.S. (2014) Do you see what I see? Perceptions of gender microaggressions in the workplace. *Psychology of Women Quarterly*, 38 (3), 340–349.

Bernardini, N. and de Poli, G. (2007) The sound and music computing field: Present and future. *Journal of New Music Research*, 36 (3), 143–148.

Boehm, C. (2007) The discipline that never was: Current developments in music technology in higher education in Britain. *Journal of Music, Technology and Education*, 1 (1), 7–21.

Boehm, C., Hepworth-Sawyer, R., Hughes, N. and Ziemba, D. (2018) The discipline that became: Developments in music technology in British higher education between 2007 and 2018. *Journal of Music, Technology and Education*, 11 (3), 251–267.

Bohnet, I. (2016) *What Works: Gender Equality by Design*. Cambridge, MA: Harvard University Press.

Born, G. and Devine, K. (2015) Music technology, gender, and class: Digitization, educational and social change in Britain. *Twentieth-Century Music*, 12 (2), 135–172.

Brereton, J.S., Daffern, H., Green, M.C., Stevens, F.S. and Hunt, A.D. (2019) Enhancing student employability through innovative programme design: A case study. *In*: M. Lopez and D. Walzer, eds. *Audio Education: Theory, Culture and Practice*. Waltham, MA: Focal Press, (Routledge).

Burrelli, J. (2008) *Info Brief: Science Resources Statistics*. Directorate for Social, Behavioral, and Economic Sciences.

Ceci, S.J. and Williams, W.M. (2011) Understanding current causes of women's underrepresentation in science. *Proceedings of the National Academy of Sciences of the United States of America*, 108 (8), 3157–3162.

Ceci, S.J., Williams, W.M. and Barnett, S.M. (2009) Women's underrepresentation in science: Sociocultural and biological considerations. *Psychological Bulletin*, 135 (2), 218–261.

The Center for WorkLife Law (2018) Bias interrupters: Small steps, big change [online]. *Bias Interrupters*. Available from: https://biasinterrupters.org/ [Accessed July 12, 2019].

Cheryan, S., Plaut, V.C., Davies, P.G. and Steele, C.M. (2009) Ambient belonging: How stereotypical cues impact gender participation in computer science. *Journal of Personality and Social Psychology*, 97 (6), 1045–1060.

Colley, A., Comber, C. and Hargreaves, D.J. (1994) Gender effects in school subject preferences: A research note. *Educational Studies*, 20 (1), 13–18.

Collins, N. (2012) Science 'girl thing' video branded offensive. *The Daily Telegraph*, 22 June.

Dasgupta, N. and Stout, J.G. (2014) Girls and women in science, technology, engineering, and mathematics: STEMing the tide and broadening participation in STEM careers. *Policy Insights from the Behavioral and Brain Sciences*, 1 (1), 21–29.

Daubney, A. and Mackrill, D. (2018) *Changes in Secondary Music Curriculum Provision over time 2016–18. Summary of the Research by Dr Ally Daubney and Duncan Mackrill*. University of Sussex.

Department for Education (2019) *Attitudes Towards STEM Subjects by Gender at KS4 Evidence from LSYPE2*. Department for Education.

Dobbin, F. and Kalev, A. (2016) Why diversity programs fail. *Harvard Business Review*. Available from: https://hbr.org/2016/07/why-diversity-programs-fail

Duffy, M. (2019) New research shows how playing music can improve your life [online]. *Fender.com*. Available from: www.fender.com/articles/play/new-research-shows-mental-physical-and-emotional-benefits-of-playing-music [Accessed July 7, 2019].

Edexcel, P. (2013) *Specification: GCE Music Technology*. Harlow, Essex: Pearson Education Limited.

Ely, K. (2015) The world is designed for men: How bias is built into our daily lives [online]. *Medium*. Available from: https://medium.com/hh-design/the-world-is-designed-for-men-d06640654491 [Accessed July 15, 2019].

Faulkner, W. (2000) The power and the pleasure? A research agenda for 'making gender stick' to engineers. *Science, Technology & Human Values*, 25 (1), 87–119.

Feldon, D.F., Peugh, J., Maher, M.A., Roksa, J. and Tofel-Grehl, C. (2017) Time-to-credit gender inequities of first-year PhD students in the biological sciences. *CBE Life Sciences Education*, 16 (1).

Gearnews (2018) Ladies, your audio interface has arrived and it's called mirror [online]. *Gearnews.com*. Available from: www.gearnews.com/ladies-your-audio-interface-has-arrived-and-its-called-mirror/ [Accessed July 7, 2019].

Gill, T. (2010) *GCSE Uptake and Results, by Gender 2002–2009 (Statistics Report Series No.22)*. Cambridge Assessment.

Gill, T. (2012) *A-Level Uptake and Results, by Gender 2002–2011 (Statistics Report Series No. 48)*. Cambridge Assessment.

Gill, T. (2016a) *GCSE Uptake and Results, by Gender 2005–2014 (Statistics Report Series No.98)*. Cambridge Assessment.

Gill, T. (2016b) *A-Level Uptake and Results, by Gender 2005–2014 (Statistics Report Series No.97)*. Cambridge Assessment.

Goode, J., Estrella, R. and Margolis, J. (2006) Lost in translation: Gender and high school computer science. *In*: J. Cohoon and W. Aspray, eds. *Women and Information Technology*. Cambridge, MA: The MIT Press, 89–114.

The 1752 Group (2017) Strategic priorities [online]. *1752Group*. Available from: https://1752group.com/strategicpriorities/ [Accessed July 15, 2019].

Guardian (2019) Head of ITV comedy drops all-male writing teams, June 18, 2019.

Guarino, C.M. and Borden, V.M.H. (2017) Faculty service loads and gender: Are women taking care of the academic family? *Research in Higher Education*, 58 (6), 672–694.

Harvard Kennedy School Women and Public Policy Program (2019) Gender action portal [online]. *Harvard Kennedy School Women and Public Policy Program*. Available from: http://gap.hks.harvard.edu/ [Accessed June 13, 2019].

Heilman, M.E. and Welle, B. (2005) Formal and informal discrimination against women at work: The role of gender stereotypes. *Center for Public Leadership Working Paper Series;05–02*, 24–40

Hello World [online] (2019) *Hello World.* Available from: https://helloworld.raspberrypi.org/ [Accessed July 15, 2019].

HESA (2019) What do HE students study? [online]. *HESA*. Available from: www.hesa.ac.uk/data-and-analysis/students/what-study [Accessed July 15, 2019].

Hewlett, S.A., Luce, C.B., Servon, L.J., Sherbin, L., Shiller, P., Sosnovich, E. and Sumberg, K. (2008) The athena factor: Reversing the brain drain in science, engineering, and technology. *Harvard Business Review*. Available from: https://store.hbr.org/product/the-athena-factor-reversing-the-brain-drain-in-science-engineering-and-technology/10094?sku=10094-PDF-ENG.

Holman, L., Stuart-Fox, D. and Hauser, C.E. (2018) The gender gap in science: How long until women are equally represented? *PLoS Biology*, 16 (4), e2004956.

Hughes, M. (2015) *Embedding Equality and Diversity in the Curriculum: A Physical Sciences Practitioner's Guide.* York: Higher Education Academy.

Institute of Physics (2019) Project Juno [online]. *IOP*. Available from: www.iop.org/policy/diversity/initiatives/juno/index.html [Accessed July 15, 2019].

Joint Council for Qualifications (2019a) Examination results: GCSEs [online]. *Examination Results: GCSEs*. Available from: www.jcq.org.uk/examination-results/gcses [Accessed June 7, 2019].

Joint Council for Qualifications (2019b) Examination results: A-Levels [online]. *Examination Results*. Available from: www.jcq.org.uk/examination-results/a-levels [Accessed June 7, 2019].

Jones, K.P., Peddie, C.I., Gilrane, V.L., King, E.B. and Gray, A.L. (2016) Not so subtle: A meta-analytic investigation of the correlates of subtle and overt discrimination. *Journal of Management*, 42 (6), 1588–1613.

Knott, J. (2018) EBacc blamed for growing gender imbalance in GCSE choices [online]. *ArtsProfessional*. Available from: www.artsprofessional.co.uk/news/ebacc-blamed-growing-gender-imbalance-gcse-choices [Accessed June 12, 2019].

Macaluso, B., Larivière, V., Sugimoto, T. and Sugimoto, C.R. (2016) Is science built on the shoulders of women? A study of gender differences in contributorship. *Academic Medicine: Journal of the Association of American Medical Colleges*, 91 (8), 1136–1142.

Mathew, M., Grossman, J. and Andreopoulou, A. (2016) Women in audio: Contributions and challenges in music technology and production. *Audio Engineering Society Convention*, 141.

McLeish, T. (2019) *The Poetry and Music of Science: Comparing Creativity in Science and Art*. Oxford: Oxford University Press.

Mills, J.E., Ayre, M.E. and Gill, J. (2008) Perceptions and understanding of gender inclusive curriculum in engineering education. In: *SEFI Annual Conference 2008*, Aalborg, Denmark, July 2–5.

Mixmag, (2018) Forget high fidelity: How women are reclaiming record stores [online]. *Mixmag*. Available from: http://mixmag.net/feature/forget-high-fidelity [Accessed May 21, 2018].

Murphy, D.T., Brereton, J.S. and Brookes, T. (2006) SpACE-Net: The spatial audio creative engineering network. In: *AES 28th International Conference, The Future of Audio Technology, Surround and Beyond*. Piteå, Sweden, 279–282.

National Centre for Universities and Business (2019) See our talent 2030 dashboard (2015) [online]. *NCUB*. Available from: www.ncub.co.uk/reports/talent-2030-dashboard-2015.html [Accessed July 12, 2019].

Onanuga, T. (2019) Virtual reality: How women are taking a leading role in the sector [online]. *The Guardian*. Available from: www.theguardian.com/careers/2019/may/28/virtual-reality-how-women-are-taking-a-leading-role-in-the-sector [Accessed July 15, 2019].

Ovseiko, P.V., Chapple, A., Edmunds, L.D. and Ziebland, S. (2017) Advancing gender equality through the Athena SWAN charter for women in science: An exploratory study of women's and men's perceptions. *Health Research Policy and Systems/BioMed Central*, 15 (1), 12.

Pearson Qualifications (2019) Results and certification: Grade statistics [online]. *Pearson Qualifications*. Available from: https://qualifications.pearson.com/en/support/support-topics/results-certification/grade-statistics.html?Qualification-Family=A-Level [Accessed July 7, 2019].

Peters, J. (2011) *Inclusive Curriculum Design in Higher Education Engineering*. London: Royal Academy of Engineering.

Prior, B., Barra, E. and Kramer, S. (2019) *Women in the U.S. Music Industry: Obstacles and Opportunities*. Boston, MA: Berklee Institute for Creative Entrepreneurship.

PRS Foundation (2018) *Key Change Manifesto: Recommendations for a Gender Balanced Music Industry*. London: PRS Foundation.

Recording Academy Task Force (2019) Recording academy task force on diversity and inclusion announces industry-wide inclusion initiative to expand opportunities for female producers and engineers [online]. *grammy.com*. Available from: www.grammy.com/press-releases/recording-academy%E2%84%A2-task-force-diversity-and-inclusion-announces-industry-wide [Accessed July 19, 2019].

Salminen-Karlsson, M. (2002) Gender-inclusive computer engineering education: Two attempts at curriculum change. *International Journal of Engineering Education*, 18 (4), 430–437.

Sánchez de Madariaga, I., Chalude, M., Suzanne, de C., and Rice, C. (2012) Science: It's a girl thing! Statement from members of the gender expert advisory group [online]. *curt-rice.com*. Available from: http://curt-rice.com/2012/06/26/science-its-a-girl-thing-statement-from-members-of-the-gender-expert-advisory-group/ [Accessed July 7, 2019].

Savonick, D. and Davidson, C.N. (2016) Gender bias in academe: An annotated bibliography of important recent studies [online]. *Impact of Social Sciences*. Available from: https://blogs.lse.ac.uk/impactofsocialsciences/2016/03/08/gender-bias-in-academe-an-annotated-bibliography/ [Accessed July 13, 2019].

Serra, X., Bresin, R. and Camurri, A. (2007) Sound and music computing: Challenges and strategies. *Journal of New Music Research*, 36 (3), 185–190.

Shepherd, I. (2009) What does a music producer do, anyway? – Production advice [online]. *Production Advice*. Available from: http://productionadvice.co.uk/what-is-a-producer/ [Accessed June 12, 2019].

Smith, S.L., Choueiti, M. and Pieper, K. (2019) *Inclusion in the Recording Studio? Gender and Race/Ethnicity of Artists, Songwriters and Producers across 700*

Popular Songs from 2012–2018. Los Angeles, CA: USC Annenberg Inclusion Initiative. Available from: http://assets.uscannenberg.org/docs/inclusion-in-the-recording-studio.pdf

Spotify (2018) Announcing the Equal (EQL) studio residency program for women – Spotify [online]. *Spotify*. Available from: https://newsroom.spotify.com/2018-08-15/announcing-the-equal-eql-studio-residency-program-for-women/ [Accessed July 15, 2019].

Sutch, T. (2014) *Uptake of GCE AS Level Subjects 2007–2013*. Research Division Assessment, Research and Development Cambridge Assessment.

This is Engineering (2018) This is engineering | Home [online]. Available from: www.thisisengineering.org.uk/ [Accessed July 15, 2019].

UK Music (2018) *Diversity Report*. UK Music.

Valian, V. (1999) *Why So Slow? The Advancement of Women*. Cambridge, MA: MIT Press.

Walton, G.M., Logel, C., Peach, J.M., Spencer, S.J. and Zanna, M.P. (2015) Two brief interventions to mitigate a 'chilly climate' transform women's experience, relationships, and achievement in engineering. *Journal of Educational Psychology*, 107 (2), 468–485.

Wolfe, P. (2019) *Women in the Studio: Creativity, Control and Gender in Popular Music Sound Production*. London: Taylor & Francis.

Young, K., Lovedee-Turner, M., Brereton, J. and Daffern, H. (2018) The impact of gender on conference authorship in audio engineering: Analysis using a new data collection method. *IEEE Transactions on Education*, 1–8.

15

The Female Music Producer and the Leveraging of Difference

Sharon Jagger and Helen Turner

INTRODUCTION: DOES GENDER MATTER IN THE RECORDING STUDIO?

Our research into women and music production has led us to the London studio of a full-time female music producer who has agreed to share her story with us. It sits in a building that houses a cluster of recording studios on a small industrial estate, and as we wait for the security door to open, we discuss whether such austere metal-and-concrete surroundings might have a phenomenological impact upon a woman's experience of working as a music producer. We resist seeing the space as masculine simply because of its industrial appearance; we resist until our interviewee tells us there are no female toilets in the building and most of them are labeled male. We laugh about the obvious semiotics. ("So, where do they put the tampon machine?" we ask.) Testing whether "music producer" and the studio space are imaginatively male is one of the aims of our inquiry, and the clue seems to be (rather prosaically) embedded within the material world of these purpose-built recording studios where references to maleness are literally nailed to the doors.

We have stumbled on a "somatic norm" (Puwar, 2004: 13), where male bodies belong in a constructed space and female bodies do not. This norm can also be said of the discourse in general: "music producer" conjures the image of the male, directing studio proceedings, in fraternity with the male engineer, who controls the (masculine) technology. If music technology and the production process have accrued masculine meaning, how does a female music engineer/producer come to develop a sense of belonging? Research suggests that when women enter a male-dominated profession in small numbers they are sometimes regarded and find themselves behaving as the honorary male to fit in with masculine norms (see Bagilhole, 2002). We want to explore whether this can be said of the music production profession,[1] or whether the "phenomenal environment" (Young, 1990) – that is, the physical and the social world of the music producer – emphasizes and reproduces sex and gender difference.[2] In other words, does gender matter in the technical and creative process of producing recorded music?

We are recording artists with decades of combined experience in the studio environment, yet we have never worked with, nor even met, a female engineer/producer. The truism that few women enter music production or music technology as a profession seems to be substantiated in our UK-based experience, and we had to reach well beyond our usual networks to find women who are full-time professional music producers. The reasons there are so few women professionally active in this part of the music industry seem to be only partially understood, a research lacuna that explains the bafflement in the industry as to why music production is dominated by men (Wolfe, 2012). The stories we have collected indicate there are potential barriers for women to overcome, but our interest also lies in the lived experience of women who have successfully developed careers as producers. There is plentiful research on how music genres reproduce and subvert gender constructions (see, for example, Whiteley, 2000; Downes, 2012; Hill, 2016) and on how music has a history of being encoded with gendered meaning (Citron, 1994), but less has been written about the gendered dynamics of the technical production of recorded music. As we have reflected on the experiences of two women who are full-time professional music producers we have found, intriguingly, that perceptions of sex and gender difference can be consciously and strategically leveraged by women producers as they deliberately cultivate *feminine* approaches to music production.

Our two interviewees are both based in London. Aubrey Whitfield is 38 years old and has been actively involved in the music industry for most of her life, and like many producers she is also a musician and a songwriter. She has produced hundreds of tracks, many of which have entered international charts. Aubrey's years of experience as both an artist and a producer are an important part of her story as she reflects on her approach to the production process:

> I think I wouldn't have been able to do this job in my 20s because I don't think I had enough life experience to deal with all these people. Some of them are so difficult. But now I've got older and I've got my confidence, I think it's really helped me.

This is a clue to the amount of relational work that characterizes Aubrey's approach to producing. Whilst she finds being a music producer fulfilling, Aubrey has experienced the profession as "brutal", requiring a toughness and resilience that she has acquired over the years to build a music production business that values emotional labor as much as the technical and creative process.

Lauren Deakin Davies is 23 years old and is proud to be the youngest female producer to have had tracks played on BBC Radio 2 (one of Britain's most prestigious music stations) and the youngest member of the Music Producers Guild. A musician and recording artist from a young age, Lauren made the transition to producer, building her own studio as well as working in other established commercial studios; she said: "I ended up being hired in [a] studio at 17. . . this was like 'music is my life'. So that's how I got into the roots of production". Lauren describes the industry in benign terms, and she feels fortunate to have "been surrounded by a lot of positive men". Social capital is a feature of Lauren's story, which she recognizes is a way of mitigating the perception that the engineer and producer are always male.

Both women tell remarkably similar stories about their experiences of the music production industry and their approach to the recording process. We do not claim these are universal experiences; however, these commonalities help to understand how gender discourses and perceptions impact on women's experience of the industry and on the production process itself. Aubrey and Lauren are both entrepreneurial, transferring their skills as musicians and songwriters to the role of producer to offer a full package to clients; they write, perform, engineer, and produce music. Both women see the music production industry as egalitarian, largely supportive of the small number of women producers who are active, and with few systemic barriers to female success. This description deepens the question of why so few women are drawn to the profession. There is, though, a leitmotif running through the stories of Aubrey and Lauren that suggests they are required to manage gender difference in the wake of sexist discourses, and both women identify situations where being female brings certain forms of precarity to their status.

We are seeking to understand how femaleness belongs in music production. In this chapter, we discuss three themes that arose out of the narratives of Lauren and Aubrey that reveal how gender impacts the music production process and the experiences of women producers. These themes are: (1) entry points into the profession, and how women might legitimize their status as music producers; (2) the significance of emotional labor and how gender difference is reproduced through social processes; and (3) the ways in which women leverage sex and gender difference as an entrepreneurial strategy.

"LEAVE IT TO THE MEN": MUSIC PRODUCTION AS A NON-TRADITIONAL OCCUPATION FOR WOMEN

Before elaborating further on entry and legitimization, emotional labor, and leveraging of difference, it would be useful to briefly examine the landscape in which we situate this discussion as feminist researchers. Understanding women's relationship to the music production process is not simply a question of counting how many women become professional producers and engineers. We have learned from previous research projects that increasing the numbers of women in a male-dominated profession does not automatically change an androcentric culture (Jagger, 2019). As part of the feminist project to understand and deconstruct the social forces that produce masculine and feminine arenas in music as a cultural process (see Cook and Tsou, 1994), we explore the ways sex and gender differences are reproduced and supported in the studio environment and in the professional world of music production that, at first glance, seems to be oriented around the male.

There are multiple feminist perspectives on what sex and gender difference means, ranging from humanist feminisms that see gender difference as a fiction and biological difference as something to be overcome, to gynocentric feminisms that seek to define and elevate sexual difference on women's own terms (see Young [1990] for a concise discussion on these perspectives). For clarity, when we refer to sex and gender difference in the world of music production, we understand such differences as socially and culturally constructed rather than as innate characteristics of men and women. Lauren and Aubrey also speak in terms of gender differences as learned behaviors rather than essentialized qualities, meaning they are conscious of the different ways they interact with male and female clients and understand that clients might expect a different approach from a *female* music producer. Both women have a sense of gender being a performance, the rules of which are laid down through a lifetime of socialization (see, for example, Butler, 2007). In other words, there are no natural reasons why women might be ill-suited to handling music technology or lack creative authority because they are women, but women may be transgressing gender norms when they enter a technological field. Equally, men are not naturally attuned (forgive the pun) to music technology, but the profession's historical alignments with masculinity create the impression that men are naturally suited to working within it. Given this context, we suggest that Lauren and Aubrey consciously and strategically manage, subvert, and leverage such constructions of femininity and masculinity as they seek belonging in the industry as women.

A woman, once she has managed to gain entrance into a male-dominated industry, is required to negotiate gendered barriers in order to develop her career. As Reimer and Bridewell comment:

> Male discrimination, differential sex role socialization, and institutionalized sexism are often mentioned reasons why women *cannot* or *will not* enter certain occupations. Yet, we know some women do, and are successful at their work. Some women have learned to deal with work related discrimination, on the job sexual harassment, and related personal frustration and exploitation.
>
> (1982: 153)

(For more detailed discussion on women entering non-traditional occupations see also Bagilhole, 2002.) What also complicates women's experience of a non-traditional occupation is the accrual of gendered meaning around words, objects, spaces, traits, and sex difference itself. Music technology makes up a significant part of the producer's toolkit, especially in the context of a small business where producers are also the engineers, as is the case with Aubrey and Lauren. Technologies are associated with gender, whereby men are traditionally considered the producers and women the consumers (Lerman, Oldenziel and Mohun, 2003). Research into the use of music technology (for example, Pegley, 2000) suggests that social and material processes reinforce its alignment with masculine traits, and the studio space is understood as male space (Negus, 1992). It is important to acknowledge these cultural alignments in the context of the morphing of the music production industry into its digital form, which allows self- and entrepreneurial producing to thrive (see Wolfe, 2012). If music production and engineering are dominated by men because of a cultural tendency to see the technical process of creating recorded music as suited to masculinized traits, then digital music technology continues to recreate the music producer as male.

Women, on the other hand, are encouraged to be artists using the "soft" technologies of creativity and emotional engagement (Lerman, Oldenziel and Mohun, 2003). Aubrey sees this unfolding around her in the industry: "I do think the industry is female-centered in terms of the singer-songwriters and artists, and they are encouraged to kind of go into those roles more than the technical roles". She wonders if women are resistant to the idea of producing because it is associated with (masculine) technological engineering in the studio. Not only this, but as Wolfe (2012) points out, the skills of the (often female) singer in the vocal booth are overshadowed by the technical skill displayed by the (usually male) engineer/producer, who has access to power in the studio setting. The notion of power being accessed differently according to the roles played out in the studio certainly chimes with our own experiences as recording artists, and where these roles are aligned to gender, this access to power is also gendered.

Women who disrupt the usual gendered roles in the studio can be subversive. The alignment of music technology and the production process to the masculine is supported by the cultural capital[3] attached to the role – the music engineer and producer have access to power because of their elevated status. As she discusses how the technical intricacies may be a barrier for women, Lauren is intent on deconstructing the mystique that surrounds music technology and production, and in doing so troubles gendered power discourses. Pulling back the curtain, she reveals how the creative magic harnessed to the power of 'cool' is illusory:

> Like, there's very few careers that are actually cooler than being a music producer in public perception. Like, so I think anything to make something cooler. Because in reality you're sitting in a dark fucking room, pressing buttons, slightly altering the snare. That's what your day actually is. . . . You're really told that it's so hard and so technical, just leave it to the men. And I think that's what women do, and when they find it's not hard, it's like. . .

Lauren is subversively diminishing the cool capital of music production and refuses to prop up gendered meanings that have accrued around technology. She is suggesting that the barrier to women is a constructed one and that a sense of masculine distinction is being generated through discourses of male technical prowess. Women, she ventures, are culturally discouraged from assuming they have the right skills to enjoy and be successful in music technology. Whilst undoubtedly there is skill and knowledge required to engineer, Lauren is deliberately uncoupling the "hard" technology from gendered alignments that direct power towards the male.

Both Lauren and Aubrey recognize their presence in the industry is juxtaposed against these masculine norms and associations. Does this context present obstacles for women once they have managed to enter the industry? A US study by Reimer and Bridwell (1982) suggests that women entering traditionally male occupations face three significant barriers to belonging: male skepticism, sexual harassment, and feelings of personal inadequacy. To varying degrees, all of these barriers are experienced by Aubrey and Lauren as they pursue their careers in music production. Whilst both women face sexism that many women generally experience, we focus on those experiences that are specifically related to their work life. The stories of Aubrey and Lauren reveal that male skepticism and feelings of inadequacy are significant issues that require the cultivation of resilience and are connected to how both women find legitimacy as professional producers.

Male skepticism can take the form of initial dismissal of a woman in the studio environment, as Lauren has noticed:

> When I first started working [in the studio] a couple of people were doing rehearsals and . . . they would assume I was a secretary or something. And I'm like, I'm literally a multi-award-winning engineer and producer, this is so frustrating. . . . Or like people ring up and they go like, "Oh, I want an engineer, can I speak to one of the men, please? Because I want to speak to an engineer". And I'm like, "I'm an engineer as well". So, it's stuff like that that still happens.

Encounters like these reveal how Lauren is required to continually re-establish her credentials, which in turn suggests that the studio space has been originally developed as space for men, containing roles that are highly gendered; the female entrant is initially perceived as auxiliary. The need to say "I'm an engineer" to reinforce Lauren's place at the desk highlights the dissonance between the female and the role in the imagination. This is enforced materially where the physical space is shaped around the male body, as we noted in the introduction with, for example, the absence of female toilets; women become "bodies out of place" (Ahmed, 2000; Puwar, 2004) when they do not fit in with the assumptions made about what a woman should be *doing* in the studio (singing, administration, cleaning, making the tea). This is a hidden skepticism that women face because they do not *look* like an engineer or producer.

For Aubrey, the male skepticism is more overt and consciously expressed in the form of hostility from male clients:

> I've had a lot of trouble with middle-aged men who are my clients . . . when I get very rude, harsh emails through . . . I thought about it and I looked at these quotes [from emails] and every single one of them was from a middle-aged man, which I thought was interesting.

The pattern Aubrey has noticed in how her work is judged can be interpreted as both a gendered and generational conflict; the middle-aged male here is potentially more likely than younger men, or women, to display hostility, condescension, and disrespect, which Aubrey suspects is related to her being female. Aubrey highlights a suspicion that gendered power asymmetries can be compounded by age difference; her experience is of the older man discrediting her status, whilst she draws confidence from her longevity in the industry. Such male skepticism implies that women producers need to outperform male counterparts to overcome (sometimes unconscious) beliefs that women are not "naturally" suited to the technical production process.

Perhaps more insidious and undermining than male skepticism are the feelings of inadequacy women experience in male-dominated fields. Such personal anxieties are clearly expressed by Lauren, who identifies the "imposter syndrome" as a typically female response to forms of skepticism. The sense of marginalization that the "imposter syndrome" fosters is fundamentally connected to how gender is framed in difference, which impacts on how women earn their credentials. Whilst both Lauren and Aubrey have chosen individualized paths and have strategically used their femaleness to carve out a space in the music production industry, the cost is a vulnerability to feelings of inadequacy that surface as Lauren and Aubrey talk about the ways in which they seek legitimization.

"THEY PICK ME BECAUSE I'M A GIRL": ENTRY POINTS, LEGITIMIZATION, AND THE IMPOSTER SYNDROME

There is not a single route into the music production profession, but it may be that not all entry points carry equal validity and may indeed be weighted according to gender. Sandstrom's (2000) discussion of US women mixing engineers in the live music scene suggests that there is an audible difference in the quality of sound that women mix in comparison to that of male engineers, and she argues this is partly to do with the gendered entry points to the profession. Men are apprenticed as roadies (the lifting of heavy equipment required for live amplification), leaving women to pursue the craft of live mixing through the classroom, grounding them in the basic science of amplified sound, which Sandstrom argues produces more nuanced listening skills. Entry points may then artificially highlight gender difference. The choice between entry through education and entry through experience is more complex for Aubrey and Lauren, who have neither obtained qualifications nor served apprenticeships as runners or assistants but have entered the profession through the leveraging of social capital and self-training. What is significant in their accounts is the hint that established routes into engineering and production may be more difficult paths for women because of the cultural conditions that masculinize the role.

In contrast to Sandstrom's commentary, Aubrey was resistant to the classroom route because, at the time she sought to enroll, such courses were dominated by men, something she observes, as a visiting lecturer, is still often the case. Lauren also eschewed the classroom route and left school as soon as she could to pursue a career as an artist before making the transition to producer. Both women disrupt the usual system of legitimacy, gained either through educational credentials or employment in learner positions. Using Bourdieu's (1984 [2010]) understanding of the process of belonging, such legitimization is understood as designed to protect the boundaries of cultural fields; to be recognized as competent or expert requires the university certificate or the completing of an apprenticeship, however informal. To belong is to have command of specialist insider language and knowledge (Porcello, 2004) and to possess official credentials that allow entry. The autodidact (a person who teaches herself), using Bourdieu's framework, may not be recognized as legitimate.

It may be that Lauren and Aubrey jeopardize their belonging in music production because of their autodidactic approach. However, they perceive that the educational entry point is weighted towards the male student, a situation that is challenging for women where they are in the minority. Aubrey's story of how she became a music producer offers a deeper insight into why female and male entry points might be differentiated. As a young person she had applied to a 'prestigious' university to do a music industry course. At the interview stage she was told by (male) lecturers that she would be the only female on the course:

> That put me off. . . . I just thought to myself, "I don't want to be the only girl on a course", so I decided not to go for that. . . . I think probably at that age I didn't want to, I liked to blend in, I wouldn't like to stand out. I thought I would get too much attention.

At the formalized entry point to a music or music technology career, where women are in an extreme minority, there is a gendered barrier. According to Young (1990), girls are often conditioned in feminine "modalities" that respond to "the threat of being seen" (p. 155) giving rise to a self-consciousness that, in Aubrey's case, prevented her younger self from entering the classroom as the only female. Again, the "body out of place" (Puwar, 2004) is generated through being a minority body, and the discomfort this brings may be a significant barrier for some women. Visibility is a burden and therefore a barrier.

Where women are resistant to the classroom route to gain credentials because of their minority status, the autodidactic route is available. However, this lack of official credentials means those who are self-taught are more vulnerable to the "imposter syndrome". Lauren sees her lack of a formal education in engineering and production as problematic at the same time as she strongly defends her individualism:

> I think the common thing with women in music is like the imposter syndrome situation, because I haven't followed a path, I have no seal of approval, I have nothing, sometimes I do feel like, I've dug really far into this hole and I have no support backing me. . . . Someone's

gonna go, "she doesn't really know what she's talking about" and I'm going to go, "I don't" [laughs]. So, definitely, like, that is a disadvantage, but at the same time, because I've always had the mindset of being different, I kind of don't care and go okay you may think that but I'm just going to carry on doing my own thing. You can judge me however much you want.

The lack of a credentialed entry into the profession allows Lauren's sense of being an imposter to undermine her confidence. At times, she feels fraudulent as she imagines that others question her status; she has no certificate that proves her abilities as an engineer and producer, yet she remains proud of her ability to carve out her individual path. She also subverts normative views towards the way in which credentials in the music business are earned, announcing on her website (Deakin Davis, 2019) that she is self-taught, giving the autodidact an elevated status and troubling how legitimacy is defined.

Our conversation with Lauren took a more interesting turn. The imposter syndrome is gendered not only because she is the precarious autodidact, but also because her femaleness is made the basis of her success and therefore simultaneously undermines her sense of deserving that success on a level playing field of skill:

> But the imposter syndrome is really evident through practically every female, even artists, and engineers, especially the engineers and producers I know. They don't feel like they deserve it, because, and this is like the crux of the situation of being given advantages for being a woman, because I feel like I've only achieved this because I'm a woman, and my skill set doesn't match the opportunities that I've been given. So that's where imposter syndrome comes in, thinking, "I shouldn't be working on this session, I don't know what I'm talking about". But they pick me because I'm a girl.

Being treated favorably because she is female, being the recipient of positive action, feeds Lauren's sense of unbelonging and compounds the feelings of inadequacy that the lack of credentials generates. What Lauren is describing is the toxic side of leveraging sex and gender difference. Whilst there are benefits to marketing the notion of the *female* producer, which we discuss shortly, chasing legitimacy becomes a perpetual task for the self-taught female, which is undermined by the suspicion that opportunities are offered "because I'm a girl" rather than being skilled enough for the job.

Being recognized publicly within the industry may provide a credentialed status that can be used as an antidote to the feelings of inadequacy, but this can be double-edged. Returning to Lauren's response to being mistaken for the secretary in the studio, she uses her award-winning status as a defense against skepticism. However, Lauren's ability to maintain confidence in her status is jeopardized by her anxieties about inadequacy: "This goes to the women's imposter syndrome. I think I won because I'm a woman". The sense of legitimacy is continually undermined by the powerful feelings that one is an outsider, someone who does not and will not belong.

As Lauren and Aubrey reflect on how they cultivate their status as producers, they identify other sources of legitimacy. Aubrey sees legitimacy in collaborating with popular artists, who accrue cultural capital from which she can borrow, becoming an unofficial talent spotter as she attempts to catch hold of an artist's upward trajectory: "You kind of have to do a bit of A&R scouting as producers as well because you need to figure out which ones are potentially going to take off hugely, and you kind of latch on to them and produce them". Both women invest significantly in the relationships they have with artists to benefit from a *quid pro quo* arrangement where legitimacy is shared.

The attainment of credentials through awards and through relationships with significant people who have already accrued social and cultural capital demands a high level of emotional labor on the part of both women. Whilst there are benefits to being able to provide continual emotional labor, particularly when leveraged by women, there is a negative side, a cost that both Aubrey and Lauren recognize they pay.

"I DEFINITELY OVERCOMPENSATED": EMOTIONAL LABOR AND THE ECONOMY OF SMILES

When Lauren began her foray into music production, she invested heavily in emotional labor to accrue the social capital she felt she needed to progress. Networking as an entry point for Lauren is gendered in nuanced ways:

> Starting out as I did, I have to proper like, "Oh my God, you're so amazing. You're so lovely". I was really, really enthusiastic. . . . I would say starting out I definitely overcompensated. And it was crucial that I did, because otherwise . . . and I'll admit some of these guys liked me to be around because I was nice, smiley, and laughy. . . . I don't think the guys have to be overly enthusiastic, smiley, laughy, to get the same level of response. . . . I had to invest so much into my emotional, social expression to get the same return.

This vignette of female emotional labor is part of the "economy of smiles" (Bartky, 1997: 135). Lauren's description of gendered expectations in interactions relates as much to her body as to her emotional work. As Bartky explains:

> Feminine faces, as well as bodies, are trained to the expression of deference. . . . Women are trained to smile more than men, too. In the economy of smiles, as elsewhere, there is evidence that women are exploited, for they give more than they receive in return.
>
> (p. 135)

Lauren reflects that to gain entrance to the field she needed to provide effervescence, especially in her contact with men. She hints at the way sexual undercurrents underpin interactions between young female neophytes to the business and established males; women in positions similar to Lauren must negotiate how their need to engage in the "economy of smiles" is sexually interpreted. Lauren is in her twenties but says she looks younger: "People do think I'm sixteen, which is, whatever, their issue". This statement acknowledges the negative side of the gendered nurturing economy, a "dark side" that is documented in other research (Ward and McMurray, 2016). At the same time as relying on the ability to provide emotional labor, Lauren must manage the male gaze that constructs her as object, even as she seeks to develop subjectivity in the industry.

Lauren describes this process as "crucial" to establishing herself because she is unable to benefit from the tacit fraternity to which men have access. For a woman to resist the "economy of smiles" may stymie a fledgling career. But, to balance this dark side to emotional labor, there is also the positive side, where women can use the gendered nurturing economy to their advantage, and this becomes part of their approach to music production, the crux of the leveraging of perceptions of femininity. We might describe this as a "difference dividend"; the expectation that a female producer will be aligned to the relational is fully embraced by Aubrey and Lauren and, we suggest, this has a significant impact on their business success, but also may change the production process and the music product.

Partaking in the nurturing economy and being willing to undertake significant amounts of emotional labor means that both Lauren and Aubrey become visible as *female* music producers. They are both seeking to bring something different and valuable to the role in ways that they perceive men are less inclined or less equipped to do. This is a more positive approach to the nurturing economy that seems to pay off, entrepreneurially. In other male-dominated professions, the emotional labor that women are expected to undertake is often devalued, invisible, and certainly unlikely to be marketable (Bagilhole, 2002).[4] Both Lauren and Aubrey see their gender difference as a way of standing out from the male crowd, contrasting with the context of the denigration of relational, pastoral, and emotional labor in other industries. Women in music production can capitalize on difference, on the idealized feminine relational skills.

Lauren senses and experiences the difference in approach to emotional labor in the studio in gendered ways amongst her clients. Whilst she is uneasy about feeding gendered stereotypes, nevertheless, there is in her experience a different dynamic with male clients and female clients:

> before any session I do with female artists, we literally will spend an hour or two talking. . . . Male artists are like "I know what I want. Let's go". I'll start recording in the first five minutes with a male artist . . . that was a hundred percent of the time. I can't deny it, . . . but I think a lot of the time it's just that the guys aren't taught to talk a lot.

The difference is not naturalized or essentialized by Lauren, but she understands tendencies to engage in talk as learned behaviors that accord with notions of masculinity and femininity. The ability and willingness to listen to the artist is framed as an opportunity to leverage perceptions that a female producer and engineer is likely to have a sensibility that changes the dynamic of the process, and therefore the sound of the product. Commenting on how mixing recorded and live music is influenced by gender, Sandstrom (2000) argues that female engineers respond more sensitively to what is asked of them by the artist, which is how both Lauren and Aubrey explain their approach. Reflecting on this phenomenon more deeply, we suggest this is a two-way process and that female artists may also undertake emotional labor in return; women working with women are able to leverage this dynamic, a process that becomes anchored in the relational.

The emotional labor offered by female music producers is something that Aubrey sees as the appeal for artists, and again sees a difference between the male and female client and their expectations:

> The majority of them will say, "this is great, I've been looking for a female producer, I spotted you working with a male producer and I wanted to work with [a female producer]". The girls say that more than the men.

The notion that male and female artists may sometimes be seeking different experiences from music producers is coupled with the understanding that women producers are better equipped – or perhaps more willing – to deliver the emotional labor sought. Both Aubrey and Lauren spend time talking and getting to know their clients, and in some cases they develop friendships. Significantly, Aubrey sees her work as enhanced by traits that are perceived as feminine:

> There's been a lot of stories that with men, they can't connect with the producer that well. So, they're not getting what they want musically, they don't feel comfortable. So, I think a woman may be more empathetic and will take the time to get to know them and build a relationship. I think that's the feedback I seem to have had.

It is the explicitly *feminine* (if constructed) ability to undertake emotional labor that becomes the basis of how both Aubrey and Lauren leverage gender difference, not only to genuinely nurture the artist through the process for the sake of musicality, but also to establish themselves as *female* producers, standing out against the male crowd.

"BEST OF THE GIRLS": THE LEVERAGING OF DIFFERENCE

Sandstrom (2000), a music engineer herself, suggests that in some cases when women enter a male-dominated world, such as music technology, there is a suspension of gender. The stories of Aubrey and Lauren suggest, however, that far from dissolving gendered distinctions, their identity as music producers relies heavily on their femaleness and on perceptions of feminine nurturing traits. This is leveraging a *strategic* difference, and both Aubrey and Lauren understand their femininity as marketable. In whatever terms difference is couched, Aubrey seeks to elevate her femaleness to establish space in which difference is positive. On her website, she writes:

> I am also proud to be one of the very few successful female Record Producers from the UK and I do all I can to promote the important role of women in music by teaching Music Production at colleges, running monthly online masterclass sessions and offering an internship scheme specifically for female students.
>
> (Whitfield, 2019)

Aubrey believes that being female in the male-dominated music production industry does not require her to be the honorary male, but that her difference, manifested through those traits of relationality that are perceived to be feminine, is a boon to her career. She is not seeking to belong in the same way as the male producer but is rather elevating the feminine and moreover encouraging other women to do the same; promoting femaleness is connected to a political sisterhood that will encourage more women to call themselves music producers through the construction of credentials that are tailored to women. Aubrey is consciously carving out female space in the industry.

Aubrey's taxonomical choice – calling herself a "female producer" – is subversive. Aubrey is troubling the notion that the female negative semantic space must always be overcome (Spender, 1985; Weatherall, 2002).[5] In other words, rather than attempting to rehabilitate the title of "music producer" to incorporate the female, she makes herself visible in being entitled as a "female music producer", which capitalizes on the currency she generates by acting out the positive gender difference in the studio. She is also eschewing the idea of being the same as the male, declaring difference as a positive attribute rather than aping the male producer. We see this in paradigmatic terms; Aubrey is establishing a feminine approach to music production, one she believes some artists seek out, and is profoundly affective because emotional labor is affective (see Ahmed, 2004; Liljeström, 2016).[6] In other words, the emotional is an intrinsic part of the production process that is given currency using attributes accorded to the feminine.

Leveraging difference through language makes Aubrey highly visible, but this is a strategy that may only pay dividends whilst women are in a significant minority. Both Aubrey and Lauren, however, are committed to trailblazing to draw more women to the industry. For the time being, though, emphasizing gender difference allows women who are producers to distinguish themselves in a competitive field. Aubrey is ambitious, but she couches her definition of success in gendered terms: "My goals are kind of high ones, you know, to be one of the biggest *female* producers of all time" (italics ours). When we asked whether this was a deliberate choice of differentiating language, she was taken aback at her own unconscious use of the term "female producer". As we discussed this use of language, Aubrey theorizes why she carries the prefix, and this reveals a profound sense of separation because of her sex:

> Ok, that's interesting that you've picked up on that. Because in my head, God, if I had any ambition it would be to be the best record producer in this country, but in my head that's completely out of reach because it's so competitive. So, the next best option is kind of best of the girls [laughs]. But if you think about it, you are, there are some really successful female producers at the moment, not a huge amount, but one or two. But it's still an open playing field. For there to be someone, if you think of the huge producers . . . they're just as famous as artists. You know, we know who they are. But there's no female producer you can say they're as big as an artist. So, there's no huge role model. I just think it's so important to fill that space, to help encourage people.

For Aubrey, entering a field that is dominated by men means the space at the top is already crowded with well-known male producers. The feeling that there is no room for the female in this elite circle indicates that Aubrey perceives women as 'outsiders' in music production. To mitigate the inability to penetrate the field on male terms, Aubrey can see the emerging space ready to be occupied by the female; it is "an open playing field." That Aubrey sees success as measured against the male elite as unreachable raises several questions about how women see themselves in the music production and engineering field. The benefit of seeing the figurative space in separated and gendered terms is that women become visible, and this is important to Aubrey, who has a passion for being a role model to other aspiring women producers and engineers; seeing the industry as carved into gendered spaces may help women raise their profile *as women* and on their own terms. This is a positive visibility that contrasts the out-of-place visibility that both Aubrey and Lauren have experienced; in this perceptual context they are visible as *producers*.

CONCLUSION: THE EMOTION OF PRODUCTION

During this research process, we have been affected by the idea that working with women producers may profoundly alter the experience of the music-making process. Reflecting on our own experiences in the studio, we speculate whether the product has an appreciably different quality under the production of a woman who values the relational and who is willing to not only listen but also to engage in emotional labor to make the artist *feel* good, feel understood. The culturally feminine attention to nurture and emotion is part of an affective process that we suggest can change the final recorded product. As recording artists, we find this proposition extremely attractive, and it seems to us, having listened to the stories of Aubrey and Lauren as well as drawing on our own experience, that producing recorded music is as much an emotional process as it is a technical one.

We have shown that visibility for women in music production is double-edged; they can be out of place, becoming visible because they are transgressing gendered expectations and because of their minority status. We have argued that women as music producers are at times constructed as "bodies out of place" (Ahmed, 2000; Puwar, 2004), being required to adapt to a masculinized space and to overcome perceptions that they do not look like a (male) music producer. But visibility is also marketable for women; Lauren and Aubrey have a genuinely nurturing attitude and harnessing traits that women are expected to possess, their approach to the production process is sought out by artists (especially female artists). However, there remains a cost to differentiation for women in music production, which is not entirely mitigated by the attempts to elevate a feminine approach. Entry points may be gendered, and Lauren's story suggests a heavy reliance on accruing social capital through gendered (and sometimes sexualized) emotional labor. Moreover, there remain historically pre-existing binary alignments between gender and technology that frame the female engineer as transgressive.

Ultimately, the success stories of Aubrey and Lauren are about leveraging strategic difference to carve out space that allows women to flourish as producers without being required to compete with, or be the same as, male producers. Of course, this leveraging of strategic difference is not simply an entrepreneurial quirk, but a way of troubling binaries that are constructed as part of a masculine paradigm that leave women alienated from technology and production. This troubling is appealing to us, as it seems to be to many other artists who wish to elevate the emotional contribution to the process of making music. As we have explored this subject, we have understood more clearly the gendered nature of our own studio experiences. Perhaps one day we will record our next album on a London industrial estate where the toilets are not labeled as exclusively male and where music is produced in an environment that supports female belonging.

NOTES

1. We focus here on women who are professional music producers, rather than women who self-produce as artists (see Wolfe, 2012).
2. We separate the terms *sex* and *gender* to emphasize that sex is the biological difference and gender is the constructed set of characteristics given to feminine and masculine. This separation of sex and gender has been an important feature of feminist thought, given early prominence in Simone de Beauvoir's *The Second Sex* (1949 [2009]).
3. 'Cultural capital' is a term coined by French sociologist Pierre Bourdieu (1984 [2010]) and describes the accumulation of specialized knowledge that enables a person to maintain distinction and belonging in a particular field.
4. Bagilhole's discussion of non-traditional occupations includes the priesthood, where pastoral skills are often seen as feminine, but also marginalized, juxtaposed against leadership qualities that are aligned to the masculine.

In academia, Bagilhole argues that women take on nurturing, pastoral, and administrative roles that are neither rewarded financially nor contribute to a career trajectory. It is an interesting and important point to raise, then, that both Lauren and Aubrey feel able to leverage the emotional labor they provide.
5. There are debates in feminist linguistic studies about how to overcome the negative semantic space of the female – that is, the way the female is invisible in language that encourages the imagination to see the male as universal. Requiring the prefix of "woman" generates a separate feminine taxonomy that many feminists would see as sexist (see Spender [1985], for an overview of this debate). For the music producer, there is clearly value in being differentiated linguistically.
6. Feminist theory has engaged with what has become known as the "affective turn", which explores the ways in which emotion and intimacy are forms of power and have an impact on social and cultural processes. Further research is needed into the affective economy in the studio.

REFERENCES

Ahmed, S. (2000). Embodying Strangers. In A. Horner and A. Keane (eds.). *Body Matters: Feminism, Textuality, Corporeality*. Manchester: Manchester University Press, pp. 85–96.

Ahmed, S. (2004). *The Cultural Politics of Emotion*. Edinburgh: Edinburgh University Press.

Bagilhole, B. (2002). *Women in Non-Traditional Occupations: Challenging Men*. Basingstoke: Palgrave Macmillan.

Bartky, S. L. (1997). Foucault, Femininity, and Patriarchal Power. In K. Conboy, N. Medina and S. Stanbury (eds.). *Writing on the Body: Female Embodiment and Feminist Theory*. New York: Columbia University Press, pp. 129–154.

Bourdieu, P. (2010). *Distinction*. Abingdon: Routledge. (Originally published by Routledge, Kegan and Paul, 1984).

Butler, J. (2007). *Gender Trouble*. London: Routledge. (Originally published by Routledge, 1990).

Citron, M. J. (1994). Feminist Approaches to Musicology. In S. C. Cook and J. S. Tsou (eds.). *Cecilia Reclaimed: Feminist Perspectives on Gender and Music*. Chicago: University of Illinois Press, pp. 15–34.

Cook, S. C. and Tsou, J. S. (eds.). (1994). *Cecilia Reclaimed: Feminist Perspectives on Gender and Music*. Chicago: University of Illinois Press.

Deakin Davis (2019). [Online]. Available at: www.laurendeakindavies.com/home

de Beauvoir, S. (2009). *The Second Sex*. Translated by C. Borde and S. Malovany-Chevallier. London: Jonathon Cape. (Originally published as *Le Dexième Sexe* by Editions Gallimard, Paris, 1949).

Downes, J. (2012). The Expansion of Punk Rock: Riot Grrrl Challenges to Gender Power Relations in British Indie Music Subcultures. *Women's Studies*, 41(2), 204–237.

Hill, R. (2016). *Gender, Metal and the Media: Women Fans and the Gendered Experience of Music*. Basingstoke: Palgrave Macmillan.

Jagger, S. (2019). *The Dialectic of Belonging: Resistances and Subversions of Women Priests in the Church of England*. Unpublished. University of York PhD.

Lerman, N. E., Oldenziel, R. and Mohun, A. P. (eds.). (2003). *Gender and Technology: A Reader*. Baltimore; London: John Hopkins University Press.

Liljeström, M. (2016). Affect. In L. Disch and M. Hawkesworth (eds.). *The Oxford Handbook of Feminist Theory*. Oxford: Oxford University Press, pp. 16–38.

Negus, K (1992). *Producing Pop: Culture and Conflict in Popular Music*. London: Hodder and Stoughton.

Pegley, K. (2000). Gender, Voice, and Place. In P. Moisala and B. Diamond (eds.). *Music and Gender*. Chicago: University of Illinois Press, pp. 306–316.

Porcello, T. (2004). Speaking of Sound: Language and the Professionalization of Sound-Recording Engineers. *Social Studies of Science*, 34(5), 733–758.

Puwar, N. (2004). *Space Invaders: Race, Gender and Bodies Out of Place*. Oxford; New York: Berg

Reimer, J. and Bridwell, L. (1982). How Women Survive in Non-Traditional Occupations. *Free Inquiry in Creative Sociology*, 10(2), 153–158.

Sandstrom, B. (2000). Women Mix Engineers. In P. Moisala and B. Diamond (eds.). *Music and Gender*. Chicago: University of Illinois Press, pp. 289–305.

Spender, D. (1985). *Man Made Language*, 2nd ed. London: Pandora Press.

Ward and McMurray (2016). *The Dark Side of Emotional Labour*. London: Routledge.

Weatherall, A. (2002). *Gender, Language and Discourse*. London: Routledge.

Whiteley, S. (2000). *Women and Popular Music: Sexuality, Identity and Subjectivity*. London; New York: Routledge.

Whitfield, A. (2019). [Online]. Available at: www.aubreywhitfield.com/

Wolfe, P. (2012). A Studio of One's Own: Music Production, Technology and Gender. *Journal on the Art of Record Production*, 7. [Online]. Available at: www.arpjournal.com/asarpwp/a-studio-of-one%E2%80%99s-own-music-production-technology-and-gender/

Young, I. M. (1990). *Throwing Like a Girl and Other Essays in Feminist Philosophy and Social Theory*. Bloomington: Indiana University Press.

16

Conversations in Berlin

Discourse on Gender, Equilibrium, and Empowerment in Audio Production

Liz Dobson

Statistics documenting gender diversity in music and audio production have shown audio domains to be overwhelmingly white and cis male across a range of audio industry sectors (Born & Devine, 2015; Dobson, 2018; Gavanas & Reitsamer, 2013; Smith, Choueiti, & Pieper, 2018, female:pressures' facts survey[1]). This situation can reinforce assumptions that associate masculinity and technology (Leonard, 2017), where women are believed to understand less while simultaneously being subjected to a higher standard – to a kind of 'super-surveillance' (Puwar, 2004: 92). Drawing from a series of 18 interviews conducted with Berlin-based womxn[2] (Gourd, 2018) music producers, this chapter explores some of the ways in which this male and homogenous space becomes a complex terrain for women.

The qualitative data is useful as it provides incentives to address this imbalance, providing a baseline by which to measure change and ammunition for new initiatives investing in women and non-majority genders, such as the PRS Foundation Keychange initiative,[3] and the Spotify and Sound Girls EQL Directory.[4] The initiatives are important for economic reasons alone, as audio professionals make an enormous contribution to our UK Creative Industries; in 2016 the creative industries contributed £27 billion value of service exports (11% of total UK services exports), and in 2018 an overall value of £101.5 billion.[5] Although projects that improve diversity in audio are key to the industry's transformation, this chapter looks more closely at qualitative data, personal experiences, and narratives that help us understand the issues women face – experiences that otherwise abstract data about gender, and provide a path to understanding how personal and systemic sexism may be contributing to a situation where women leave audio engineering professions, as articulated briefly by Beth O'Leary and Kim Watson in *The Guardian* (2019).

Other sources of such insight include the personal narratives of gender-based violence and discrimination faced by women, including Tarana Burke's "Me Too" movement and Alyssa Milano's subsequent #metoo. But also a growing body of scholarship including Tara Rogers' *Pink Noises: Women on Electronic Music and Sound* offer a substantial portfolio of qualitative materials that include nuanced discussion around the impact of sexual violence, sexism, and prejudice in music (Bannister, 2017; Fuller, 2016; Gavanas & Reitsamer, 2013; Hill, 2016; Hill & Savigny, 2019; Leonard, 2017; Rodgers, 2010; Théberge, 1997; Thornton, 1996; Walser, 1993; Wolfe, 2019 to cite a few). Paula Wolfe's *Women in the Studio: Creativity, Control and Gender in Popular Music Production* (Wolfe, 2019) offers a significantly deeper dive into this topic, drawing on her analysis of 93 (conference/industry) speakers, six informal conversations, 31 formal primary and 10 secondary interviews. Wolfe is also careful to emphasize that gender-based prejudice is not so simply attributed to individual people, but that it is a consequence of "the historically marginalised status of women practitioners, combined with the gendering of the field that has ... resulted in gendered assumptions about the profession" (Wolfe, 2019, loc 2218). Concerned with Electronic Dance Music culture more specifically, Gavanas and Reitsamer draw on 75 face-to-face interviews with DJs in London, Stockholm and Vienna, and suggest that the informality of this scene is a factor:

> We argue that this scarcity of women artists originates partly in the gendered social construction of technology and partly in the informal character of working environments and social networks in electronic dance music cultures, dominated by images of male artists/musicians/producer/entrepreneur and the sexualised images of (young) women.
> (Gavanas & Reitsamer, 2013: 54)

Grace Banks' article in *The Quietus* refers to an "unconscious bias [that] plays out under the radar" (Banks, 2017); providing a range of examples, she explains that "[t]hey had an insult for anyone who fucked up: 'girl'. It was banter, of course, but, it made me feel bad each time I heard it". She cites examples of objectification, and the underestimation of her technical agency, explaining that "[e]nvironmental factors [such as images of male musicians and a naked women] leading to such feelings can be unquantifiable and subliminal but they are, nevertheless, there" (Ibid.). These contributions bring us closer to exploring a pervasive undercurrent of misogyny, and perhaps often the subtle but damaging unconscious bias still faced by women in audio. The most powerful stories of sexism, racism, homophobia, and ableism have necessary impact, but they often receive more attention than the subtle everyday actions – that might appear to be relatively harmless when viewed in isolation, while eroding self-belief, self-confidence, and resilience, doing slow permanent damage, like water lapping a coastal cliff until even the strongest foundations have crumbled and dissolved into the sea.

I am personally concerned with access points: once we have inspired girls and young women to pursue an interest in audio, the spaces where they can explore this interest. As a much younger woman interested in music technology, I was dissuaded by male targeted music tech adverts (featuring men in studios, and female models used to sell magazines and equipment): "these publications routinely associate electronic music machines with seductive, female sexuality" (Sterne & Rodgers, 2011: 37) (see this Flickr album[6]). I wondered how I could fit in, and now I wonder how this and these spaces are perceived by the next generation.

Following my 16th university induction week welcoming a predominantly male undergraduate cohort, I visited Berlin to interview womxn music producers. I conducted 18 interviews, asking about these producers' creative practice, how they came to use technology, and if or when they have become conscious about their gender when working in this field. Those interviews provided a springboard for exploring further insights, particularly on activism, inclusion, and diversity. However, at that time my selection process for participants lacked attention to diversity amongst women around race, socio-economic background, sexuality, and disability. This is one significant shortcoming of the work, as the narrative represents the experiences of some, but not all women, because I did not prioritize this due to my own blindness and lack of preparation, or understanding about intersectional feminism.

ARTISTS INTERVIEWED

Artist A

Composer, producer and electro-pop artist, their work spans from glitchy, abstract, multichannel electroacoustic experimental pieces, live electronics and audio-reactive visuals to the electro basslines and synth pop melodies of the solo project 'Strip Down'. They studied composition at Birmingham Conservatoire and the "Hanns Eisler" Hochschule Für Musik in Berlin.

Kaltès

Musician, DJ, curator with a background in jazz and improvisation, and an active member of female:pressure, she released music on prominent labels such as Eotrax and MORD.[7]

Sol Rezza

She is a sound artist, radio producer, and sound designer for media and an electroacoustic music composer.[8]

Miruna (Boru) Borusiade

A DJ with a classical music education, whose "DJ sets combine bold and obscure sounds and genres fluctuating mostly in the field of Dark Disco, EBM, tropical mutants with an Acid touch".[9]

Flora Könemann

Works at the interface of experimental music, sound, and performance:[10]

> Currently, she has a keen interest in combining sound performance, dance/choreography and shamanism. In her sound performances she

uses everyday materials and "broken" and/or sorted out instruments/ objects. Often these are objects and instruments that are "found" and recorded locally. For example, she was invited to a sound performance on the Canadian clothesline and let the listeners look at a shopping street, while in the background they sewed a loudspeaker noisily using a yarn and needle.[11]

Adrienne Teicher

Sound designer, composer, and movement artist known for collaborative work HYENAZ[12] with Fischer:

> The techno soundscapes, performative installations and a/v works they produce are based on the sonic shapeshifting of field recordings gathered everywhere from refugee camps to amphibian habitats to intentional communities. These works ask the question: What happens when the body becomes a foreign object, an unknown territory even to itself, a thing to be feared, managed even annihilated?[13]

Katheryn Fischer

Electronic music producer and

> creator of multimedia and immersive live performances polypartner Teicher for our project HYENAZ. We are scavengers, shapeshifters and androgynous creatures whose soundscapes spring from the surplus and detritus of our cultural and consumerist material. We are determined to create the fabric of our music entirely from found sounds that relate closely to the conceptual content of the song, so that process relates to product. We are learning and conversing with the world around us, a world that is increasingly harder to survive in without amassing money.[14]

I conducted a thematic analysis (Braun & Clarke, 2006) of the interview transcripts, seven of which feature directly in this chapter.[15] The salient themes are presented with further annotations that connect them with relevant cultural literatures, musicology, sociology, and education research.

'THE PREROGATIVE OF BOYS'

These interviews presented many stories of bias around technical competence and gender. These first examples illustrate perceptions of how this sub-cultural capital in audio is still "depicted in gender-free or masculine terms and remains the prerogative of boys" (Thornton, 1996,: 163):

> in my everyday life and my profession, it's extremely obvious that every time I have a technical problem, I'm a woman having a technical problem. Whatever I do. I don't even think I can reach a point of

knowledge so extreme that I will stop being judged as a woman having a technical problem.

Kaltès

there's always that feeling of being outnumbered and having to not make mistakes, be better than other people when you need to be.

Könemann

I mean it often happens that there's some kind of say gender comment, or they don't take you so serious, and to be really like a – in some ways you're confronted with this issue even if you don't ask for it. In that way it always – it's the gender issue it always comes up.

Könemann

Interviewer: Can you give me any examples of more specific situations? You don't have to name anybody, it's just to help somebody who doesn't understand what you mean by these encounters that you experience. Könemann: Yes, I mean it's also like in this situation like, for example, you come to a concert and then what happened to me is that the technician doesn't really recognise you. You say what you need. Like all the time I always have my own mixer so what I only need is – so I modulate the sound myself.... Sometimes they don't like it because it's their job ... once I was playing with a friend and the technician he was sticking to him, just talking to him, and he was like I've no idea about the techniques I only play the guitar, you have to ask [Könemann] because she's the one who is doing the more [electroacoustic] – electronic things. He said it three times and this guy still didn't look at me. So this kind of thing is like – I mean it's really annoying.

Könemann

Teicher works and performs with Fischer.

If a sound engineer that we're working with understands me as a man, then, often, they will just talk to me about various aspects – different aspects of our performance. I find that annoying, to be begin with. We then actively make [Fischer] more of a contact person when we enter a club for technical aspects, so that it's – just to break that whole thing a little bit.

Teicher

So many of these interviews showed how women working in audio are assumed not to have knowledge about audio: "Because women are not expected to have certain abilities, there is always an element of doubt, even if it is temporary, concerning their capacity to do the job well" (Puwar, 2004: 91). These interviews mirror many extensive studies that show how women in audio are underestimated but simultaneously held to higher expectations, required to work harder in order to be taken seriously, while also often experiencing sexual intimidation and teasing in good humor (Wolfe, 2019: loc 2325–2296). Puwar suggests that this is because "the minority position of women makes women much more visible and therefore they are noticed much more easily" (Puwar, 2004: 92).

Fischer's reflection on her experience of working closely with Teicher moves towards a more inward-facing interrogation of her actions. This self-questioning was the next prominent theme:

> One thing I feel is that [Teicher] having grown up as male, gave her a different experience growing up of power and owning power, especially even around technology. So, yeah, even with us, and we're really close and we really agree with each other and really are passionate about queer politics and feminism, but even the two of us, I sometimes feel like she can come at her knowledge or present her knowledge in a way that can sometimes make me feel like disempowered. It's not something that I would point fingers at her or point fingers at me, I think it's about the way that we interact and being aware of that.
>
> Also for myself, being aware of why I feel disempowered or why I don't take control. Why am I not like, no, get away, get away from the computer? I'm doing this thing now. Why do I sit back more? What is my learning style? Because I certainly feel that I explore; I am able to go even more deeply when I'm completely alone. Once I have the tools that I need and I'm completely alone but the thing, of course, that makes that not ideal is that I think collaborations do make things better and collaborations are great places for learning and knowledge sharing and stuff. So I wouldn't say that my goal is to just work alone. I want to work with other people and I want those other people to not just be women too because I know that that is the reality of the world and also because I don't think that fundamentally I shouldn't work with men. But how do we go at that so we all feel really good in the collaborative process? It's a complex question.
>
> **Fischer**

Fischer considers where power is situated or taken, examines her actions, and even learning style. It begins to show how this self-checking, combined with a normatively masculine backdrop, unconscious bias, and everyday sexism from others erodes self-belief, sometimes to such an extent that some ultimately simply withdraw from the discipline. One crucial insight Fischer offers though here is that this is not a gender or indeed binary gender issue, but one that relates to power. For example, Kaltès explains how a female promoter held her to a higher standard specifically because of her gender:

> It's so deeply rooted in minds that I even sometimes had comments from other women – I remember I had a gig and my sound card crashed just before it. I was like, okay, big problem. So I found another sound card, someone brought it to the club but I didn't have the right cables. So I went to the sound guy and said, I don't have the right cables. I need this and that.

> He told me that, no, I could directly "plug my computer into the mixer". I said no, that my material/setting wasn't allowing that but he said "just do it". He was extremely frustrated and he disappeared for half an hour without a word. I had to start my gig and he didn't want to help for a reason I don't understand.
>
> I went to him again and said, you didn't give me these cables, it's really basic cables, and after another half hour I just found the cables on my desk. He didn't give them directly to me nor informed me he got them even though he could see I was running around to try to find a solution. The promoter who wasn't helping either told me, it's your job, you have to know your gears, especially as a woman. You have to know better because what are these guys thinking now? (the night focused on the representation of female-identified artists). . . . I actually felt guilty. I felt responsible. I felt that my knowledge was not good enough, and it took me a couple of days to be like, fuck this [laughs]. I know my gear and the problem has nothing to do with it.
> **Kaltès**

Kaltès concluded, after several days of reflection, that her request was entirely reasonable; however, in this situation she felt unsupported but under pressure to perform to a higher standard than the others, specifically because of her gender. I asked Kaltès what effect this kind of experience has on her as an artist:

> I think the first direct effect is that I questioned myself, I questioned my work. I lost my balance and doubt myself, that was the first thing. Then I felt angry. Also at myself, which is an ugly feeling. But eventually I learned from such experiences to trust myself more. To identify more clearly my own emotion or frustration and not let others burden me. You constantly have to keep some distance with that or you end up being crippled with self doubt, bitterness and anger. It's work.
> **Kaltès**

Experiences like this make the possibility of such discrimination feel plausible, and they can become anticipated; experiences of gender-based discrimination and elevated expectation are difficult to set aside.

In the predominantly white (cis) male space then, it is important to be aware, to be cognizant of the situation, and supportive (while not explicitly drawing attention to it). Within education specifically, this 'super-surveillance' (Puwar, 2004: 92) affects the individual's freedom to speak, take risks, explore ideas/equipment; some enjoy a greater privilege of being vulnerable in the classroom. I make this connection because personal development, creativity, and learning is dependent on the social circumstance of learning. My own research in creativity and learning in the collaborative situation is framed by developmental Psychologist Lev Vygotsky's (1978) emphasis on language being a psychological tool for higher mental development. So there is enormous value in creating spaces where risk can be felt less acutely (Dobson & Littleton, 2016;

Mahn & John-Steiner, 2002). While there is always a *degree* of risk and vulnerability, not all students possess the sub-cultural capital (Thornton, 1996), some lacking the socio-economic capital that enables them to take the same creative or technical risks. The best audio production, and audio education environments in particular, provide: detailed information about technical equipment and support available, and actively promote professional conduct practices based on an anti-harassment policy that is fit for purpose, and proactively enforced by those responsible for a space.

THE BENEFIT OF A NETWORK

One might argue that audio digitization and availability of free or relatively inexpensive music software, combined with online tuition and YouTube, is a kind of audio democratization – that this kind of access should result in greater diversity in audio. Research shows that the socio-cultural context has a much greater influence. For example, Patrick Bell investigates an idea that there has been a "great democratization of digital recording technologies such as DAWs" (Bell, 2015) through his survey of New York based–music, n = 104 (n = 89 male and 70% white). He explains that "[t]he putative democratization of recording technology is tethered to a basic economic argument – affordability equals access – while analyses of the sociocultural conditions of this mode of music-making have been largely overlooked" (Ibid.: 133). The transition from vinyl to digital audio was also a potentially significant step towards DJ democracy, but members of this scene still need to navigate social capital and masculine DJ cultures. The most significant factor contributing to the low proportion of women working in audio is community (Wolfe, 2019; King, 2018; Armstrong, 2016; Comber, Hargreaves, & Colley, 1993). This raises questions about entry points. Learners and early career professionals need access to equipment (economic capital), a community of supportive peers (social capital), plus information about the domain (scene capital), and opportunities to learn through low-risk exploration of new knowledge and skills. At some point any engineer, artist, or explorer with professional ambition needs to bring their work out into the world; however, the possibility of this happening is connected with how isolated or integrated a person is socially. Through a tested mastery of, and confidence with, professional audio equipment comes increased autonomy and agency as a practitioner. Ultimately, this can lead to elevate the symbolic capital of more diverse people, which begins with empowerment. Fischer explains the benefit of being in environments that include other women or at least feminist people who understand the impact of power dynamics:

> I don't think that all men make women feel disempowered or can't be those teachers, but to some extent there can be a divide along those lines. So I think it is important for women to be around female people, or female identified people or feminist people or people at least who recognise this power dynamic because, again, to me that's where it

gets back to power. It's not necessarily about gender but it's about power and the feeling of power in the world.

Fischer

All-women environments can (but don't necessarily) offer this insight about power and: "The very real exclusion from scene networks is the central motivation for women to launch their own forms of local, translocal and virtual networking, separate from the existing male-dominated scene networks" (Gavanas & Reitsamer, 2013: 69). Further discussion of such feminist networks can be found elsewhere (Gavanas & Reitsamer, 2013; Fitzpatrick & Thompson, 2015; Dobson, 2018), though my own view is that groups like our Yorkshire Sound Women Network[16] are valuable in terms of building social, economic, cultural, and symbolic capital amongst women, benefiting women's creative practice, collaborative learning, and enterprise creativities through the increased availability of facilities and opportunities to develop domain specific knowledge and skill (Dobson, 2018). For many it is also a pathway into intersectional feminism as women meet others with relatable and also diverse experiences and priorities. For those who find this too jarring, I suggest that what we effectively have had (for some time) are all-male environments with all of the benefits that I have already outlined.

WHERE ARE THE SITUATIONS FOR POWER TAKING?

It is enormously problematic to suggest that power is only given; power can of course be taken when conditions support a person's capacity to take. Drawing on two Higher Education experiences, Artist A explains how their[17] technical support encouraged students to really exploit the help offered. Conversely, the subsequent absence of technical support as a postgraduate influenced their learning and also their perspective on seeking help:

> I really feel like I let men run around and do things for me that I should have really been doing and learning myself. That definitely happened. So I think I came away not quite as technically able as I should have been, and then I learned it all on my own, after I actually graduated. . . . They have a full-time technician who sorts everything out. I didn't really have to do anything there ever. Whenever I had a problem, I'd just knock on the door of one of the two technicians. Either the computer technician or the sound technician, and they'd come and sort it out for me. Whereas going to the Hanns Eisler, it was really good because it's just one professor who does all of the electronic music in the whole place, and there was a 32-speaker Lautsprecher Orchester with speakers all over the place doing completely different things. It's his design, but obviously he can't do that all himself, so it was up to us basically to put it all together, following instructions. It was a bit of a mess with 10 composers running around not really knowing who's going to do what or what's already done and what still needs to be

done or who needs to go and get what. But it was really good because it was, no, that's how it should be. It should be hands on because people are here to learn and they're here to learn how to be electronic composers, and if you want to be an electronic composer, you have to be a technician.[18]

Artist A

In the end Artist A's point is not explicitly about gender, but against the socio-cultural norms already described gender becomes relevant. I have often witnessed situations where students benevolently support others by taking control of the equipment (mouse, desk, speakers) to solve a problem in a way that is mostly disempowering to a point that does more to reinforce their helplessness.

ASSUMING POWER

When assuming that someone either has knowledge (or a least a capacity to understand), we reinforce this belief. Assuming power is more empowering than benevolently taking over. Fischer explained that empowerment comes from working with mentors and teachers who have power:

> [When] women want to learn more about electronic music, feel more empowered or feel like they can be more outspoken when they work with other women because they don't feel like sometimes men or people – and this is, again, I'm trying to step lightly around it because I don't think it's just about men, but there is a certain style of importing knowledge or being a teacher or having knowledge that some men have, which may create problems for some women who are trying learn – in that learning environment they really need someone who is treating them in a way that assumes that they have power, that they can do it, that they're not just to be following. It really creates a supportive environment for them.

Fischer

For me the key point here is the value of a teacher who is attuned to this idea of power and empowerment. Fischer also touches on the idea of a traditional education discourse: "a certain style of importing knowledge or being a teacher or having knowledge that some men have"; a style that echoes some of my earlier points about the relationships between technology, masculinity, and sub-cultural capital. The discourse of traditional Western education is often based on a more knowledgeable person imparting knowledge in a way that elevates the teacher's status, power, and agency, particularly in Higher Education where the traditional discourses of learning are instructor led, promoting ideas of authority and power more typically associated with the men historically responsible for established academic institutions. It is even built into the very fabric of the building with large, fixed seating lecture halls designed for larger cohorts, to enable – dare I suggest – a more economic mode of delivery.

But symbolic capital, the currency of credibility around who we assume to be in command, can be given:

> If you can be at least with other people who are mutually empowering and not competitive or not about throwing knowledge around, or throwing weight around, that doesn't necessarily have to be an all-female space, but it does have to be understood that that is what we're going after. It's easier in some ways to be, okay, let's just make it an all-female or female identified space because then to some degree we know what we're after and what some of our mutual values are.
>
> **Fischer**

... and empowerment is separate from gender:

> how we deal with gender in society and, moreover, how we deal with power in society and where our feelings of being empowered come from or disempowered come from, and that it's not necessarily a one on one relation to male/female binary but has something to do with a lot of other elements.
>
> **Fischer**

Indeed, Goh and Thompson's Cyberfeminist (Haraway, 1990) inspired development of sonic cyberfeminism (Goh, 2014; Thompson, 2018) asks us to deconstruct our very conceptual framing of society as a precursor to system change, to address "the very categories in which we think and speak, the very categories which create the gendered subjects we commonly call 'male' and 'female'" (Goh, 2014: 57). Sonic Cyberfeminism offers "a way of understanding better the logics of inclusion and exclusion which are at play" (Goh, 2018: 86), referring not only to gender but also our conceptions of technology that further embed and reinforce norms around class, race, gender, and disability.

I suggest that the more empowering audio education situations include better opportunities for learning because staff understand the learner's priorities and interests, create more appropriate spaces (socially and physically) by exploring less hierarchical models of education, and encourage confidential feedback around inclusion, the learning environment, and the curriculum design, as much as the delivery. As educators I suggest that we consider who we are intentionally and unintentionally empowering through our course design and delivery.

POLICING SELF

My conversation with Rezza required a translator. She narrates her experience of difference in the electroacoustic performance domain:

> ... when she came here [to the festival we both attended] and needed to prepare the presentation and to deal with other people in the festival, especially the technicians, they all were men, all of them. So in

that context they were somehow discussing technical issues or technicalities about the sound and whatever, she doesn't feel that because they are evil or mean in any way, but only because they are men, they just put her aside somehow and did not include her by themselves in the conversation. One of the things that she was learning during this process of the [festival name] was to deal in a context in which she needed to negotiate with almost 12 men at the same time in the same room. She learnt that she needed to let them discuss first and to have this thing of who knows more about this and who knows more about this. After that she just put herself in the middle and started giving her points of view. She said I [the translator who is male] was so blinded about that, so using the situations we were sharing during this process, she made me truly, I don't know, acknowledge the fact that these things happen and that it was happening in front of my eyes and I wasn't really aware of that. I was like why are you so affected about this meeting and she said no, you didn't realise that this and this and this happened. Then I started paying more attention and yes, it was happening actually, it was happening all of the time. She says that she was able to kind of put in context a lot of things that I wasn't, as a man, really aware of in this kind of situation.

She says that it was actually a very enjoyable process and a fun process, because she had this fun way of dealing with it and kind of understanding that dealing with it in a fun way could be better than dealing in a negative, ominous way.

[Rezza] thinks that if she hasn't her ideas as clear as she has, perhaps they will always override her or somehow ignore her. But as long as she was very minimal in what she was saying but at the same time very clear about what she wanted and how she wanted it, they were very receptive at the end. Men in a group is the problem. Men separately it's just okay, they can speak at the same level.

Rezza had to find her way in to navigate a lot of male energy. There are power dynamics amongst men, described as male "homosociality, a hierarchical structure in which men compete together 'as a team', whilst simultaneously policing themselves and others against and by accusations of effeminacy (Sedgwick, 1985; Hawkins, 2017: 36).

GENDER FEEDBACK LOOP

One magazine I love is called [Name removed] Magazine, I think it's great. They produce great content, but early on I was reading it and I wrote a piece about . . . I went through and I noted all of the male names that were the author, the editor or mentioned in the entire magazine [to] female. I noticed that the ratio was . . . something like 200 to four. So, I was thinking about this in relation to music, like electronic music and the idea of the remix, and so I was thinking part of the problem is just that if all of your references are male, then you're going to remix those same males. Who are your samples? Your samples are men and then you sample from those.

Fischer

> If all your tools are already men and your references are already men, mostly, then mathematically you're just going to keep having more and more and more of them. So the less women that we see, it obviously is going to produce that in the way that we write and the way that we talk about it. I think a lot of it is about trend and who we are talking about actually. This online, just names that are thrown around as who is important and who to listen to. There is a lot of elitism in that and there is a lot of who you know, name your call kind of stuff.
>
> It's sad but also understandable in the sense that we're so saturated with new artists, new sounds, new music, you can't possibly listen to everything. So if your best friend, who you respect tells you listen to this person, then you do. I think a lot of things operate that way to some extent.
>
> **Fischer**

The individual and systemic issues that shape this homogenous male majority in audio remains unchallenged, may be connected with a long and unchallenged precedent of 'self'-amplification, possibly shaped by confirmation bias that constantly brings us back to the same narratives and people. In popular culture, people who have established greater prominence on the scene ultimately reinforce the greater prominence of a particular white male demographic in audio.

Still, in 2019, the majority of computer scientists developing systems that touch every aspect of our individual lives are men. Research approved for publication in computer science (including digital signal processing) prioritizes certain people and topics (Alessandrini, 2007), but it is also

> a cautionary example for our own technological field [computer music, but also broader areas of audio systems development] of how discourses contributing to discrimination are re- produced – consciously or unconsciously – in institutions, ultimately determining not only who performs research, but what research is performed.
>
> (Alessandrini, 2018: 6)

Alessandrini asks how this research could be different "if the relevant fields were more inclusive in terms of gender and race?" (Alessandrini, 2018: 8). It is not enough to create better spaces and check our bias; we must diversify our awareness, and engage with the writers, thinkers, innovators, and explorers who bring diverse lives and contributions. It is not sufficient to make women and racialized minorities more visible, to tick a diversity box, or be seen supporting diverse people, because diversity takes us to much more diverse places, and this makes me feel so optimistic for the future of audio.

I believe this begins with education. Patrick Bell suggests that we need to:

> foster a culture in which learners go beyond simply using music technologies and retroactively navigating their pre-programmed biases to avoid perpetuating a simplistic user mentality. Instead,

music educators must engage their students in activities of iterative technological tinkering that nurture a design mentality. Then, not only will we play the studio, but we will design it too.

(Bell, 2015: 140)

I urge you to watch this stunning short film from the Yorkshire Sound Women Network (https://tinyurl.com/YSWNsolder), because this captures Bell's sentiment perfectly. Not only are the girls inspired by audio electronics, but they realize that with this knowledge they can change the world.

To accomplish this though, we must also acknowledge and deal with issues of sexism, racism, unconscious bias, discrimination, and tokenism within STEM, music technology, and music cultures.

KEY RESOURCES

Questions for Reflection and Debate

1. How can inclusive and welcoming audio practices be fostered and communicated beyond the immediate circle of people who already 'belong'?
2. Could experiences of sexism, gender-based unconscious bias, tokenism, and other forms of explicit discrimination (such as racism, homophobia, and ableism) inform the creation of an 'inclusion risk assessment' pro forma?
3. Is it possible to train a community to consider microaggressions relating to power and motivate investment in empowerment mindsets in audio?

ACKNOWLEDGMENTS

With my thanks to Dr Rosemary Lucy Hill for feedback, challenges, and insights on this work.

NOTES

1. https://femalepressure.wordpress.com/facts/facts-2017-discussion/
2. Womxn is used here to include all women, genderqueer, non-binary, and gender non-conforming people; people who are typically masked behind a cis male majority in audio and therefore of interest in my work, avoiding heteronormative gender norm/definitions.
3. https://keychange.eu/
4. https://makeiteql.com/
5. www.thecreativeindustries.co.uk/resources/infographics
6. www.flickr.com/photos/bdu/albums/72157621935462430
7. https://soundcloud.com/kaltes
8. https://solrezza.com/
9. https://soundcloud.com/borusiade

10. www.there-is-something-wrong-with-the-view.net/
11. http://stadtbesetzung.de/kuenstler/flora-koenemann/flora-koenemann/
12. www.hyenaz.com/about/
13. https://at.freudianslit.com/
14. http://alfabus.us/about/
15. The participants have read these transcripts, withdraw, or update their meaning/position, and also approved the presentation of their words in this chapter.
16. https://yorkshiresoundwomen.com/
17. Pronouns: them, their, they.
18. As an aside I find this interesting because UK HEIs are measured according to the National Student Satisfaction Survey and Teaching Excellence Framework, so the idea of not providing sufficient technical support is in complete contradiction with the notion that students may benefit from facing certain challenges. Those challenges can easily be presented in other ways, but in music technology agency comes from understanding which equipment to choose, how to set it up, solve problems, imagine and also explore unfamiliar technologies.

FURTHER READING

Donna Haraway's Cyborg Manifesto. Available at: http://people.oregonstate.edu/~vanlondp/wgss320/articles/haraway-cyborg-manifesto.pdf

Gavanas, A., & Reitsamer, R. (2013). DJ technologies, social networks and gendered trajectories in European DJ cultures. In *DJ Culture in the Mix: Power, Technology and Social Change in Electronic Dance Music*. New York: Bloomsbury Publishing (pp. 51–78).

Goh, A. (2018). Why Sonic Cyberfeminisms? In Alessandrini, P., & Knotts, S. (Eds.), In *Array*. The International Computer Music Association 2017–2018 (pp. 83–89). Available at: http://computermusic.org/media/documents/array/Array-2018-special.pdf

Nicholas, L., & Agius, C. (2017). The Persistence of Global Masculinism: Discourse, Gender and Neo-colonial Re-articulations of Violence. Cham: Palgrave Macmillan, Springer International Publishing.

Plant, S., & Sadie, Z. (1998). *Ones: Digital Women and the New Technoculture*. London: Fourth Estate.

Women's night safety charter (2019). Available at: London.gov website. https://www.london.gov.uk/what-we-do/arts-and-culture/24-hour-london/womens-night-safety-charter

REFERENCES

Alessandrini, P. (2007). Not all ideas are the same: Challenging dominant discourses and re-imaging computer music research. In P. Alessandrini & S. Knotts (Eds.), *Array. The International Computer Music Association*

2017–2018 (pp. 8–13). Available at: http://computermusic.org/media/documents/array/Array-2018-special.pdf

Armstrong, V. (2016). *Technology and the Gendering of Music Education.* London: Routledge.

Banks, G (2017). *Off the Record: How Studios Subliminally Silence Women.* Available at: https://thequietus.com/articles/22436-recording-studios-sexism-music-industriy (Accessed 17 January 2020)

Bannister, M. (2017). *White Boys, White Noise: Masculinities and 1980s Indie Guitar Rock.* Routledge.

Bell, A. P. (2015). DAW democracy? The dearth of diversity in 'Playing the Studio'. *Journal of Music, Technology & Education, 8*(2), 129–146.

Born, G., & Devine, K. (2015). Music technology, gender, and class: Digitization, educational and social change in Britain. *Twentieth-Century Music, 12*(2), 135–172.

Braun, V., & Clarke, V. (2006). Using thematic analysis in psychology. *Qualitative Research in Psychology, 3*(2), 77–101.

Comber, C., Hargreaves, D. J., & Colley, A. (1993). Girls, boys and technology in music education. *British Journal of Music Education, 10*(2), 123–134.

Dobson, E. (2018). Digital Audio Ecofeminism (DA'EF): The glocal impact of all-female communities on learning and sound creativities. In *Creativities in Arts Education, Research and Practice* (pp. 201–220). Brill Sense.

Dobson, E., & Littleton, K. (2016). Digital technologies and the mediation of undergraduate students' collaborative music compositional practices. *Learning, Media and Technology, 41*(2), 330–350.

Fitzpatrick, S., & Thompson, M. (2015). Making space: An exchange about women and the performance of free noise. *Women & Performance: A Journal of Feminist Theory, 25*(2), 237–248.

Fuller, S. (2016). 'Something revolting': Women, creativity and music after 50. In *Gender, Age and Musical Creativity* (pp. 21–38). Routledge.

Gavanas, A., & Reitsamer, R. (2013). DJ technologies, social networks and gendered trajectories in European DJ cultures. In *DJ Culture in the Mix: Power, Technology and Social Change in Electronic Dance Music* (pp. 51–78). New York: Bloomsbury Publishing.

Goh, A. (2014). Sonic cyberfeminism and its discontents. In CTM Festival & J. Rohlf (Eds.), *CTM 2014: Dis Continuity Magazine* (pp. 56–59). Berlin: CTM.

Goh, A. (2018). Why sonic cyberfeminisms? In P. Alessandrini & S. Knotts (Eds.), *Array. The International Computer Music Association 2017–2018* (pp. 83–89). Available at: http://computermusic.org/media/documents/array/Array-2018-special.pdf

Gourd, A. (2018). Being, fxminist. *Summit to Salish Sea: Inquiries and Essays, 3*(1), 14.

Haraway, D. (1990). A manifesto for cyborgs: Science, technology, and socialist feminism in the 1980s. In Linda J. Nicholson (Ed.) *Feminism/Postmodernism* (pp. 190–233). New York and London: Routledge.

Hawkins, S. (Ed.). (2017). *The Routledge Research Companion to Popular Music and Gender.* London: Taylor & Francis.

Hill, R. L. (2016). *Gender, Metal and the Media: Women Fans and the Gendered Experience of Music*. Springer.

Hill, R. L., & Savigny, H. (2019). Sexual violence and free speech in popular music. *Popular Music, 38*(2), 237–251.

King, A. (2018). The student prince: Music-making with technology. In *Creativities, Technologies, and Media in Music Learning and Teaching: An Oxford Handbook of Music Education* (p. 162). Oxford: Oxford University Press.

Leonard, M. (2017). *Gender in the Music Industry: Rock, Discourse and Girl Power*. Routledge.

Mahn, H., & John-Steiner, V. (2002). The gift of confidence: A Vygotskian view of emotions. In *Learning for Life in the 21st Century* (pp. 46–58). Oxford: Blackwell.

Puwar, N. (2004). *Space Invaders: Race, Gender and Bodies Out of Place*. New York: Berg.

Rodgers, T. (2010). *Pink Noises: Women on Electronic Music and Sound*. Durham: Duke University Press.

Sedgwick, E. K. (1985). *Between Men: Male Homosocial Desire and English Literature*. New York: Columbia University Press.

Smith, S. L., Choueiti, M., & Pieper, K. (2018). Inclusion in the recording studio? Gender and race/ethnicity of artists, songwriters and producers across 600 popular songs from 2012–2017. *Report from USC Annenberg, Annenberg Inclusion Initiative*. Available at: http://assets.uscannenberg.org/docs/aii-inclusion-recording-studio-2019.pdf

Sterne, J., & Rodgers, T. (2011). The poetics of signal processing. *Differences, 22*(2–3), 31–53.

Théberge, P. (1997). *Any Sound you Can Imagine: Making Music/Consuming Technology*. Wesleyan University Press.

Thompson, M. (2018). Sonic (cyber) feminisms: Questions, strategies and contestations. *Paper presented at MUTEK Festival*, Montreal.

Thornton, S. (1996). *Club Cultures: Music, Media, and Subcultural Capital*. Wesleyan University Press.

Vygotsky, L. S. (1978). *Mind in Society: The Development of Higher Psychological Processes*. Cambridge, MA: Harvard University Press.

Walser, R. (1993). *Running with the Devil: Power, Gender, and Madness in Heavy Metal Music*. Wesleyan University Press.

Wolfe, P. (2019). *Women in the Studio: Creativity, Control and Gender in Popular Music Sound Production*. Routledge.

Index

Note: Page locators in *italics* indicate a figure.

Aakær, Annika 134
Abbey Road Studios 19, 41
acoustic: engineer 224, *225*; era, as early sound 11, 12; masculine, as not 130; properties, as desired sound 17, 20, 26, 45, 208
Acuña, María 101, 102, 115, 119
Agirre, Jozean 53
Aguirre, Rafael 52 53, 102
alboka 52, 53, 79n10
Alembic 145, 148, 149, 154n3
Alessandrini, Patricia 280
Algueró, Augusto 37–38, 49n2
Alonso, Celsa 35
Alvarez, Dunia 110, 115, 119
Amuriza, Xabier 71, 73–77
Anderson, Ruth 98
Angel Records, classical division of Capitol Records 24
audio: engineering 7, 23, 187, 194, 219–223, 229–235; industry 223, 227–229, 232, 234, 268; production 7, 145, 153, 168, 268, 275; quality of 17; technology 15, 175; women in 6, 7, 162, 175, 187, 269, 272
Audio Engineering Society 15, 233, 236
Audio Record 15
Avalon (Ballroom) 145–147
award: Grammy 24, 25, 104, 182; Juno 27; Oram 195; Producer of the Year 228; Woman of the Year 196
Azoff, Irving 159

bagpiper, female 3, 97–98, 99, 102, 104, 111
Bailey, Alice 88
bands: Baby Zebras, the 134; Beatles, the 2, 38, 40–41, 43–48, 130; Cinco en Zocas (Five in clogs . . .) 111, 113, 118; Formula V, 39, 41, 48; Galician Trio 104, 105; Gene Loves Jezebel 187, 188; Gose 56, 60, 66–68, 70–71; Grateful Dead, the 145; Huntza 56, 60, 66–68, 71; Ironing Maidens, the 202–204, 206, 208, 210–211; Koban 56, 60, 66–68, 71; Los Brincos 38–41, 44, *46*, *47*, 48; Maraghotas *112*, 112–114, 116n15; marstal:lidell 134, 140; Os Rosales 99
Barron, Bebe 16
Basque: culture, complexity of 51, 78n2; dance tunes of 54; female artists 55; society, vision of 52, 53, 55, 59, 70, 77
Batlle, José María 40, 41, *46*, *47*, 48
Bayton, Mavis 136
Beatles, the 38, 40–41, 43, 46, 48, 130
Bell, Patrick 275, 280
Beltzza, Esne 56, 59, 61–62, 69, 77
Berguices, Aingeru 52
bertsolaris 55, 57, 71, 78n8
bertsos 52, 55, 57, 73
Björk 5, 187, 193–194, 196
Born, Georgina 222, 223, 232, 242, 243
Boulton, Laura 13, 30n4
boundaries: blurred, as being 22, 23, 190; breaking through, as in gender 123, 124, 239
Boy Band, culture of 1
Boyd, Liona 25
Brereton, Jude 6, 223
Brightman, Candace 149
Bro Code, the 135
bromance 134, 136
Bromhill, Olive 18–19, 24, 28, 30n16
Burgess, Richard 3, 38, 39
Bush, Kate 5, 177, 184, 193–194, 196

285

Index

Callejo, Maryní (Nieves Callejo Martinez-Losa, Maria de las): early life and career 36–37; music producer, career begins 38–42, *46*, *47*
Cantor-Jackson, Betty 4; Betty Boards 150, 154n8; interview with 145–153
Chiesa, Plinio 41, 48
Cixous, Hélène 6, 170–171, 200
Cohen, Bob 146
Cohen, Sara 129, 130
Columbia-Princeton Electronic Music Center 22
composer: Alguero, Augusto 37; Cianni, Susanne 22; Fisher, Linda 22; Golijov, Osvalso 104; Ives, Charles 15; Oliveros, Pauline 22; Osborn Joujon-Roche, Aynee 156; Peón Mosterio, Mercedes 105; Prince 120; Purcell, Henry 19; Rezza, Sol 270; Teicher, Adrienne 271; Wood, Catharine 203
concert promoter 14
Costa, Marco Antonio 115, 119
counterculture movement 154
couplets, poetic verse 51–53, 57, 62–68, 71–76
Cozart-Fine team 17, 20
Cozart, Wilma 16–20, 25, 27
Craft-Union 14, 15, 19
Culshaw, John 21, 27
Cunningham Fletcher, Alice 12

Dahl, Linda 23
Das Klagende Lied 20
Dasgupta, Nilanjana 236, 237
De Lauretis, Teresa 6, 201
dead studio 26, 27
Deakin Davies, Lauren 252–265
Delia Derbyshire Archive 229
Densmore, Frances 12, 13
Derbyshire, Delia 5, 22, 187, 194–195, 196
Des Knaben Wunderhorn 20, 21
Devine, Kevin 222, 223, 232, 242, 243
Dewey, Warren 162, 163
discrimination: bagpiper, as to 97; free from, as bands 229; gender, female 2, 82, 94, 106, 154, 228, 268, 274; homophobia 269, 281; male 254; positive, as 100–101, 189; public, as to spaces 136–137
DJs (Disc Jockey): abuse, verbal 91 barriers, as female 92–93; beginning, in the 84–85, 88; changeover, issues with 90; difficulties of 89–90; gender bias 83, 89; harassment 90–92; LGBTQ, non-binary, as focus 88; male dominated 82–83; Melodie, Miss 88; music obsession, as primary 88–89; promoters 89, 92, 94; rewarding, as personal 95; training program 88, 93
Dobbin, Frank 234, 236
Dobson, Liz 7
documentary film recording: *Alalá* 108; *Asi es Galicia* (This is Galicia) 102; *España Insólita* (Amazing Spain) 102; Native American Indian song 12; pioneers of 12, 102; *Saudade, Retrato en Si Bemol* (Saudade, portrait in B flat) 102
dreampop alternative rock subgenre 140, 194

electroacoustic 270, 272, 278
electronic music: feminine approach 6, 16, 22, 269, 277, 279; producers, female 187, 202, 268, 271
engineer: audio, sexism in 7, 14, 21, 130; colleagues, as predominately male 23, 25; Craft-Union mode 14, 15, 19; female role, as increasing 187, 194, 196, 214n6, 219; gender imbalance, dead shows 148; recording 2, 12, 14–15, 147, *225*
Equalizer, talent promoter 3, 88, 93, 94, 95
equality, diversity, and inclusion (ED&I): diversity 65, 232, 234, 236, 240–244, 268, 275; equality 91, 156–157, 190, 219; inclusion 232, 236–242
Esne Beltza 56, 59–62, 65–66, 69, 77
Estudios Celada 37
Evans, Bill 23

Facebook is like disco, Twitter is like punk 3, 86–88
Family Dog 145, 146, 147, 154n1
femaleness 102, *112*, 114, 253, 259, 263
feminine: experience 200, 205, 211; music, approach to 6, 202–206, 210, 263, 265
feminism: cyberfeminism 7, 278; ecofeminism 101; equality is 170; gender identity and 254, 273; second-wave 157, 170
feminist: male, activist (pro-feminist) 135; movement 22, 69, 77, 100, 102; networks 276; philanthropic, as mindset

Index

132; postfeminist 91, 94; project 3, 253; research 199, 253; style, aesthetics and 207; theory 128, 266n6
Fernández, Marina 111, 115, 119
fidelity, as in sound, defined 17
Fine, C. Robert 16–17
Firth, Simon 87, 93, 139
Fischer, Katheryn 271, 273, 275, 277
Fjeldsted, Marie (Penny Police) 131
Fonit Cetra studios 41, 48
Fórmula V 39, 41, 48
Frey, Glenn 159–159

Gabriel, Ethel 23, 24, 27
Gadir, Tami 91, 93, 94
Gaisberg, Fred 12, 13, 19, 27
Galicia: culture, as to pipers 97–100, 111, 114; glocal, cultural production 105, 109, 114; piper band, as all women 98, 101; piper band, Saudade 102–103, *103*; piping, popular music 110–112, 113–115; touristic image of 100, *100*
García Morencos, Esteban 38, 49
Garcia, Jerry 151, 154, 154n4
Gavanas, Anna 82, 83
gender: allies, change from within 5, 187, 190–192, 196; balance 93–94, 143, 197, 223, 231–233, 240; bias 228–229, 235, 239; cultural capital 93, 98, 109, 255, 260; deconstructing 123, 157–158; difference, male dominate 6, 84, 156, 171, 251–254, 257, 260–263; discrimination of 2, 14, 82, 100, 238; inclusive, as curriculum 243–244; inequality 91, 190, 227, 237, 242, 244; issues 102, 106, 111, 114, 128, 135, 227; lyrics regarding 132–133; modalities, of music production 199–200, 207, 210–212; neutral, 170–171; perspective, 51–52, 57, 73, 78, 128, 141; relationships 128–129, 141, 156; rock, pop association of 130–131; roles, contradicting 56, 63–65, 69–70, 76–77, 175, 206; social capital 253, 257, 260, 265, 275; stereotypes 56, 60, 77, 171, 192, 233, 243; sub-cultural capital 271, 275, 277; transgender and 66, 77, 88; violence, gender-based 57, *58*, 69, 154, 269, 274, 281; workplace, as in the 1, 6, 135, 153, 203, 228–229, 235
gendering 29, 61, 128, 130, 141, 219, 269
Gil Calvo, Enrique 36

González, Manolo 41, 42, 45–46
González, Nacho 98, 115, 119
Gose (Iñaki Bengoa, Ines Osinaga, Jon Iñaki Zubiaga, members) 56, 60, 66–68, 70, 71, 77
gramophone 18–19, 21
Gramophone 21
Graphophone 12
Grateful Dead, the 4, 145–153
Green, Lucy 130
Grosz, Elizabeth 6, 200
Grzesik, Ania 205, 209–211, 213, 215n9
Guðmundsdóttir, Björk *see* Björk

Harana, Mielanjel 53, 55
Harding, Phil 1, 242
Hathaway, Thomas 27
Healy, Dan 147, 148, 152
Hedkandi 88, 90
Hello World 241
Hepworth-Sawyer, Russ 1
Her Noise 299
High Fidelity 18
Hirdman, Yvonne 128
Hodgson, Jay 1
homosociality 93, 141, 279
Howard, Mary Shipman 14–16, 27, 30n6
Huntza 56, 60, 66–68, 71
Hurk, Fieke van den 202, 203, 205, 208–211

Ibsen, Henrik 132, 133, 142n3
irrintzi 52, 79n9

Jagger, Sharon 6
Jepson, Barbara 27, 28
Joint Audio Media Education Support (JAMES) 242
Junkera, Kepa 56, 59–60, 62–63, 65, 68, 73

Kahn, John 141
Kalev, Alexandra 234, 236
Keane, Helen, 23, 27
Kearney, Mary Celeste 208, 210
Kepa Junkera 56, 60, 62, 65, 68, 70
Keychange 3, 94, 227, 268
Koban 56, 58, 60, 66, 68, 71

label: Allegro 16; Angel 24; Aquitaine 25; Blue Note 42; Boot 25; Brunswick 14, 42; Camden 24; Capitol Records 24,

25; Columbia 36, 38, 54; Contemporary 42; Coral 42; Deadhead 150; Decca 20, 21, 195; Delysé 19, 20; Deutsche Gramophon 27; DGG 42; Elektra Records 160; Elite Recordings 25; EMI 18, 19, 20; Eotrax 270; Estudios Kirios 37; Mary Howard Recordings (MHR) 15; MORD 270; Musicraft 16; Nonesuch 25, 26, 31n23; Novola 38, 39, 46; Phillips 39, 41–42; Polydor 41, 42; RCA Victor 13, 15, 23; Riverside 42; Rudy Records 122; Spanish Music Center 16; sub pop 85; Variety 14; Vox 25; Zafiro 36, 38, 40, 41, 48
Lane, Lilith 192
Laursen, Patti 24–25, 27, 28
Legge, Walter 19
Lidell, Anna 134
LinTon Taun 56, 61
Living Presence 17–18, 30n11
Lockett, Lucy 88, 90, 93
Lont, Cynthia M. 21, 22, 23
Los Brincos, 38–41, 44, *46*, *47*, 48
Los Brujos 37
Lössl, Astrid Nora 128, 131–133, 135–141
Lössl, reflections of 137–140
Lovehearts, Hattie 88, 90

Ma, Yo-yo 103, 104, 105
Macarthur, Sally 200–212
machism 69, 97
Mahlar, Gustav 20, 21
Manchado, Marisa 57
Marrington, Mark 2, 6
marstal:lidell, dreampop duo 134, 140
Martin, George 38, 40
masculinity: bagpipes, associated as 98, 109; culture of, as in recording 130, 157; embracing 170–171, 200, 202; hegemonic, defined as 122–123, 254, 268; muted, as 187; rock music, as rooted in 134; semantic, musical reference 58, 82; toxic 196
Matthews, Bob 147, 148, 149, 152, 154n2
McRobbie, Angela 83, 139, 210
Meniñas de Saudade (Nostalgic Girls) 98, 101–102, 113, 115, 118
microphone, *43*; close miking 26, 44; minimal miking 17, 26; positioning 17, 20, 42, *45*, 185; Telefunken, as brand 17, 41, 44

Milhaud, Alain 36, 48
Mills, Irving 14
misogyny 84, 90, 94, 95, 148 269
Miss Melodie DJ, 88
Morris, Wyn 20, 21
music genre: alternative 4, 127; American minimalism 185; Anglo-Saxon 35, 48; classical 5, 16–20, 24–26; dance 82–83, 91, 269; Dark Disco 270; disco 87; EBM (electronic body music) 270; ethnic idioms 185; German classicism 185; hip hop 83; indie rock 83, 134; Italian polyphony 185; jazz 5, 14, 42, 105, 158, 182; jotas 116n15; modern 37; pop, 35, 37, 39, 48, 187; punk American post-punk 84; rock and roll 28, 40, 89, 149, 194; Russian romanticism 185; Triki pop 55, 71, 77; trikitixa 3, 52, 73; ye-ye, 38
music producers, female: Björk 5, 187, 193–194, 196; Botta, Sophie 202, 214n2; Bromhall, Olive 18, 28; Bukvich, Svjetlana 5; Bush, Kate 5, 177, 184, 193–194, 196; Callejo, Maryní 2–3, 36; Deakin Davies, Lauren 252–265; Grzesik, Ania 205, 209–211, 213, 215n9; A Hundred Drums 205–206, 208, 210–212, 214n8; Hurk, Fieke van den 202, 203, 205, 208–211, 213n1; Kearney, Mary Celeste 208, 210; McRobbie, Angela 83, 139, 210; Missy Thang 202, 204, 208–210, 214n3; Peón Mosterio, Mercedes 98, 102, 105–106, *107*, 107–109, 113; Preece, Patty 202–204, 206, 208, 210–213, 214n5; Rivero, Karina 203–205, 208–213, 214n7; Wallich, Isabella 18–21, 25, 28; Warren, Amelia 202, 204–206, 209–211, 214n4; Whitfield, Aubrey 252–264; Wood, Catharine 203–205, 208, 212, 214n6
music producers, male: Marstal, Henrik 4; Milhaud, Alain 36, 48; Rasmussen, Kasper 128, 131, 132, 135–140
music production: Do-It-Yourself (DIY) 210, 212; feminine approach to 6; gender bias, change as necessary 235; studio based 13
music, evolution of 18, 49

NBC (National Broadcasting Company) 14–16, 29
Nelson, Prince Rogers *see* Prince

Index

Nickrenz, Joanna 24, 25–28
non binary 3, 84, 88, 92–94

Oakley Dance, Helen 14, 27
Osborn Joujon-Roche, Aynee 156–158; interview with 158–162, 168–170
Oubiña, Leticia 110, 115, 119

pandero 52, 68
Parkening, Christopher 24–25
Parker, Ramona *see* Miss Melodie, DJ
Pato, Cristina 98, 100, 102–103, *103*, 104–106, 114
Peón Mosterio, Mercedes 98, 102, 105–106, *107*, 107–109, 113
Perspectivees on Music Production (POMP) 1
Pettinger, Peter 23
Piercy, Marge 157, 171
Pink Noises 229, 268
Plunkett, Donald 15, 30n8
Preece, Patty 202, 214n5
Prince: career, early 123; icon, musically 120; masculinity, challenged concept of 4, 120–123; Meniñas de Saudade (Nostalgic Girls) 98, 101–103, 113–115, 118; Milladoiro 100, 108; musicians, female support of 121; Paradanda 111; persona, as sexually suggestive 122; Restless Blues Band 156; Rolling Stones, the 40; Tapia eta Leturia 55–56, 60–65, 69–71; White Flag Society 134; work ethic 124; Xironsa 110
producers: executive 17, 184; Great Britain, in 18; jazz, first female 14
Producing Music 2
promoters: concert 14, 145; cultural 101; DJs, of, 92
PRS Foundation 94, 195, 268

Rasmussen, Kasper 128, 131, 132, 135–140
rave 85
record production: commercial 16; feminist approach to 6, 17, 20, 111, 137–138, 190–191; historical gaps in 11; holistic approach 15, 17, 191; musicology 82, 128, 178, 271; studio based 4, 13, 22, 220
recording: electronic 16, 22, 185, 220; high fidelity 16; Living Presence, classical 17; process of 17, 20, 23–25, *45*, 130, 141, 253
recording techniques, 15, 26
recordist 12–16
Reed, Vanessa 94, 95
Reitsamer, Rosa 82, 83
Rezza, Sol 270, 278, 279
Rich, Adrienne 5, 157
Rivero, Karina 203–205, 208–213, 214n7
Rodríguez, Aurea 98
Rodríguez, Theresa 101
Rogers, Susan 4, 121, 122–124
Rogers, Tara *(Pink Noises: Women on Electric Music and Sound)* 268
Rothchild, Dan 158, 159
Rothchild, Paul 158, 159

Saar studios 41, 48
Saudade *see* Meniñas de Saudade
Sayang 88, 94
Seivane, Susana 98, 101, 102, 105, 108–109, 113
self-production 13, 22, 29, 208
Sevilla, Carmen 37
sexist 92, 106, 148, 253
sexual: diversity 65; expression 3, 66; freedom 77, 108; harassment 90, 94, 192, 239–240, 254; identity 3, 67, 157; relations 61–63, 65, 67
sexuality: freedom to express 66, 67, 68, 88; heterosexual 61–63; music, as in 57, *58*, 64
She is the Music 229, 240
Sniderman, Eleanor 24, 25–26, 27, 28
social network 7, 229, 237, 240, 275
Socialist Jukebox 85, 86
sound: analogue 162, 188; dead studio 26–27; digitized 162, 164, 168, 180; Dolby 183; echo 24, 26, 179; fidelity 16–17; mixing process 17, 39, 44, 47, 128, 150–152, *225*; musician placement and 21
Spanish music 35–37, 40
Spotify 168, 229, 240, 268
Stagg, Allen E. 20–21, 30n19, 31n20
STEM (science technology engineering and maths) 226, 228–229, 231–337, 240
Stout, Jane 236, 237
Straw, Will 93, 130
studios: A&M Records 159; Abbey Road studios 19, 41; CEA studios 41; Columbia 38, 54; Fonogram, owned by

Polydor 41; IBC, British Independent 20; Larribee 159; male dominance within 128–129; Mikroskopet 128; Ocean Way 159; Pacific Recordings 147; Regal Record Company (purchased by Columbia) 53; Studio 2, 43

Tapia en Leturia 55–56, 59–60, 62–65, 69–71
technology: audio, sound and 14, 145n3, 175, 224; degree 222–223, 230, 232, 242; recording 22, 122, 188, 275; track, 16, 145
Teicher, Adrienne 271, 272, 273
Thompson, Louise M. 6, 7, 278
Thornton, Sarah 93
tokenism 281
Toscanini, Arturo 15, 24
trade literature: *Audio Record* 15; *Billboard* 42; *High Fidelity* 18; *Newsweek* 15
trikitixa: accordion, diatonic 53–55, 79n18; history of 53; onomatopoeia, as defined 3, 51, 52; players 68, 71; tributes to women 68
Turner, Helen 6
twenty four seven (24/7) 3, 85–86

Urteaga, Miguel 53–54

Viñuela, Laura 57, 58
vocalist 105, 130, 139, 197

Wallich, Isabella 18–21, 25, 28
Warren, Amelia 202, 204–206, 209–211, 214n4
webography (websites) 118
Where Did Nora Go 127–129; *see also* Lössl, Astrid Nora
Williams, Allistair 128, 133
Williams, John 20, 21
Wolfe, Paula 13, 28, 47, 128, 135, 139, 199, 219, 255, 269
Women in Music 229
Women in Science and Engineering (WISE) 229
Women's Engineering Society (WES) 229
Wood, Catharine 203, 204, 205, 208, 212, 214n6

XR (virtual reality, augmented reality, mixed reality) 233

Yorkshire Sound Women's Network 229, 276, 281
Yurchenco, Henrietta 13, 30n5

Zafiro, record company 36, 38, 40, 41, 48

Printed in Great Britain
by Amazon